AF388914

Sustainability Assessment in Higher Education Institutions

Sustainability Assessment in Higher Education Institutions

Special Issue Editors

Sandra Caeiro
Ulisses Miranda Azeiteiro

MDPI • Basel • Beijing • Wuhan • Barcelona • Belgrade • Manchester • Tokyo • Cluj • Tianjin

Special Issue Editors

Sandra Caeiro
Universidade Aberta
Portugal

Ulisses Miranda Azeiteiro
University of Aveiro
Portugal

Editorial Office
MDPI
St. Alban-Anlage 66
4052 Basel, Switzerland

This is a reprint of articles from the Special Issue published online in the open access journal *Sustainability* (ISSN 2071-1050) (available at: https://www.mdpi.com/journal/sustainability/special_issues/sus_assess_edu).

For citation purposes, cite each article independently as indicated on the article page online and as indicated below:

LastName, A.A.; LastName, B.B.; LastName, C.C. Article Title. *Journal Name* **Year**, *Article Number*, Page Range.

ISBN 978-3-03936-535-7 (Hbk)
ISBN 978-3-03936-536-4 (PDF)

Contents

About the Special Issue Editors

Sandra Caeiro, Prof. Dr., Department of Science and Technology, Universidade Aberta, Rua Escola Politecnica, n. 147, 1269-001 Lisboa, Portugal. CENSE, Center for Environmental and Sustainability Research, Department of Environmental Sciences and Engineering, NOVA School of Science and Technology, NOVA University Lisbon, Lisbon, Portugal. Interests: sustainability assessment; indicators; higher education; campus sustainability.

Ulisses Miranda Azeiteiro, Prof. Dr., Department of Biology & CESAM Centre for Environmental and Marine Studies, University of Aveiro, 3810-193 Aveiro, Portugal. Interests: global change biology and ecology; marine biology and ecology; zooplankton; larval fish; climate change and sustainability.

 *sustainability*

Editorial

Sustainability Assessment in Higher Education Institutions

Sandra Caeiro [1,2] and **Ulisses M. Azeiteiro** [3,*]

[1] Department of Science and Technology, Universidade Aberta, Rua Escola Politecnica, n. 147, 1269-001 Lisbon, Portugal; scaeiro@uab.pt
[2] Department of Environmental Sciences and Engineering, CENSE, Center for Environmental and Sustainability Research, NOVA School of Science and Technology, NOVA University Lisbon, 2829-516 Lisbon, Portugal
[3] Department of Biology & CESAM Centre for Environmental and Marine Studies, University of Aveiro, 3810-193 Aveiro, Portugal
* Correspondence: ulisses@ua.pt

Received: 3 March 2020; Accepted: 9 April 2020; Published: 23 April 2020

Abstract: This Special Issue "Sustainability Assessment in Higher Education Institutions" provides peer-reviewed research from several geographies and institutions and covering various topics with the broad objective of achieving an assessment of the effectiveness and impact of different implementation dimensions measuring and evaluating how sustainability is being applied in practice. A set of nine papers, covering sustainability education, interdisciplinary teaching, sustainable assessment, governance strategies, commitments and practices, and social responsibility at Higher Education Institutions, contribute significantly to this area of knowledge.

Keywords: education for sustainable development; higher education institutions; commitments; practices

---

Higher Education Institutions (HEIs), within their mission and activities, have an important responsibility in the transformation of societies and, in particular, in contributing to the development of a more sustainable society. These institutions can implement sustainable development in different dimensions, according or not to a holistic approach, from education and curricula, campus operation, organizational management, external community and research, to assessment and communication. Ideally, these implementations should be based on a holistic/integrated approach that the United Nations Educational, Scientific and Cultural Organization (UNESCO, Paris, France) calls the "Whole-School Approach". Assessment of the effectiveness and impact of these different implementation dimensions allows measuring and evaluating how sustainability is being applied in practice, and highlighting weakness, strengths, and improvements needed.

This Special Issue aims to share knowledge and stimulate innovation within the larger theme of Sustainability Assessment in Higher Education Institutions. It gathers nine articles from the USA, China, Taiwan, and Europe (Portugal, Spain, Germany, and Slovakia). Readers can find research outputs and theoretical discussions about knowledge, perceptions, and motivation toward sustainability education, interdisciplinary teaching for sustainability, performance indicators and sustainable assessment, governance sustainability strategies, commitments and practices, and social responsibility at Higher Education Institutions.

Please find below a brief summary of each article, organized by chronological order of acceptance.

Pompeii et al. (2019) [1] analyzed student and faculty knowledge and perceptions toward sustainability education at undergraduate level in a USA University. Findings identified diverse levels of sustainability knowledge within the student body and among faculty and revealed barriers in pursuing interdisciplinary sustainability curricula across disciplines. A common pattern showed a denial of personal responsibility when addressing sustainability challenges.

Liu, Z. et al. (2019) [2] analyzed three global ranking indices, the Academic Ranking of World Universities, the Quacquarelli Symonds World University Ranking, and the Times Higher Education World University Rankings in newly formed world-class universities. The analysis aimed to discuss, based on the sustainability indicators of the ranked indices, what the common shared sustainability indicators, their variations, and contributions in the future and the research productivity and government initiatives of the universities are. The authors concluded that for the sustainability of universities, it is necessary to have an increasing emphasis on the effectiveness and efficiency of government-supported research, stability of investments, and more approaches to employ international initiatives, allowing outstanding educational programs and comprehensive internationalization. Nevertheless, the authors highlighted the criticism and cautions with regard to the used indicators, the institutions being measured, and the diversity of features to make comparisons using this type of ranking indices.

Farinha et al. (2019) [3] aimed to identify to what extent the integration of sustainability in universities in Portugal has been achieved through an analysis of their strategic and activity plans and sustainability reports. This paper highlighted the importance of analyzing the content of plans and reports from higher education institutions when intending to assess and define a country profile for the implementation of sustainability in the educational sector. According to the authors, this research may also be helpful in sharing and encouraging best practices of sustainability implementation in these types of institutions and ways of improvement.

Weng et al. (2019) [4] developed and tested an integrated model for the evaluation and improvement of university teachers based on the official teacher evaluation criteria of China's International Scholarly Exchange Curriculum program. A multiple-criteria decision-making methodology was used. Authors concluded that the developed model can be a support tool for decision makers to improve their current evaluations of teachers and to provide a cause–effect improvement strategy for education reform committees and higher education institutions, namely within sustainable development.

Filippo et al. (2019) [5] aimed to evaluate Spanish universities performance and sustainability research, development and innovation, based on indicators of scientific activity. Scientometric techniques to analyze the journal (Web of Science) and European project databases, along with reports issued by Spanish institutions, were used. The authors concluded that Spanish universities' research sustainability projects within sustainability are still insufficient, with a gap between policies and results. Nevertheless, the use of this type of analysis can be important in terms of transparency and accountability to help to promote measures that encourage information on the impact of university sustainability actions on society.

Kucharcikova et al. (2019) [6] aimed to identify factors affecting the motivation of students in a university at the Slovak Republic so they can actively engage in the education process. It also aimed to define recommendations for the increase of this motivation and contribution to the sustainability of education at the universities. Through a questionnaire survey, authors found that motivation factors are mainly related with future job expectations (friendly working team, working conditions, meaningful work, and the opportunity for self-fulfillment). Motivation is also related with the quality of education with new, progressive, and participative education methods and updated content of the study programs connected with the actual requirements of the labor market.

Marqués-Sánchez et al. (2019) [7] aimed to analyze university students' behavior in their networks following a cooperative interdisciplinary educational intervention (with the participation of students from different undergraduate programs) and the association with academic performance, resilience, and engagement. According to the authors, this way of understanding, through collaborative work between different faculties and how the university context approaches social reality, is a useful way to propose innovative and sustainable solutions in teaching–learning.

Roos (2019) [8] conducted a systematic review on how Higher Education Institutions (HEI) assume their responsibilities as social institutions, analyzing their social performance. According to the authors, social matters, namely, responsible management, strategy implementation, and leadership, as well as the measurement of outcomes at HEIs, are a recent interest of HEIs. Also according to the authors, ecological sustainability prevails within the scientific discourse and reporting, whereas social performance plays a minor role. Furthermore, the existing assessment tools for sustainability at HEIs are not measuring this performance well. HEIs are strongly determined by their mission on research and teaching and so far have not focused on other external demands from outside the organization, so future work is needed in this field.

Caeiro et al. (2020) [9] aimed to critically reflect the existing tools to assess and benchmark education for sustainable development implementation at Higher Education Institutions and to discuss their applicability in two case studies in Portugal and Spain. The authors concluded that there is a need to define a common objective of the assessment tools and continuous improvements on their development, namely, the need to integrate the external impact of HEI on sustainability, to integrate participatory processes, and to assess nontraditional aspects of sustainability.

In conclusion this Special Issue on "Sustainability Assessment in Higher Education Institutions" presents an overview of ongoing research on updated and holistic strategies and initiatives, integrated and collaborative learning, engagement of personal and institutional responsibility, and long-term performance assessment for sustainability implementation at Higher Education Institutions (inside and outside impact) that is essential for the achievement of Sustainable Development Goals and quality education. A long pathway is still needed, but HEIs are on their way.

The Special Issue Editors would like to thank the authors and reviewers, without whom such a high-quality publication would be impossible to achieve.

Author Contributions: Both authors were Guest Editors for this Special Issue and contributed equally to the Editorial. All authors have read and agreed to the published version of the manuscript.

Funding: This research received no external funding.

Acknowledgments: Thanks are due to CENSE that is financed by Fundação para a Ciência e Tecnologia, I.P., Portugal (UID/AMB/04085/2019). Thanks are due to FCT/MCTES for the financial support to CESAM (UIDP/50017/2020+UIDB/50017/2020), through national funds.

Conflicts of Interest: The author declares no conflict of interest.

References

1. Pompeii, B.; Chiu, Y.; Neill, D.; Braun, D.; Fiegel, G.; Oulton, R.; Ragsdale, J.; Kylee Singh, K. Identifying and Overcoming Barriers to Integrating Sustainability across the Curriculum at a Teaching-Oriented University. *Sustainability* **2019**, *11*, 2652. [CrossRef]
2. Liu, Z.; Moshi, G.J.; Awuor, C.M. Sustainability and Indicators of Newly Formed World-Class Universities (NFWCUs) between 2010 and 2018: Empirical Analysis from the Rankings of ARWU, QSWUR and THEWUR. *Sustainability* **2019**, *11*, 2745. [CrossRef]
3. Farinha, C.; Caeiro, S.; Azeiteiro, U. Sustainability Strategies in Portuguese Higher Education Institutions: Commitments and Practices from Internal Insights. *Sustainability* **2019**, *11*, 3227. [CrossRef]
4. Weng, S.S.; Liu, Y.; Yen-Ching Chuang, Y.C. Reform of Chinese Universities in the Context of Sustainable Development: Teacher Evaluation and Improvement Based on Hybrid Multiple Criteria Decision-Making Model. *Sustainability* **2019**, *11*, 5471. [CrossRef]
5. Filippo, D.; Sandoval-Hamón, A.L.; Casani, F.; Sanz-Casado, E. Spanish Universities' Sustainability Performance and Sustainability-Related R&D+I. *Sustainability* **2019**, *11*, 5570. [CrossRef]
6. Kucharcikova, A.; Miciak, M.; Malichova, E.; Durisova, M.; Tokarcikova, E. The Motivation of Students at Universities as a Prerequisite of the Education's Sustainability within the Business Value Generation Context. *Sustainability* **2019**, *11*, 5577. [CrossRef]

7.   Marqués-Sánchez, P.; García-Rodríguez, I.; Benítez-Andrades, J.A.; Portillo, M.C.; Pérez-Paniagua, J.; Reguera-García, M.M. with Undergraduate Nursing and Computer Engineering Students. *Sustainability* **2019**, *11*, 6325. [CrossRef]
8.   Roos, N.A. Matter of Responsible Management from Higher Education Institutions. *Sustainability* **2019**, *11*, 6502. [CrossRef]
9.   Caeiro, S.; Hamón, L.A.S.; Martins, R.; Aldaz, C.E.B. Sustainability Assessment and Benchmarking in Higher Education Institutions—A Critical Reflection. *Sustainability* **2020**, *12*, 543. [CrossRef]

 *sustainability*

Article

Identifying and Overcoming Barriers to Integrating Sustainability across the Curriculum at a Teaching-Oriented University

Brian Pompeii [1,*]**, Yi-Wen Chiu** [2,*]**, Dawn Neill** [3]**, David Braun** [4]**, Gregg Fiegel** [5]**, Rebekah Oulton** [5]**, Joseph Ragsdale** [6] **and Kylee Singh** [7]

[1] Sociology, Social Work & Anthropology, Christopher Newport University, Newport News, VA 23606, USA
[2] Natural Resources Management & Environmental Sciences, California Polytechnic State University, San Luis Obispo, CA 93407, USA
[3] Social Sciences, California Polytechnic State University, San Luis Obispo, CA 93407, USA; dbneill@calpoly.edu
[4] Electrical Engineering, California Polytechnic State University, San Luis Obispo, CA 93407, USA; dbraun@calpoly.edu
[5] Civil and Environmental Engineering, California Polytechnic State University, San Luis Obispo, CA 93407, USA; gfiegel@calpoly.edu (G.F.); roulton@calpoly.edu (R.O.)
[6] Landscape Architecture, California Polytechnic State University, San Luis Obispo, CA 93407, USA; jragsdal@calpoly.edu
[7] Facilities Energy, Utilities, and Sustainability, California Polytechnic State University, San Luis Obispo, CA 93407, USA; klsingh@calpoly.edu
* Correspondence: brian.pompeii@cnu.edu (B.P.); yichiu@calpoly.edu (Y.-W.C.)

Received: 17 March 2019; Accepted: 30 April 2019; Published: 9 May 2019

Abstract: This research collects and analyzes student and faculty knowledge and perceptions toward sustainability education at a predominately undergraduate, teaching-oriented university. In-depth, qualitative methods distinguish low- and high-knowledge student and faculty cohorts, identify perceived barriers to sustainability education in each cohort, and recognize strategies to overcome the barriers identified by each cohort. Data collected from recorded and transcribed semi-structured interviews of student and faculty subjects underwent analysis via repeated readings to uncover key themes. Results required developing metrics for student and faculty sustainability knowledge and attitudes across disciplines, determining discipline-specific gaps in sustainability knowledge and differences in attitudes, and relating implementation barriers to general or specific knowledge gaps and attitudes. Findings identified low and high levels of sustainability knowledge within the student and faculty subject population and revealed barriers in pursuing interdisciplinary sustainability curricula across disciplines and among both students and faculty at the study university. Overall, higher sustainability knowledge participants tend to identify barriers related to institutional accountability while lower sustainability knowledge participants tend to identify barriers related to personal responsibility. Distributing barriers and solutions along a continuum from personal responsibility to educational institution responsibility reveals more recognition of barriers at the personal level and more solutions proposed at the institutional level. This result may reflect a common tendency to deny personal responsibility when addressing sustainability challenges.

Keywords: sustainability education; qualitative research; interviews; implementation barriers

---

1. Introduction

1.1. Context and Background

This research contributes to the broadening understanding of impediments to integrating sustainability education into higher education. Prior studies have investigated structural conditions ranging from educational priorities to disciplinary silos to competing values [1,2]. This study seeks to understand the relationship between a level of sustainability knowledge and perceived barriers to integrating sustainability-based instruction in higher education. Specifically, this project explores the perceptions of students and faculty regarding issues of sustainability education and identifies potential barriers to implementing the teaching and learning of sustainability at the university. As such, this study identifies barriers and solutions to the implementation of sustainability among different sustainability knowledge groups of faculty and students.

The focus institution is California Polytechnic State University, San Luis Obispo (herein referred to as Cal Poly). As a non-PhD granting and predominantly undergraduate university, Cal Poly enrolls approximately 22,000 students in six colleges with an emphasis on hands-on pedagogy to prepare students for the job market and "success in a global economy" [3]. Work aimed at advancing sustainability education and curricula at the university accelerated with the university's signing of the Talloires Declaration in 2004 [4,5]. The resulting action plan committed Cal Poly to "sustainability and environmental literacy in teaching, theory, and practice". The university took steps to advance this plan with the establishment of the Sustainability Learning Objectives (SLOs). The SLOs promote the idea that all graduating students should have some knowledge of fundamental sustainability principles. The Academic Senate Resolution 688-09 establishing the SLOs states [6]:

"Cal Poly defines sustainability as the ability of the natural and social systems to survive and thrive together to meet current and future needs. In order to consider sustainability when making reasoned decisions, all graduating students should be able to:

(1)   Define and apply sustainability principles within their academic programs,
(2)   Explain how natural, economic, and social systems interact to foster or prevent sustainability,
(3)   Analyze and explain local, national, and global sustainability using a multidisciplinary approach, and
(4)   Consider sustainability principles while developing personal and professional values."

In 2014, the California State University (CSU) sought to further advance sustainability education for all its campuses (including Cal Poly) when it updated its sustainability policy [7]. The policy states that the "CSU will seek to further integrate sustainability into the academic curriculum working within the normal campus consultative process." Cal Poly more recently signed the Second Nature Climate Commitment, stating that "Cal Poly is committed to achieving carbon neutrality and climate resilience as soon as possible, and is infusing this work into curriculum, research, and student experience."

To support the advancement of sustainability education on campus, the Center for Teaching, Learning, and Technology at Cal Poly formed an interdisciplinary faculty learning community in 2016 focused on "Teaching Sustainability Across the Curriculum." This faculty group, representing four of six academic colleges, works to improve students' sustainability learning through the creation and promotion of educational experiences based on current best practices. Within the group discussions, anecdotal evidence and faculty experiences pointed to a consensus that implementation of sustainability goals was at best limited in the current campus climate, despite ongoing institutional efforts. Therefore, a campus-wide survey was proposed to assess student and faculty sustainability knowledge and awareness in order to make more informed future decisions.

Concurrent with the development of the survey, Cal Poly applied for certification through AASHE/STARS (Association for the Advancement of Sustainability in Higher Education/Sustainability Tracking, Assessment and Rating System) receiving a silver rating (62.57 of 100 possible points) in February 2017. This rating considers six domains of university sustainability: Institutional characteristics, curriculum and research, engagement, operations, planning and administration, and innovation and leadership. Cal Poly received only 28.13 of 40 possible points in the curriculum section, with two notable curricular areas contributing to this result—the lack of sustainability-focused and -related academic courses available (6.13 of 14 points) and the absence of assessment of sustainability literacy (0 of 4 points). The results indicate that only 4.9% of courses at Cal Poly are considered sustainability course offerings. Zero points were scored in the category of sustainability literacy assessment, because, at the time of submission, an annual assessment of students' sustainability knowledge did not exist. These scores reveal that while Cal Poly has theoretically dedicated itself to sustainability education, it is unclear how related policies and commitments materialize within the curriculum.

This study seeks to understand how the perception of barriers to and solutions for the integration of sustainability in teaching and learning correlates with sustainability knowledge, in order to identify opportunities for improving sustainability education. To achieve this goal, students and faculty from across the six colleges were assessed using qualitative methods to determine in-depth understanding of both sustainability knowledge and the identification and overcoming of barriers to integrating sustainability in higher education curriculum.

1.2. Literature Review

Multiple studies reported in sustainability education literature contribute to the integration of sustainability in the curriculum [8–10]. Although the need to assess sustainability across campus has been emphasized [11–13], former studies fall short either at pointing to a precise method of assessment or taking into account the context of sustainability knowledge. The literature does, however, reveal that sustainability learning outcomes can vary greatly even within environmental based courses and suggest further research on disciplines and majors that have historically been on the periphery of sustainability education [14]. An immense survey-based, quantitative study in European higher-education institutions also investigated the relationship between different pedagogical approaches and learning outcomes or competences. Results found that none of the competences examined were likely to address sustainability in any three of its dimensions (economic, social, or environmental) [15].

The literature identifies barriers internal to universities that prevent infusing sustainability: Financial constraints, lack of understanding and awareness of sustainability, resistance to change, and difficulty achieving a "coherent institutional approach, where operations, teaching, research, and outreach are synergized" [16]. The literature contains several examples of how silos in academia tend to act against infusing sustainability. According to Miller et al. [17], academic institutions typically organized around scholarly disciplines lack the "epistemological pluralism and reflexivity" required producing sustainability knowledge characterized by "social robustness, recognition of system complexity and uncertainty, acknowledgement of multiple ways of knowing and the incorporation of normative and ethical premises." Others also state that academic silos represent the most insidious barrier, because specialization helps to isolate faculty and "prevents the systems-level integration required to embed sustainability" [16].

Beyond silo-ing, other institutional level barriers have been identified, including institutional priorities and external pressures [18]. For example, perceptual barriers include the competition for funds on campus, the commodification of education, and the exclusion from any faculty evaluation criteria [2]. Institutional barriers to the comprehensive adoption of sustainability in higher education curriculum also include differences in understanding of the concept of sustainability and challenges of working across all areas of university structure [19]. An evaluation of faculty participation in the University of Vermont's Sustainability Faculty Fellows program examined the impact of a funded faculty learning community focused on enhancing sustainability curricula across disciplines [20]. Results identified the largest barriers for faculty included: A packed curriculum, lack of planning time, lack of department support, difficult to integrate into content, lack of content knowledge, lack of learning activity resources, and class size [20].

Arizona State University's School of Sustainability provides an example of an approach where an institution successfully applied an adaptive cycle to create a sustainability program emphasizing "interdisciplinary collaboration and community engagement" [17]. The literature offers several approaches to distinguish individual from institutional responsibilities towards infusing sustainability. A proposed sustainability compass depicts five axes of individual and institutional elements required to foster sustainability knowledge [17]. Similarly, Sterling's model for integrating sustainability in education distinguishes "bolting-on" by adding separate sustainability courses from the deeper level of integration via "building-in", which educates for sustainability by teaching sustainability issues in discipline-specific courses [21,22].

Our research is built on broad based projects like Lozano et al. [15] with an in-depth textured analysis of student and faculty experiences, in order to examine a level of sustainability knowledge in relation to the identification of barriers and solutions to further integrate sustainability into the curriculum. This approach involves categorizing interview participants' responses based on their level of knowledge in sustainability.

2. Materials and Methods

Given the lack of existing data on sustainability knowledge among Cal Poly students and faculty, qualitative methods were deemed the most appropriate for data collection and analysis. Data were collected using semistructured interviews [23], in which a set of open-ended questions were prepared to guide the interview process but might be asked in a particular order or format. Interview questions were designed to gauge each participant's general sustainability knowledge and behaviors, to assess how sustainability is approached as a learning objective across disciplines, and to identify potential barriers to teaching sustainability across the curriculum. A total of 17 faculty and 39 student interviewees from six colleges at Cal Poly (i.e., agriculture, architecture, business, engineering, liberal arts, and science and math) voluntarily participated in this survey. Students were recruited from large general education (GE) courses within a variety of disciplines and provided minimal assignment extra credit incentives for participation. The large GE courses chosen were defined as courses with over 125 students where all academic departments were represented in the possible student pool. Recruitment announcements were made in four such classes. Third-year and fourth-year students were specifically targeted as they would have more class experience to draw upon.

There are several qualitative data collection practices for conducting interviews based on what type of data the researcher wants to collect [24,25]. This project used a purposeful interview sampling technique, which has been recognized as a powerful tool to capture empirical relationships between different groups of the data [26]. In qualitative research, sample size has been shown to be less important when the participants have personal experience with the project subject, when small numbers of participants are studied intensively, and when the type of participants are chosen purposefully [27]. Moreover, this is not a hypothesis-based study, and the selected method does not aim for deriving statistical significance to test any predeveloped hypothesis. The responses from the semistructured interviews provided considerable data for analysis, including over 10 h of recorded transcripts, which serves the purpose of the study despite the small sample size for both students and faculty.

2.1. Interview Design and Implementation

Interviews were conducted by a small team of student researchers. Prior to commencing data collection, all student researchers participated in an in-depth training session with faculty researchers to ensure interviewer consistency. The same faculty researchers were present during all interviews to further ensure consistency and maintain rigorous oversight of data collection. Each interview took approximately 10–20 min to complete. All interviews were audio-recorded and transcribed. The transcripts were individually coded for emergent themes using a grounded theory approach [28].

This approach allows the researchers to determine patterns on how interviewees perceive sustainability in academia. Coding and analysis relied primarily on assessment by three faculty researchers with experience in qualitative methods to ensure inter-rater reliability. The semistructured interviews were designed to assess each participant's knowledge of, perceived importance of, and exposure to sustainability concepts and practices, with the following questions guiding that conversation:

- How do you gauge your own knowledge on sustainability?
- How do you define sustainability?
- How important do you think sustainability is? Why do you think that?
- Do you think sustainability learning is important to include in the Cal Poly curriculum?
- How does Cal Poly teach sustainability?
- What courses have you taken that discuss sustainability or focus on sustainability? (Students).
- What courses have you taught that present information on sustainability? (Faculty)
- What prevents you from receiving more sustainability instruction at Cal Poly? (Students).
- What prevents you from providing more sustainability instruction at Cal Poly? (Faculty).
- What are some ways to make sustainability education more accessible at Cal Poly?

2.2. Transcript Analysis

The stage of analysis in this study was conducted by utilizing several established techniques. Ryan and Bernard (2003) list several techniques for identifying themes when analyzing qualitative data [29]. Interview transcripts were analyzed for the following themes: Repetitions, indigenous typologies or categories, similarities and differences, missing data, and theory-related material. Recognizing repetitions is one of the most commonly used procedures for identifying themes in interviews [30–32]. Multiple, collaborative readings of the transcripts allowed for the identification and marking of statements that succinctly characterized the repeated themes.

Data analysis relied on coding, an iterative methodology identifying text "that captures and signals what is going on in a piece of data in a way that links it to some more general analysis issue" [33]. Coding schemes provided a framework for identifying emergent themes linking specific data points to the broader concepts under investigation. Following the development of a coding scheme, analyses were then incorporated to identify emergent themes, derive explanations, and actionable responses related to main research objectives [33,34]. In this study, data analysis was conducted by multiple researchers in order to avoid interpretive bias from a single researcher in the coding process, thereby gauging inter-rater reliability and establishing qualitative rigor [35,36].

Transcription analysis consisted of three phased readings. The entire interdisciplinary research team carried out an initial reading to develop a tentative, emergent coding scheme based on the repetition of certain ideas. A second reading was carried out with a smaller group of three researchers, each with expertise in qualitative methodologies. During the second reading, each researcher first coded each transcript for level of sustainability knowledge. These researchers then engaged in group discussions that gauged and normalized transcripts for either high or low sustainability knowledge. The same three qualitative researchers then completed a third reading, individually coding the text according to the coding scheme developed by the entire research team, then analyzing codes for emergent themes related to barriers or solutions. The researchers then engaged in group discussions to reach consensus on the key actionable emergent themes. Data saturation was achieved, indicating that further interviews would have produced similar results [37].

3. Results

All participant responses were reviewed and analyzed for determining high or low level of sustainability knowledge through analysis of the introductory questions "How do you gauge your own knowledge of sustainability" and "How do you define sustainability?" A high or low level of sustainability knowledge was determined through phased readings and defined through researcher congruence. Researchers referenced common definitions of sustainability including: Cal Poly's definition of sustainability "the ability of the natural and social systems to survive and thrive together to meet current and future needs", the Brundtland Commission's statement on sustainable development "meeting the needs of the present without compromising the ability of future generations to meet their needs", and references to the 'three Es—Environment, Equity, Economy'. High knowledge had a relatively low threshold for connection with agreed-upon definitions. Any mention of a broad understanding of sustainability was rated as high. When identified according to the structure of observed learning outcomes (SOLO) taxonomy, high knowledge responses contain multistructural, relational, or extended abstract statements, whereas low knowledge responses operate at the prestructural or unistructural levels [38]. Thus, participants responding with general or greater information implying broader or more comprehensive perception to the question "How do you define sustainability?" were defined as "high". Responses indicating a high level of sustainability knowledge included:

"Meeting the needs of the present without compromising the ability to meet the needs of the future"

" … it's the practice or philosophy that resources should not be used up so that any kind of practice or any materials that are used, should be used in such a way that the resource doesn't get depleted for the foreseeable future or for infinity."

"Sustainability has to do with making sure that the way that humans live, the resources we use … the inputs and outputs of our society are things that could continue for thousands of years without a problem."

Responses demonstrating a low level of sustainability knowledge were those that did not recognize a larger philosophy or were simply unrelated to the question asked. For example, if responses simply eluded to activities such as recycling or driving hybrid cars, these would be classified as low knowledge. Of the 39 student responses, 22 were noted as having a high level of sustainability knowledge, and 17 were noted as having a low level of sustainability knowledge. Of the 17 faculty responses, 10 were noted as having a high level of sustainability knowledge, and 7 were noted as having a low level of sustainability knowledge.

Participants were grouped in this way in order to develop a deeper understanding of how their prior interest and/or knowledge regarding sustainability might impact identification of issues associated with sustainability in the Cal Poly curriculum. An a priori assumption was that sustainability 'adherents' (i.e., those students and faculty with prior or continued exposure to sustainability education) would represent a qualitatively different subset of responses with a generally more positive attitude toward sustainability education due to their understanding of the importance of sustainability practices. Given this a priori assumption, the analysis sought to identify whether the barriers and solutions identified by students and faculty were similar regardless of their adherence to or knowledge about sustainability, or whether those with more knowledge about or adherence to sustainability practices would identify different types of barriers for curriculum development.

3.1. Student Responses

3.1.1. Student-Identified Barriers

Table 1 summarizes student responses identifying barriers to sustainability-based education at Cal Poly, including frequencies (total number of student interviewees N = 39). Any statement wherein a student identified a relevant barrier inhibiting their participation in sustainability-based education was coded as a Barrier. A single transcript could contain multiple coded barriers.

Though there is little variation between the "high" and "low" knowledge groups for the most frequently stated barriers in student responses, some interesting key results can be seen in Figure 1, which depicts the data graphically. The top three most frequently stated barriers in both "low" and "high" student groups are *Accessibility*, *Time Constraints*, and *Neglect*. *Accessibility* identifies barriers wherein students note they are unable to access sustainability-related courses. For example, as this student (second-year child development major, with low sustainability knowledge) shares:

Table 1. Student-identified barriers to sustainability-based education at Cal Poly and corresponding frequencies (Total number of student interviewees N = 39).

Barriers	Examples	LSK [1] (N = 17) [3]	HSK [2] (N = 22) [3]
Neglect	Lack of interest or care in topic	12	6
Time constraints	No opportunities in schedule or curriculum	12	8
Major & background	No connection or relation with discipline	1	2
Personal attitude	Insignificant subject, not important	0	4
Conflicts with goals	Concepts not aligned with career goals	1	0
Personal priority	No incentive	5	6
Accessibility	Courses not offered	9	12
Lack of resources	To make courses available	3	1
Approach of promotion	Over advertising and integrating	0	7
Professor motivation	Concern not expressed by faculty	2	3
Professor not equipped	Faculty lack competence	0	1
Lack of institutional investment	Funds unavailable to develop courses, initiatives	1	0
Institutional priorities	Not an emphasis or strategic goal for campus	1	2

[1] LSK = Low Sustainability Knowledge; [2] HSK = High Sustainability Knowledge; [3] Values in columns three and four give the response frequency.

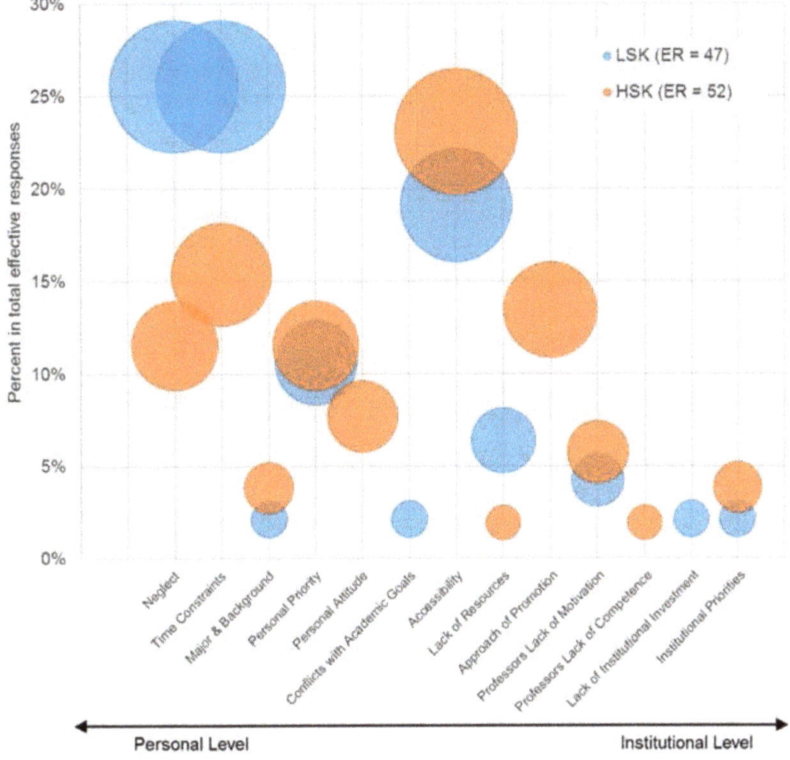

Figure 1. Distribution of percent effective responses corresponding to student-identified barriers to sustainability-based education at Cal Poly. Size of bubbles represent percentage of responses in each knowledge group. (Number of student interviewees N = 39; Total effective responses ER = 99. LSK and HSK indicate low and high sustainability knowledge, respectively).

"I haven't really seen a lot of classes that focus on it, so I'm not enrolling in those classes because I don't know what they're about."

Time Constraints represent a grouping of responses that represent an inability for the student to schedule sustainability-focused courses. As this student (second-year landscape architecture major with high sustainability knowledge) shares:

"Time. I could actively seek out a place to educate myself if I had the time, but with classes and stuff, but if it were a GE, I think we would have time to go, if it was in an actual class that I had to take, then I would be thrilled to have to go to it."

Neglect represents responses where students share attitudes that represent a disinterest in the idea of sustainability, in general. For example, as this student (second-year electrical engineering major, with low sustainability knowledge) shares:

"It's not really on my mind since it's not a problem that's currently affecting me to a great degree..."

Though the top three 'barriers' overlap between groups, those students with 'low' sustainability knowledge most frequently list *Neglect* as a barrier. In fact, *Neglect* is listed twice as often by 'low' as 'high' students. This indicates that *Time Constraints* and *Accessibility* are key barriers across student respondents, and those who lack sustainability-related knowledge may face a self-selection barrier beyond *Accessibility*.

Populating Table 1 data into Figure 1 suggests that most of the student participants appear to have a tendency to address barriers from personal experience, and few student participants can address barriers beyond the personal level. Figure 1 orients barriers on a continuum from personal to institutional level of experience. The bubble diameters display the percentage of responses in each knowledge group describing each barrier. The plot shows that responses derived from participants with low sustainability knowledge significantly skew toward personal level of experience and perception, whereas high-knowledge participant responses lean toward institutional observations.

It is not surprising that students with low sustainability knowledge also appear to neglect this subject, and vice versa. Other common barriers across student groups were mentioned much less frequently. These responses include statements that reinforce the general ideas that *Time Constraints* and competing priorities limit individual ability and/or desire to pursue sustainability-related education. These statements relate to lack of institutional or personal priorities, lack of motivation, and resource constraints (e.g., time, money, available electives). Although the result does not display significant variation, Figure 1 shows that students with better knowledge in sustainability tend to envision barriers from institutional aspects, whereas those with lower knowledge addressed sustainability from a personal perspective.

3.1.2. Student Identified Solutions

Table 2 and Figure 2 summarize student responses identifying solutions to address barriers to sustainability-based education at Cal Poly and the corresponding frequencies. These were generally coded as part of responses to the interview question: "What are some ways to make sustainability education more accessible at Cal Poly?" Any statement wherein a student identified a relevant strategy to enhance participation in sustainability-based education was coded as a Solution. A single transcript could result in multiple solutions.

Despite the fact that several barriers were recognized based on student interviewees' personal perception (Table 1 and Figure 1), all solutions proposed suggested how Cal Poly should tackle the challenges from an institutional level (Table 2 and Figure 2). This discrepancy implies student participants collectively recognize institutional opportunities to promote sustainability yet are less willing to act or make a commitment at a personal level.

Under the "low" and "high" knowledge categories, the top three solution responses in terms of frequency of occurrence are: *Promotion*, *Integration*, and adding a *General Education Option*. Statements coded as *Promotion* include responses identifying the use of fliers, booths, or events to promote sustainability curricula. This result seems ironic, because Cal Poly organized and heavily promoted an Earth Week event during the week prior to our interviews. There is a significant disengagement between ignorance and the suggestion to "promote" sustainability. As a result, the authors are skeptical about the effectiveness of event promotion in overcoming sustainability barriers on campus.

Table 2. Student-identified solutions to improve sustainability-based education at Cal Poly and corresponding frequencies (Total number of student interviewees N = 39).

Solutions	Examples	LSK [1] (N = 17) [3]	HSK [2] (N = 22) [3]
Promotion	Increase awareness, advertising	8	12
Integration	Add material to existing, disciplinary courses instead of creating new courses or requiring added courses	5	11
Ge option	Add material or courses to existing GE requirements	5	9
Ge required	Add and require a new GE course focused on sustainability	2	2
More classes	Add/schedule additional classes	2	1
Link to major/job	Connect importance and benefit of topic to finding a job or disciplinary knowledge	2	6
Activities & events	Create extracurricular events and activities to increase awareness	2	5
Early awareness	Include information in freshman orientation programs	1	5
Smaller class size	Reduce enrollment to promote discussion, inclusion of subject	1	0
Institutional responsibility	Make priority/goal for campus	1	1

[1] LSK = Low sustainability knowledge; [2] HSK = High sustainability knowledge; [3] Values in columns three and four give the response frequency.

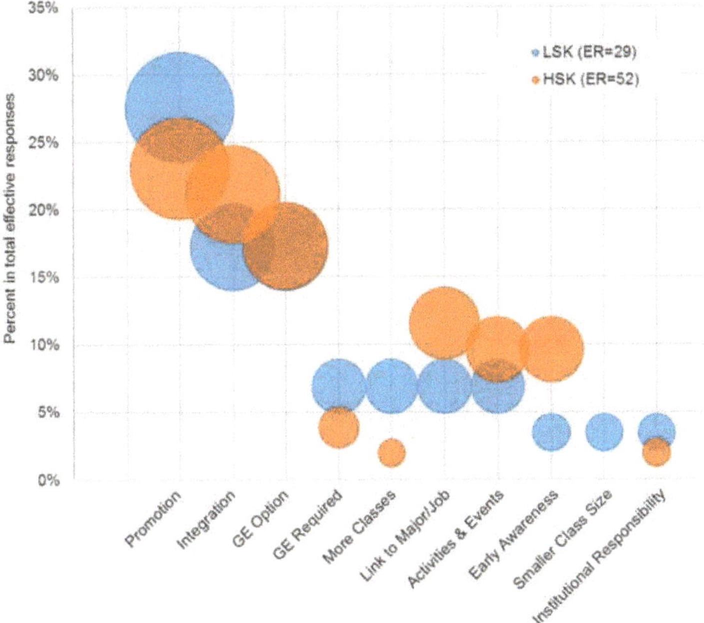

Figure 2. Distribution of percent effective responses corresponding to student-identified solutions to improve sustainability-based education at Cal Poly among low- and high-knowledge (LSK and HSK) student participants. Size of bubbles represents number of responses (Number of student interviewees N = 39; total effective responses ER = 81).

Though the top solutions from the 'low' and 'high' groups once again overlap, the 'high' group lists a greater number of solutions than the 'low' group. Additional responses include *Link to Major/Job*, creating *Activities and Events*, *Early Awareness*, and making it a *General Education Requirement*. The 'low' group also mentioned *Link to Major/Job* and Making it a *General Education Requirement*.

The ideas of *Integration* and *Linking to Major/Job* pose interesting and nuanced solutions derived from the transcript data. *Linking to Major/Job* represents responses where students suggest the importance of sustainability should explicitly link to how it will benefit their future job prospects and careers. For example, as this student (first-year political science major, with high sustainability knowledge) shares:

" . . . if you did it in the curriculum, not like literally taking a sustainability class, but it could be integrated into certain classes, because it's very applicable to different things."

Integration captures responses where students suggest that instead of creating new courses or new requirements, sustainability education should integrate into already existing curricula, as this student (second-year engineering major, with low sustainability knowledge) notes:

"I'm not feeling taking a required class just for sustainability. Maybe incorporate into classes that are already . . . GEs that are already required."

Required general education (GE) courses comprise approximately one-third of the total units for each degree at Cal Poly. Subjects include lower- and upper-division courses in: Communications, sciences and mathematics, arts and humanities, society, and technology. All references to the *General Education Option* category captures student responses suggesting that a viable solution would be adding additional sustainability-related courses to the curriculum as options for completing general education (GE) requirements. This is slightly more popular (judging from the responses) than the solution of adding an additional GE Requirement. A *GE requirement* would modify the curriculum across the campus to ensure all students complete a sustainability course, whereas the *GE Option* solution would provide students with the ability to fulfill a broad GE requirement by choosing to take a sustainability-related course. The requirement is a more rigid, yet broader ranging solution.

3.1.3. Student Identified Barriers and Solutions by College

Although student responses for barriers and solutions were similar regardless of 'low' or 'high' sustainability knowledge, we note some deviation when examining responses by the six academic college units on campus (Table 3). Respondents span all colleges across campus in similar (though not identical) proportion to the make-up of the university. Due to the intensive nature of qualitative data analysis, sample sizes are small. Though our study reveals important and actionable data on barriers and solutions in sustainability education, the sample size falls short in making comparisons across different colleges. The sample has a lower representation of Science and Math students with high sustainability knowledge and generally few participants from the college of Business. Despite the small sample size, data did reach data saturation, which indicates validity of the overall findings [37]. However, we suggest only drawing tentative and university-specific conclusions from these data.

Table 3. Summary of most frequent student-identified barriers and solutions by sustainability knowledge (high "HSK" or low "LSK") and college.

College	Most Frequently Identified Barrier		Most Frequently Identified Solution	
	LSK	HSK	LSK	HSK
All (39)	Neglect (11)	Accessibility (12)	Promotion (8)	Promotion (12)
Agriculture (7)	Neglect (2)	Accessibility (4)	Promotion (2)	Promotion (4)
Architecture (6)	Neglect (2)	Time Constraints (3)	Promotion (1)	Promotion (3)
Business (3)	n/a	n/a	GE Option (1)	n/a
Engineering (7)	Neglect (3) Time Constraints (3)	Accessibility (2) Time Constraints (2)	GE Option (2) Integration (2)	GE Option (3)
Liberal Arts (13)	Accessibility (2) Time Constraints (2)	Accessibility (4)	Integration (2)	Integration (2)
Science and Math (3)	Neglect (3) Accessibility (3)	n/a	Promotion (3)	n/a

Despite these limitations, responses from students do reflect the characteristics of collegiate curriculum. For instance, Engineering students often identify *Time* and *General Education Options* as barriers and solutions because of discipline-specific constraints that limit the freedom of engineering students to pursue elective units outside their professional curriculum requirements. This has undoubtedly contributed to an institutional attitude of efficiency. Thus, one feasible approach to increasing sustainability education might arise by adding formal options to an already restricted curriculum.

Meanwhile, Liberal Arts students most frequently identify *Accessibility* and *Integrating* as barriers and solutions. This observation is interesting, because it may reflect the College of Liberal Arts' approach to a more integrated and holistic liberal arts education, despite the major within the college. Again, all conclusions are tentative given the sample size. However, results may indicate that discipline-specific solutions are needed to promote and improve sustainability education on campus.

3.2. Faculty Responses

3.2.1. Faculty-Identified Barriers

Table 4 summarizes faculty responses to identifying barriers to sustainability-based education at Cal poly, including frequencies (N = 17 faculty interviews). As apparent in Table 4 and Figure 3, the barriers noted by faculty differ considerably from those identified by students. In addition, faculty responses show variation depending on sustainability knowledge.

Table 4. Faculty-identified solutions to improve sustainability-based education at Cal Poly and corresponding frequencies (Total number of faculty interviewees N = 17).

Barriers	Examples	LSK [1] (N = 7) [3]	HSK [2] (N = 10) [3]
Lack of awareness	Unsure if applies to courses taught, unsure how to incorporate	4	3
Lack of competence	No knowledge of subject	4	2
Instructor philosophy	Subject is controversial	1	2
Personal priority	Other concerns or competing requirements have greater importance	3	5
Time constraints	Quarter system or class meeting pattern doesn't allow for additional topics/information	2	2
Accessibility	No courses offered, available	2	2
Discipline restrictive	Course topic restricts opportunity to integrate	3	7
Lack of guidance	No training, support, directions for including in teaching	1	0
No incentive	No personal benefit, no recognition for incorporating into teaching	0	1
Lack of resources	Funding not available for new courses, electives	0	1
Institutional priority	Not listed as an institutional priority/goal	0	1

[1] LSK = Low sustainability knowledge; [2] HSK = High sustainability knowledge; [3] Values in columns three and four give the response frequency.

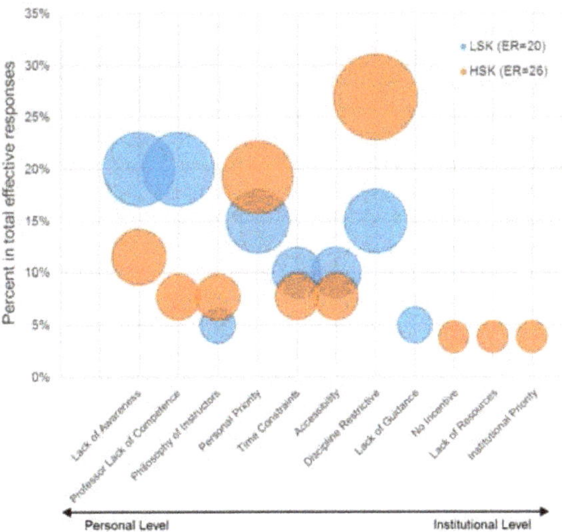

Figure 3. Frequency distribution of faculty-identified barriers to sustainability-based education at Cal Poly among groups with different sustainability knowledge levels (Number of faculty participants: N = 17; effective total responses: ER = 46).

Similar to the pattern observed in Figure 1, faculty members with low knowledge level describe barriers from a personal level, whereas high-knowledge faculty groups address the challenges across the wider spectrum of aspects (Figure 3). Faculty members with 'low' sustainability knowledge cite a *Lack of Knowledge* or *Lack of Awareness* as major barriers towards advancing sustainability across the curriculum. The attitude of neglect or carelessness in sustainability appears to be a noticeable driving force determining the perception of teaching sustainability among faculty participants with 'low' sustainability knowledge. Less frequently mentioned by 'low' knowledge faculty are *Discipline Restrictive, Priority, Accessibility*, and *Time*. 'High' knowledge faculty emphasized *Discipline Restrictive* and *Priority*. The theme of *Discipline Restrictive* intends to capture responses from individuals who note that the subject matter of a class or discipline can restrict the integration of sustainability themes. For example, as this faculty member (associate professor in the Mathematics Department, 12 years, with high sustainability knowledge) notes:

"In some courses, it's more natural to fit in than others. When you're teaching students how to do calculus, you don't need to know about sustainability to do the technical thing, but you can include those topics … It's important to include in topics. Some topics are easier to include than others."

The theme of *Priority* generally refers to statements indicating that sustainability is not prioritized across certain curricula. This theme echoes student concerns regarding their own time (e.g., time to graduate, time for additional units). From a faculty perspective, this theme refers to institutional priorities for curriculum development. The *Priority* theme differs from those responses coded as *Time*, which refers to the ways in which faculty prioritize their own time in light of competing priorities. For example, a professor whose response was coded for *Time* might not choose to prioritize spending her time developing sustainability curricula. This observation is captured in the following response from a faculty participant (professor, Statistics Department, 18 years, with low sustainability knowledge):

"Especially me who's been here forever, I've been here a long time. I don't always change".

3.2.2. Faculty Identified Solutions

Faculty ideas for barrier-specific solutions to improving sustainability-based education were revealed during the interviews when the faculty were asked how identified barriers might be overcome. Table 5 and Figure 4 summarize the faculty-identified, barrier-specific solutions by 'high' and 'low' sustainability knowledge.

Table 5. Faculty-identified solutions to improve sustainability-based education at Cal Poly (N = 17 faculty responses).

Solutions	Examples	LSK [1] (N = 7) [3]	HSK [2] (N = 10) [3]
Hold students accountable	Require assessment of all students	0	1
Clear definition	Provide shared definition and concepts	0	1
More faculty training	Provide instruction, class support	3	0
Promotion	Increase awareness of activities, events, courses	2	1
Integration	Include information in existing courses, make coursework relevant to sustainability	2	4
Systems thinking in teaching	Add additional information on systems to courses	1	0
Link to major/job	Connect/emphasize relevance to finding a job or disciplinary importance	0	3
Early awareness	Include information in freshman orientation programs	0	3
Promote minors	Advertise existing programs on campus that are focused on sustainability	1	2
Ge option	Add materials or courses to existing GE requirements	0	2
Interdisciplinary solutions	Provide courses, opportunities for faculty from other disciplines to teach together	1	1
Institutional responsibility	Make campus priority, strategic goal	1	5
More resources	Provide funding for additional courses/electives	0	2

[1] LSK = low sustainability knowledge; [2] HSK = high sustainability knowledge; [3] values in columns three and four give the response frequency.

As is apparent in Table 5, faculty solutions varied depending on sustainability knowledge. For example, those faculty members with 'low' sustainability knowledge suggested *More Faculty Training* to help to incorporate sustainability themes in the classroom. Other solutions include *Promotion* and *Integration*. Those faculty members with 'high' sustainability knowledge suggest that solutions or improvements are the responsibility of the institution, (*Institutional Responsibility*), which might also include a responsibility for providing increased training to identify those classes that could most easily integrate sustainability-related themes and/or course buyouts to allow for curriculum development.

Similar to students, faculty in the 'high' knowledge group also frequently express *Integration*, *Link to Major/Job*, and *Early Awareness* as a potential solution for overcoming barriers. Considering together the faculty solutions of *More Faculty Training*, *Institutional Responsibility*, and *Integration* into curriculum could provide a blueprint for how universities might advance sustainability education across the curriculum in a way that not only achieves institutional goals related to sustainability education but does so in a way that enhances the desire to teach and learn about sustainability. Moreover, similar to the discrepant pattern between identified barriers and solutions, personal commitment to take action remains questionable.

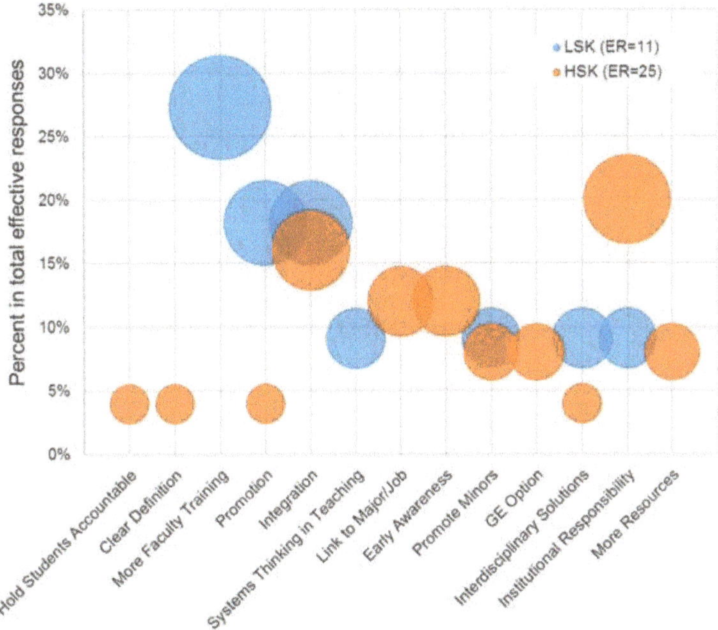

Figure 4. Distribution of percent effective responses corresponding to faculty-identified solutions to improve sustainability-based education at Cal Poly among low- and high-knowledge (LSK and HSK) faculty groups. Size of bubbles represents number of responses (Number of faculty interviewees N = 17; Total effective responses ER = 36).

4. Discussion—Designing Barrier-Specific Solutions

It is well established that interdisciplinary and non-siloed approaches are crucial to the incorporation of sustainability science into the curriculum [10]. However, details of how disciplines work as a system to contribute to sustainably curriculum are less apparent. These results reveal that gaps in the higher education system limit the expansion of an interdisciplinary sustainability curriculum. Findings from this study indicate that both students and faculty with higher sustainability knowledge have the tendency to identify barriers related to institutional accountability. A former study can well support the importance of this aspect, in which its authors found that an institution's internal rules and setting can be the driving force shaping people's behavior and decision-making processes [39]. This is particularly important for the development of solutions, as an institution will need to initiate a holistic strategy to incorporate sustainability into its core values. Being part of the ongoing effort, Cal Poly has incorporated several institutional measures, such as the creation of sustainability learning outcomes for all graduating students. It is important to assess the alignment between institutional measures and actual barriers that need to be addressed.

Noticeably, certain solutions for a singular barrier might be confronted by additional barriers. This is expected because these solutions reflected interviewees' intuitive rationalization, rather than those proposed based on thorough and systematic assessment and reasoning. Similarly, this is also why not all the proposed solutions were ranked from individual to institutional levels (Figures 2 and 4) as was done to identify barriers (Figures 1 and 3). In these cases, all the participants demanded institutional actions, rather than identifying what actions they would be willing to take. This can imply future challenges to make substantial impacts at an individual level even if Cal Poly can implement campus-wide solutions to encourage practicing sustainability. Recognizing personal responsibility to engage in learning sustainability needs to be addressed as one of the key steps to either promote a top–down or bottom–up approach to facilitate learning and teaching in sustainability. In the meantime, we also view this challenge as a new opportunity for establishing the momentum to promote sustainability education. A former study states that one of the possible tactics to improve personal awareness in sustainability is to create a "cognitive dissonance between individuals' values and behaviors" [40]. Therefore, findings from our study can provide ready input to address the dissonance between recognition of challenges and willingness to engage in making changes at a personal level, especially among the groups with lower sustainability knowledge.

Other studies note that expanding sustainability education and behavior must also address personal concerns and take into account increasing awareness of issues, messaging solutions, increasing convenience, and incentivizing change [41]. This resonates with the study's initial understanding of barriers, that the design of specific solutions may be best supported by institutional initiatives and availability to resources. Future areas of study will seek to better understand the role of faculty programs in the design and incorporation of barrier-specific solutions in higher education. As part of an effort to develop solutions to faculty-identified barriers, the authors will pursue a program of solutions, many of which could provide incentives in the form of modest stipends and opportunities for professional development and publication that begin to address identified barriers and solutions. These include:

(1)   Developing workshops that provide space, training, and education focused on the development of new course material for existing courses. Workshops would promote identified solutions "integrating" and "linking" with the goal of increasing the number of sustainability-related courses. Workshop activities would include the review of learning outcomes and course structure, while designing additional course materials.

(2)   Developing a sustainability learning community to inform faculty from a wide range of disciplines with low sustainability knowledge. Faculty learning communities are established solutions for infusing sustainability concepts into the curriculum and increasing the number of sustainability-related courses [20]. This focused community will align with the barriers and solutions of "discipline restrictive", "more training", and "lack of knowledge". A focus on overcoming seemingly discipline-specific barriers and including sustainability would be addressed through interdisciplinary approaches. The learning community participants would pair faculty with low and high sustainability knowledge as a means to address the identified "lack of knowledge" barrier with "more training" solution.

(3)   Developing a year-long "Community of Practice" consisting of a group of interdisciplinary educators with high sustainability knowledge and a shared interest in sustainability in higher education. The community will serve as a platform to exchange ideas, insights, and practices regarding sustainability in education. Meanwhile, it can facilitate the assessment and improvement of sustainability education across university curriculum.

5. Conclusions

A well-known core principle of sustainability education holds that interdisciplinary approaches are crucial, but the details of how these disciplines work as a system to contribute to an overarching sustainability curriculum are less apparent. These results reveal the gaps in Can Poly that limit the expansion of an interdisciplinary sustainability curriculum, which can resonate with institutes with similar academic setting. Cal Poly's experience indicates that the implementation of a sustainability-related curriculum must rely on multidimensional strategies and approaches. Recognition of barriers ranging from the personal to the institutional level will aid in the design and implementation of any expanded sustainability-related curriculum or program initiatives. Research results confirm that barriers to sustainability education exist across disciplines, participants with varying levels of sustainability knowledge, and among student and faculty groups.

The results from both students and faculty show that participants with higher sustainability knowledge have the tendency to identify barriers to sustainability education by holding organizations accountable, while participants with lower sustainability knowledge have the tendency to identify barriers to sustainability education by holding themselves accountable. This finding well depicts the mentality of students and faculty in supporting sustainability education. The finding also indicates that a campus seeks change to include sustainability education must initiate a holistic strategy to incorporate sustainability into its core values from the institutional level, with which individual awareness can be better promoted. This study's results confirm both approaches will be required to address perceived barriers to implementation. Continued research and understanding of the factors impeding the implementation of sustainability education could help students, faculty, and institutions to develop those holistic strategies.

Author Contributions: Conceptualization, all authors; Methodology: B.P., Y.-W.C., and D.N.; Writing—original draft preparation, B.P., Y.-W.C., D.N., and R.O.; Writing—review and editing, all authors; Visualization, Y.-W.C.; Green Campus Team supervision, K.S.; Funding acquisition, B.P., R.O., and Y.-W.C.

Funding: The APC was funded by USDA NIFA, Award No: 2017-70003-26380.

Acknowledgments: We thank Patrick Sullivan, Director, Center for Teaching, Learning & Technology, Cal Poly State University San Luis Obispo for his role in organizing the learning community where this work originated and continued support throughout the project. We acknowledge the Green Campus Team's participation in data collection via focused interviews.

Conflicts of Interest: The authors declare no conflict of interest. The funders had no role in the design of the study; in the collection, analyses, or interpretation of data; in the writing of the manuscript, or in the decision to publish the results.

References

1. Gale, F.; Davison, A.; Wood, G.; Williams, S.; Towle, N. Four impediments to embedding education for sustainability in higher education. *Aust. J. Environ. Educ.* **2015**, *31*, 248–263. [CrossRef]
2. Moore, J. Barriers and pathways to creating sustainability education programs: policy, rhetoric and reality. *Environ. Educ. Res.* **2005**, *11*, 537–555. [CrossRef]
3. California Polytechnic State University. Cal Poly Strategic Plan Draft. Available online: http://vision2022planning.calpoly.edu/wp-content/uploads/2016/10/vision-2022-strategic-plan.pdf (accessed on 30 December 2017).
4. Academic Senate California Polytechnic State University. AS-598-03 Resolution in Support of Signing the Talloires Declaration. Available online: http://digitalcommons.calpoly.edu/senateresolutions/599/ (accessed on 27 September 2017).
5. Sterling, S. *Sustainability Education: Perspectives and Practice Across Higher Education*; Taylor & Francis: New York, NY, USA, 2010.
6. Academic Senate California Polytechnic State University. AS-688-09 Resolution on Sustainability Learning Objectives. Available online: http://digitalcommons.calpoly.edu/senateresolutions/689/ (accessed on 27 September 2017).
7. California State Universities. California State University Sustainability Policy (RJEP/CPBG 05-14-01). Available online: http://www.calstate.edu/cpdc/sustainability/policies-reports/documents/JointMeeting-CPBG-ED.pdf (accessed on 27 September 2017).
8. Dmochowski, J.E.; Garofalo, D.; Fisher, S.; Greene, A.; Gambogi, D. Integrating sustainability across the university curriculum. *Int. J. Sustain. Higher Educ.* **2016**, *17*, 652–670. [CrossRef]

9.   Ferrer-Balas, D.; Adachi, J.; Banas, S.; Davidson, C.; Hoshikoshi, A.; Mishra, A.; Motodoa, Y.; Onga, M.; Ostwald, M. An international comparative analysis of sustainability transformation across seven universities. *Int. J. Sustain. Higher Educ.* **2008**, *9*, 295–316. [CrossRef]

10.   Fisher, P.B.; McAdams, E. Gaps in sustainability education: The impact of higher education coursework on perceptions of sustainability. *Int. J. Sustain. Higher Educ.* **2015**, *16*, 407–423. [CrossRef]

11.   Alghamdi, N.; den Heijer, A.; de Jonge, H. Assessment tools' indicators for sustainability in universities: An analytical overview. *Int. J. Sustain. Higher Educ.* **2017**, *18*, 84–115. [CrossRef]

12.   Lundquist, L.L.; Lucero, K.; Cox, H. Sustainability Knowledge Assessment at a Large, Regional, Minority-Serving Institution. *J. Sustain. Educ.* **2018**, *19*, 1–6.

13.   Horvath, N.; Stewart, M.; Shea, M. Toward Instruments of Assessing Sustainability Knowledge: Assessment development, process, and results from a pilot survey at the University of Maryland. *J. Sustain. Educ.* **2013**, *5*, 27.

14.   Mintz, K.; Tal, T. The place of content and pedagogy in shaping sustainability learning outcomes in higher education. *Environ. Educ. Res.* **2018**, *24*, 207–229. [CrossRef]

15.   Lozano, R.; Barreiro-Gen, M.; Lozano, F.J.; Sammalisto, K. Teaching Sustainability in European Higher Education Institutions: Assessing the Connections between Competences and Pedagogical Approaches. *Sustainability* **2019**, *11*, 1602. [CrossRef]

16.   Ralph, M.; Stubbs, W. Integrating environmental sustainability into universities. *Higher Educ.* **2014**, *67*, 71–90. [CrossRef]

17.   Miller, T.R.; Muñoz-Erickson, T.; Redman, C.L. Transforming knowledge for sustainability: towards adaptive academic institutions. *Int. J. Sustain. Higher Educ.* **2011**, *12*, 177–192. [CrossRef]

18.   Blanco-Portela, N.; Benayas, J.; Pertierra, L.R.; Lozano, R. Towards the integration of sustainability in Higher Education Institutions: A review of drivers of and barriers to organisational change and their comparison against those found of companies. *J. Clean. Prod.* **2017**, *166*, 563–578. [CrossRef]

19.   Hooey, C.; Mason, A.; Triplett, J. Beyond greening: Challenges to adopting sustainability in institutions of higher education. *Midwest Q.* **2017**, *58*, 280.

20.   Natkin, L.W.; Kolbe, T. Enhancing sustainability curricula through faculty learning communities. *Int. J. Sustain. Higher Educ.* **2016**, *17*, 540–558. [CrossRef]

21.   Sammalisto, K.; Lindhqvist, T. Integration of sustainability in higher education: A study with international perspectives. *Innov. High. Educ.* **2008**, *32*, 221–233. [CrossRef]

22.   Sterling, S. Higher Education, Sustainability, and the Role of Systemic Learning. In *Higher Education and the Challenge of Sustainability*; Springer: Berlin, Germany, 2004; pp. 49–70.

23.   Harrell, M.C.; Bradley, M.A. *Data Collection Methods. Semi-Structured Interviews and Focus Groups*; RAND National Defense Research Institute: Santa Monica, CA, USA, 2009.

24.   Agar, M. Stories, background knowledge and themes: Problems in the analysis of life history narrative. *Am. Ethnol.* **1980**, *7*, 223–239. [CrossRef]

25.   Crang, M.; Cook, I. *Doing Ethnographies*; SAGE Publications: Thousand Oaks, CA, USA, 2007.

26.   Dey, I. *Qualitative Data Analysis: A User Friendly Guide for Social Scientists*; Routledge: London, UK; New York, NY, USA, 1993; p. 62.

27.   Cleary, M.; Horsfall, J.; Hayter, M. Data collection and sampling in qualitative research: does size matter? *J. Adv. Nurs.* **2014**, *70*, 473–475. [CrossRef] [PubMed]

28.   Ellis, D. Modeling the information-seeking patterns of academic researchers: A grounded theory approach. *Libr. Q.* **1993**, *63*, 469–486. [CrossRef]

29.   Ryan, G.W.; Bernard, H.R. Techniques to identify themes. *Field Methods* **2003**, *15*, 85–109. [CrossRef]

30.   Guba, E.G. *Toward a Methodology of Naturalistic Inquiry in Educational Evaluation*; CSE Monograph Series in Evaluation 8; University of California: Los Angeles, CA, USA, 1978.

31.   Bogdan, R.; Biklen, S. *Qualitative Research for Education: An Introduction to Theory and Practice*; Allyn and Bacon, Inc.: New York, NY, USA, 1982.

32.   Strauss, A.; Corbin, J. *Basics of Qualitative Research: Grounded Theory Procedures and Techniques*; Sage Publications: Thousand Oaks, CA, USA, 1990.

33.   Rossman, G.B.; Rallis, S.F. *Learning in the Field: An Introduction to Qualitative Research*, 3rd ed.; Sage Publications: New York, NY, USA, 2011.

34. Creswell, J.W. *Qualitative Inquiry and Research Design: Choosing Among Five Approaches*; Sage publications: Los Angeles, CA, USA, 2013.
35. Syed, M.; Nelson, S.C. Guidelines for establishing reliability when coding narrative data. *Emerg. Adulthood* **2015**, *3*, 375–387. [CrossRef]
36. Taylor, J.; Gilligan, C.; Sullivan, A.M. Missing voices, changing meanings: Developing a voice-centered, relational method and creating an interpretative community. In *Feminist Social Psychologies: International Perspectives*; Wilkinson, S., Ed.; Open University Press: Buckingham, UK, 1996; pp. 233–257.
37. Fusch, P.I.; Ness, L.R. Are we there yet? Data saturation in qualitative research. *Qual. Rep.* **2015**, *20*, 1408–1416.
38. Carew, A.L.; Mitchell, C.L. Characterizing undergraduate engineering students' understanding of sustainability. *Eur. J. Eng. Educ.* **2002**, *27*, 349–361. [CrossRef]
39. Sterman, J.D. Modeling managerial behavior: Misperceptions of feedback in a dynamic decision making experiment. *Manag. Sci.* **1989**, *35*, 321–339. [CrossRef]
40. Pappas, J.B.; Pappas, E.C. The Sustainable Personality: Values and Behaviors in Individual Sustainability. *Int. J. High. Educ.* **2015**, *4*, 12–21. [CrossRef]
41. Perrault, E.K.; Clark, S.K. Sustainability attitudes and behavioral motivations of college students: Testing the extended parallel process model. *Int. J. Sustain. High. Educ.* **2018**, *19*, 32–47. [CrossRef]

Article

Sustainability and Indicators of Newly Formed World-Class Universities (NFWCUs) between 2010 and 2018: Empirical Analysis from the Rankings of ARWU, QSWUR and THEWUR

Zhimin Liu [1], Goodluck Jacob Moshi [2,*] and Cynthia Mwonya Awuor [2]

[1] Higher Education Institute, Nanjing Agricultural University, Nanjing 210095, China; liuzhimin@njau.edu.cn
[2] College of Public Administration, Nanjing Agricultural University, Nanjing 210095, China; awuormwonya@gmail.com
* Correspondence: goodluck.jacob@gmail.com or 2017209033@njau.edu.cn

Received: 25 February 2019; Accepted: 10 April 2019; Published: 14 May 2019

Abstract: In the 21st century, sustainability and indicators of world-class universities have come within the scope of an academic cottage industry. The complex problem of university sustainability implies a big challenge for countries and educators to implement important strategies in an integrated and comprehensive way. This paper highlights and analyzes the sustainability indicators of universities included as newly formed world-class universities (NFWCUs) in the top 100 from 2010 and 2018. The integration of three global ranking scales—the Academic Ranking of World Universities (ARWU), the Quacquarelli–Symonds World University Ranking (QS) and the Times Higher Education World University Rankings (THEs)—allows us to minimize the impact of the methodology used. This study integrates regression analysis by using statistical grouping, case studies and normative analysis. Our principal findings are as follows: among the commonly ranked top 100 universities in 2018, the ARWU, QS and THE counted 57, compared with 47 in 2010. Thus, comparing 2010 and 2018 shows that 44 of the universities appeared simultaneously in ARWU, QS and THE rankings and maintained a sustainable position in any ranking system in the family of top 100 groups. Three lower-ranked NFWCUs in the hybrid list for 2010 lost their ranking and did not appear in the group of top 100 universities in 2018, which are covered by some catch-up and young universities. The NFWCUs were from US, Australia, China, Singapore, Germany and Belgium. By systematic comparison, the US and UK continued to dominate the stability of NFWCUs in 2010 and 2018. The key sustainability indicators include a high concentration of talent, abundant resources to offer a rich learning environment and conduct advanced research. Generally, the factors were negatively associated with ranking suggesting that a higher score result in top ranking and vice versa. Teaching, research, citation and international outlook were negatively correlated with THE ranking in 2018. Similarly, Alumni and PUB were negatively associated with ARWU ranking in 2018. All factors except international student ratio were significantly correlated in QS ranking either in 2010 or 2018, where negative association was observed. The significant contribution of our study is to highlight that for the sustainability of universities, it is necessary to have an increasing emphasis on the effectiveness and efficiency of government-supported research, stability of investments and more approaches to employ international initiatives. The results also confirm the appropriate governance, developing global students and place emphasis on science and technology as additional factors in the approaches of pathways to NFWCUs, with delivery of outstanding educational programs and comprehensive internationalization as a key indicator for performance improvement and global university ranking systems.

Keywords: world-class universities; newly-formed; sustainability; rankings; indicators; initiatives

---

1. Introduction

Over the past few decades, the concept of world-class universities (WCU), also called globally competitive, elite, early formed or traditional universities, has emerged. At present, this term has become a catchphrase, not only in terms of improving the quality of learning and research in higher education but also for sustainability and developing the capacity to compete explicitly in the global higher education marketplace [1–5]. Consequently, since the appearance of global university rankings in 2003, the goal of establishing and changing universities to WCU is to be able to compete in the global knowledge economy [6,7]. Furthermore, the aim is to train creative human resources and advance national development through the acquisition, adaptation and creation of an advanced knowledge-based economy. However, in many ways the development of higher education has facing international and domestic pressures. In today's globalized world, several countries have removed hurdles to join a global battle to establish their institutions as WCU and to promote the knowledge-economy society [8,9]. At the same time, establishing world leading position has been ambition of both government and top universities.

Meanwhile, countries in the western world, particularly the United States (US) and the United Kingdom (UK) and several Asian countries are launching programs for university authority, structure, system and organizational goals to enhance the sustainability and competitiveness of their universities. According to Hezelkorn and Li, assuring world-class university status is the wish for every nation [8]. Subsequently, a variety of reforms and development strategies at both the national and institutional level have been defined and observed. These reforms have also been strengthened and intensified with the propagation of international league tables [7,10–13]. According to Mok, to help their universities achieve this exclusive status, many countries and regions have also implemented a number of funding initiatives.

Generally speaking, high-profile university excellence initiatives are becoming increasingly popular research and development policies globally. These initiatives were institutionalized in more than 28 nations from 1995 to 2017, including in Europe, Latin America and Asia. Such initiatives include Germany's Excellence Initiative; Japan's Centers of Excellence for the 21st Century; South Korea's BK21; China's 211, 985 and 2.0 Projects; Singapore's Research Centers of Excellence; and Australia's ARC Centre of Excellence. The aim is to help their countries and elite universities achieve global competitiveness and higher-quality colleges and universities [14,15]. Again, according to Mok and Salmi, higher education worldwide has been experiencing a continuous trend of transformation shaped by different types of international drivers which operates in a constant flux of globalization. With the strong intention to rank highly in global university leagues, governments are exerting serious efforts to boost their universities global competitiveness. The goal for establishing WCU and changing existing universities to WCU is to be able to compete in the global knowledge economy [16]. Thus, it is important not only to improve the quality of research and teaching but also more significant to develop the competency to compete in the global market economy [17,18]. The term "excellence" has gained in relevance in recent years to sustainability of higher education which are oriented to international ranking criteria. Given this, the quest for WCUs can be understood as an institution that transcends culture and education [19]. They are a "point of pride and comparison among nations that view their status in relation to other nations" [18,20]. In this sense, developing WCU is both a national and local strategy in a worldwide context and sensitive to universal referents and objectives. Several reports confirm that there is no clear definition and statement about the context of what constitutes a world-class university [9,14,21]. Three complementary sets of dimensions of action are in play and can be found in most NFWCUs in the last eight years: favorable governance features that encourage strategic vision, innovation and flexibility; abundant resources to offer a rich learning environment and conduct advanced research; and a high concentration of talent [22,23].

Indeed, all universities are desirous of achieving world-class status; however, the paradox of WCU, as Altbach has succinctly and accurately observed, is that "everyone wants one, no one knows what it is and no one knows how to get one." Thus, while the goal of WCU status is clear the definition of world-class status is not. Therefore, a set of complementary factors are needed to define the sustainability of NFWCUs [24,25]. In line with the work of these authors, the aim in paths of governance to transform WCU is rooted not just in thoughtful reflections but also in the symbolic role of such higher education institutions (HEIs). The facts tell us that the path for WCU is not something that we can fix overnight. In addition, it is not something that universities can proclaim themselves, as Salmi point out: "becoming a member of the exclusive group of WCU is not achieved by self-declaration; rather, elite status is conferred by the outside world on the basis of international recognition" [5,11]. We have to learn from the institutional characteristics of both early and newly formed WCU as a starting point [26,27]. The global university rankings make competition between states visible and thus are commonly recognized as indicators of success due to excellence-driven policy goals [28]. Hence, strong engagement toward becoming WCU is also a trend nowadays.

Since the 2000s, a new era in HEIs is characterized by global competition in which university ranking systems have assumed importance. Since the emergence of global university rankings in 2003, the interrelated connection between WCU and university rankings has been a heated topic around the world. Thus, global university rankings have become a useful measure of university competitiveness and are commonly used to undertake quality assurance of higher education systems, their strength and weakness, internal analysis, academic performance of top-ranked higher education and their variations [29,30]. It has been found that ranking systems often serve as a proxy for world- class status and despite of many issues with rankings, most are driven by the purpose and concept of university quality. Various methodologies are used in these ranking systems and their influence is striking. Ranking has become unavoidable and it will remain part of academic life.

The ranking of universities has affected HEIs, as ranking appears to strengthen or grant visibility to some universities and expose perceived challenges at the system and institutional level [19,31–33]. Ultimately, the governments of several countries have made a strategic choice for relative targets and formed new visions and missions, university structures and functions, thus eventually turned to lead HEI policy change from the national system level to form top-ranked universities [6,34,35]. According to Teodoro (2019) the compilation and use of university rankings is a widely debated issue. There was some previous interest in this field: in 2003 the Academic Ranking of World Universities (ARWU) by Shanghai Jiao University was published and then the National Taiwan University Ranking (NTU), the University Ranking of Academic Performance (URAP), the Centre for Science and Technology Studies, Leiden Ranking and Scimago Institutions Rankings. Other rankings take into account dimensions not exclusively related to research data, as they incorporate opinion surveys, such as the Times Higher Education World University Rankings (THE) and the Quacquarelli–Symonds World University Ranking (QS). The Multidimensional Global University Ranking (UMultitank) endeavors to resolve some of the criticism commonly leveled at the above rankings.

Rankings have been the object of much criticism [16,19,36]. Composite rankings have numerous weak points, as indicated by several authors. There are numerous major global and national ranking systems that describe the performance of universities in the world; among them, we chose three prominent systems: the Academic Ranking of World Universities (ARWU), the Quacquarelli–Symonds World University Ranking (QS) and the Times Higher Education World University Rankings (THE), which are the most cited and commonly used [18,37]. World rankings aspire to include the most relevant universities worldwide. Thus, composite rankings summarize several weighted indicators and assign one score, which is then used to offer a classification. Rankings should be used with caution, taking into account how they are developed [38].

Approaches to Common Composite of World University Ranking Systems

Rankings provide great insight into the strength and shifting fortunes of individual research universities, although the compilation and use of university rankings is currently a widely debated issue. University rankings, despite their limitations, have been considered to be important by many stakeholders [37]. It is evident that many, though not all, institutions of higher education take rankings seriously for the purpose of accountability, evaluation and strategic planning. At the individual level, rankings are potentially useful in providing a comparable and clear summary of information for students to select appropriate universities and destinations for their education. At the institution level, global rankings have become a marketing feature in university activities [39]. Rankings strongly influence the behavior of universities, as their presence in ranking tables alone heightens their profile and reputation, encouraging the collection and publication of reliable national data on higher education. Internationally, rankings provide an informal measure of a country's ability to compete in a knowledge-based world economy due to the emphasis on research output. Therefore, several authorities and universities have put forward policies for creating WCU and take university rankings seriously. According to Pusser and Marginson, rankings are an essential sustainability instrument for the exercise of power in the service of dominant norms in global higher education.

It is clear that rankings continue to grow in popularity and gain interest by policy makers. While the world is obsessed with rankings, there are several cautions voiced against them. Rankings have been the object of much criticism with regard to the measurement indicators, the institutions being measured and the diversity of features to make comparisons [40]. A consensus has emerged and there are plenty of discussions in the literature that despite the wide diffusion of university rankings in recent years, their methodology and usefulness are not exempt from critics, especially the validity of the results and suitability of the variables used. Several studies have also criticized that the selection of indicators and their weighting, data processing and transparency are strongly questioned. Again, every ranking is based on the availability of comparable data and is built on the subjective judgment (over indicators and weightings) of its compilers. In addition, the indicators used promote the presence of a minimal number of research institutions from peripheral countries and neglect some other types of college and universities. [7,29]

University ranking systems provide a comparative analysis of university performance and characteristics at a global level, regional level (e.g., Europe, Latin America), national level (e.g., US), development level (e.g., developing countries) and particular group level but several ranking systems are available for several purposes and only a few attract many and are commonly recognized. In line with the above, this study considers the three rankings of ARWU, QS and THE due to their reputation, characteristics, influence and limitations. There are more than 28,077 (http://www.webometrics.info/en/node/54) HEIs all over the world and the top 100 universities account to 0.004%, which should be recognized worldwide due to their sustainable performance indicators and academic excellence (www.topuniversities.com). This paper addresses the gap in sustainability indicators of NFWCUs in the last eight years. Then, if we agree that any universities listed on the top 100 in the global ranking systems of ARWU, QS and THE as world-class, this research will define those that have fresh appeared as newly formed (or progressive-type) world-class universities (NFWCUs).

2. Research Objectives and Goals

The general goal of this study is to examine sustainability indicators and establish present evidence on the empirical distribution of and contributions across NFWCUs in the last eight years according to the top world ranking systems of ARWU, QS and THE. In today's globalized world, higher education institutions worldwide are in a period of difficult transition and transformation shaped by different types of internal drives. This situation is mostly affected by globalization, the advent of massification, undefined relationships between the state and universities and current technologies, among other factors [41]. In this view, an increasingly pressing agenda and priority of many governments is to make sure that their elite universities are operating at the cutting edge of intellectual and scientific development [42]. In the 2000s, several countries joined the global pursuit of building WCU in the knowledge society [43,44] and currently, several countries are creating WCU as an essential part of their higher education reform [45]. In recent years, building a sustainable world-class university has been the dream of some nations and emerging approaches have attracted attention from scholars. However, up to now few countries have had the possibility to turn the dream into reality.

According to Peter Senge, within the unpredictable future business environment and accelerated development of the knowledge economy, universities need to increase their knowledge generation and knowledge transfer in society. Universities should strive to become learning organizations [46–48]. They must enlarge their focus on research and their traditional mission of teaching and learning. Universities should adopt a new paradigm to monitor the needs of different stakeholders. This means they should create adaptation knowledge to produce generative knowledge and to become learning organizations and at this stage the government should become a strategic driving force [3,48] for universities and a powerful integrator. Since all main functions of a university are related to knowledge transformation and distribution, the university becomes a knowledge-intensive organization dominated by intellectual capital [49].

Since the beginning, the university has always been a cultural and moral symbol of social communities. The perspectives, preoccupations, activities and goals of universities have significantly changed over time, as have their roles and strategies. Nowadays, universities are viewed as knowledge providers, innovation facilitators, promotors of entrepreneurial talent, economic and civic leaders and mostly knowledge pioneers of the creative commons [49]. Countries around the world have faced and continue to face challenges in global university ranking competitiveness and the increased importance of having top listed universities.

Higher education governance continues to grapple with increased competition under pressure from the curriculums they offer and the university ranking systems used. In developing countries, higher education faces several challenges, such as inadequate funding, outdated curriculums and governance structure. Whereas many developed countries have made changes in their higher education system to deal with the growing pressure of global competitiveness of universities across the world, developing countries continue to lag behind [50]. In many developing countries, the lack of reformation and innovative changes in management practices and governance models in higher education is stifling high-quality teaching, reputation, productive faculty, excellent students, flexible administration, full funding and internal engagement.

The literature indicates that WCU involves a complementary set of three factors. These factors can be observed in most top-ranked universities. Typically, these factors are a high concentration of talent among both faculty and students; abundant resources to support a rich learning environment and conduct advanced research; and constructive governance features that encourage strategic vision, innovation and flexibility to enable institutions to make autonomous decisions and manage resources without being burdened by bureaucracy [27].

In the rest of the world, world-class universities have in general kept pace with reforms in governance structures and academic systems. A number of researchers [46,51,52] have analyzed the contributions of university governance and educational systems to a world-class reputation. Most handful of recent scholars indicate that the key determinants that have a strong influence on the development of world-class universities include excellence in research, productive faculty, excellent students, flexible administration, adequate facilities, full funding and international engagement. According to the definition of world-class universities by Philip G. Altbach, the criteria include governance structure and productive faculty. Therefore, when we try to understand the paths of governance for NFWCUs, we must keep in mind that each nation is unique, and its universities can survive and prosper if they meet the needs of the society.

In the 21st century, the top universities are under tremendous pressure and have upgraded to world research once in a relatively short time. They have made many policies at unprecedented speed. Other universities and their governing agencies have also set such an aim. The movement for NFWCUs has become a national role, with the investment of many resources. The question is, can universities transform into NFWCUs only with a strong will and high financial investment? The answer is unknown. In line with the criteria above, the central objective of this study is to assess the driving indicators for variations and contributions of NFWCUs based on characteristics and distributions among countries. Further, the study suggests common measurable strategies applied by these universities to achieve success.

3. Research Questions and Methodology

To guide our inquiry related to this paper and reach the main findings, this study integrates research methods using statistical groupings, case studies and normative analysis. The following research questions were used to guide our study:

RQ1: What are the newly formed world-class universities (NFWCUs)?
RQ2: What are the common shared sustainability indicators of NFWCUs?
RQ3: What are their variations and contributions in the future?
RQ4: What are the research productivity and government initiatives? Specifically, how have these governments and HEIs construct and transformed to become globally competitive.

Although there still are no generally accepted definitions or clear standards for quantified assessment of world-class universities in the world, there are various indexing systems for assessing universities. Each has its own motives and purpose as well as shortcomings and limitations. It is therefore a focus on being world-class universities has become synonymous with relative efforts by universities to move upwards in the international league tables. Thus, league tables are taken as symbolic and powerful indicator to prove striving to NFWCUs.

To achieve the aims of the research questions and develop explanations of NWFCUs as a social phenomenon, according to Liu (2015) and Teodoro (2018) this study adopted a qualitative research design. Qualitative research entails opinions, feelings and experiences. It aims at describing social aspects as they occur and taking a holistic approach, thereby extricating critical experiences from NFWCUs and facilitating the optimization of excellence sustainability and indicators in the last eight years by providing suitable suggestions. According to Pusser and Marginson (2013), rankings are an important element in global higher education. In our current study, we use three known rankings with their relative indicators. By synthesizing these rankings, NFWCUs are obtained. This reduces the heterogeneity of single rankings and weaknesses as indicated by Soh (2017b). Keeping in mind that every ranking has some statistical problems, this study implements and integrates four research methods: case studies, statistical groupings, comparative analysis and normative analysis. The integration of these four research methods with the global ranking systems of ARWU, QS and THE provides an opportunity to build a greater assortment of divergent views on the issue studied, increase the credibility of the study and offer stronger inferences. Taking this into consideration, this is the important reason why this study integrates four research methods.

3.1. Methods of Data Collection

Our study retrieved data from three databases that are most publicly visible and well recognized globally for university rankings: ARWU, QS and THE, in which the factors used to determine university rankings are provided. It was evident that each ranking system has unique factors, hence this study compared university rankings among the three databases. To be included in the study, a university must appear among the top 100 universities in all three ranking systems. The universities were selected from 2010 and 2018.

The first step is to obtain data from ARWU, QS and THE using the values of indicators they use and global scores; after that, we input the top 100 universities that appear in all three ranking systems. Finally, universities with all characteristics are identified and analyzed by a regression model.

We selected NFWCUs according to a set of relative criteria to find current NFWCUs from the last eight years (2010–2018) based on the ARWU, QS and THE ranking systems. We then combined the current data from ARWU, QS and THE on the top 100 universities ranked previously by using their summation and total mean so as to know the order of overall top-ranked universities and a hybrid list of NFWCUs. Finally, the overlapping part was synchronized to drive classifications, disparities and distributions among countries.

It should be noted that all university ranking information related to institutional practices are available from the web pages of the three ranking systems: http://www.shanghairanking.com/, https://en.wikipedia.org/wiki/ and the NFWCUs' home pages. By considering the adopted system, the four methods were used to integrate the three ranking systems to produce a hybrid list of current NFWCUs in the last eight years. The timeframe appeared to be a period during which the governance of several universities-initiated excellence initiatives. Moreover, such an approach allowed us to deal with the criticism of inconsistent ranking results because of different methodologies used by the three selected systems; if we simply look a single ranking list, it is not enough to fully judge the quality and level of a university, so it seemed that integrating the top three ranking systems to evaluate the sustainability indicators of NFWCUs would be more comprehensive for this study. Following the above-explained methods, a summary of the process and the population of the study and sample design are illustrated as follows.

3.1.1. Academic Ranking of World-Universities (ARWU)

The Academic Ranking of World Class University (ARWU; www.arwu.org) undertaken by Centre for World-Class Universities of the Institute of Higher Education of Shanghai Jiao Tong University, China (hence often known as Shanghai Jiao Tong Ranking). It was firstly started in 2003 as the first world global ranking of universities. Thus, the origin was to establish the global standing of top Chinese universities but soon attracted world attention. This ranking has attracted much interest from around the world as it ranks the best world's 500 top universities from 41 different countries annually. Since 2009, the Academic Ranking of World Universities (ARWU) has been published and copyrighted by Shanghai Ranking Consultancy, a fully independent organization on higher education intelligence. Although the initial purpose of ARWU was to find the global strength standing of top Chinese universities, it has attracted a great deal of attention from universities, governments and public media worldwide. Moreover, a survey on higher education published by The Economist in 2005 commented ARWU as the most widely used annual ranking of the world's research universities. The significant influence of ARWU is that its methodology is scientifically sounds, stable and transparent. Its content is widely cited and employed as a starting point for identifying national strengths and weakness. [36,45,53]. Being as the first digital instrument of its kind, it is one of the causes of some of the key features of global academic competition as we know it today.

3.1.2. Quacquarelli-Symonds World-University Ranking (QS-WUR)

The Quacquarelli-Symonds World-University Ranking (QS; www.topuniversities.com) is an annual publication of university rankings. Previously known as Times Higher Education-OS World University Rankings, being similar in many ways; this is a global career & education company specializing in education and study abroad. From 2004 to 2009 THE and QS jointly published the same rankings. The fundamental criteria used were research, employability, teaching & internationalization. However, in 2010, after separating from THE, QS continued with virtually the similar criteria for its annual rankings, on the whole with some changes in the weight of the requirements ranks about 700 of the world's top universities [33,54]. Being the only international ranking to have received International Ranking Expert Group(IREG) approval, the QS is viewed as one of the three most-widely ready university rankings in the world [55,56].

3.1.3. Times Higher Education World University Ranking (THE-WUR)

Times Higher Education World University Ranking (THEs; www.timeshighereducation.com/). Is an annual publication of university rankings. Actually, as early as 2004 Times Higher Education (THE) partnered with Quacquerelli Symonds (QS)-QSWUR to publish a new set of world university rankings. However, by in 2010, THE & QS ended their partnership, as each one deciding to release its ranking with two independent programs. In 2010, THE using new data supplied by Thomson Reuters (a business data provider headquartered in New York) published its rankings using a different methodology. THEs currently uses 13 performance indicators grouped in 5 areas of indicators, most of the data that are being provided by the institutions. Thus, ranks world's 400 top universities annually comprise of the world's overall subjects and reputational rankings. THE is often considered as one of the most widely observed university rankings and praised for having a new, improved ranking methodology. THE ranking is criticized however on having and relying on subjective reputation survey. [53]. Table 1 provides details on the indicators and weight for each three ranking methodology used in this study.

Table 1. University ranking methodology and sustainability indicators.

Ranking Name	Publisher/Commencing YEAR	Indicator and Weight	Website
ARWU	Institute of Higher Education of Shanghai Jiao Tong University, 2003	Alumni (Alumni with Nobel and Field Medals), 10% Award (Nobel and Field Medal winners), 20% HiCi (Researchers cited by Thomson Scientific), 20% N&S (articles published in *Nature* and *Science*), 20% PUB (articles indexed in SCI and SSCI), 20% PCP (faculty average score in above 5 items), 10%	http://www.shanghairanking.com/aboutarwu.html
QS-WUR	Quacquarelli Symonds, 2004	Reputation, 40% Employer reputation, 10% Student-to-faculty ratio, 20% Citations per faculty, 20% International faculty ratio, 5% International student ratio, 5%	http://www.topuniversities.com/university-rankings-articles/world-university-rankings/qs-world-university-rankings-
THE-WUR	Times Higher Education, 2010	Teaching (30%), Research (30%), Citations (32.5%), Industry income (2.5%) International outlook (5%)	http://www.timeshighereducation.com/

4. The Conceptual NFWCUs Context and Analytical Framework

References to the sustainability of NFWCUs immediately imply allusion to catch up, the most progressed type, the prominent among the easily formed within the state or national higher education institutions. The NFWCUs constitute almost universally within the English-speaking countries, post-secondary institutions that constitute aiming the pinnacle of either a state or national higher education system, of those have overlapped to excel, among others. This understanding usually depicts the largest and most elite huger regarded universities within the countries and a more substantial set of commonly acknowledged universities [57]. These can be seen as, designation, visibility, differentiation, validity and the various discourses around NFWCUs in Asia and Latin American higher education institution context were vocal critics analyzed in Reference [2]. These can be perceived as several Asian countries drive to the formation of WCUs.

Expounding on different phrases of "prestigious institutions" as a research university, "world class" and "Newly Formed World-Class Universities" (NFWCUs): Altbach, [2] articulate world-class universities as leading universities in the countries, largest producer of graduate students, research and publications. They are held high national esteem and play an important role in national capacity building and innovation effort and commonly included in the top-ranked universities. In century ago, only well-deserved universities such as Harvard Universities and Oxford University that were recognized as "Early Formed World-Class Universities" "EFWCUs." However, recently, several universities are catching up to future the quest and catch up in of top 100 global university rankings. Invariably, in this study, the word "NFWCUs" used to mean, universities which are "fresh-appeared" "catching up" or "progressive type" are described based on the commonly ranking list of top 100 in the world's top three ranking systems. They are universities newly listed in the current top 100 ranked by the global university ranking of ARWU, QS and THEs in the past eight years (2010-2018). This broad conceptual framework of NFWCUs as enunciated here is deemed relevant for this study and adopted for the study of global NFWCUs. Therefore, early universities and NFWCUs shed some light on the sustainable worldwide movement to create WCU. By tracing the university ranking systems, the NFWCUs has a global phenomenon of strong reputation and influence on the development of higher education including productive faculty, excellent students, flexible administration, international engagement and plentiful funding mechanisms.

Although, the discussion as regards to developing WCUs involves integrated part assessment of teaching and research in global world rankings. In this paper, variation refers to the sequence with changes divergence in the range of sequences in global rankings of the WCUs regarding characteristics, data and functions. Since there are over 28,077 distribution and variations of colleges and universities ranked all over the world but the top 100 worldwide listed institutions in the three global ranking system are very few (cover around 0.004%). Therefore, there are certain limitations in the rankings of various universities and they are widely criticized, whether the top universities in the rankings must be the world's top universities still have doubts. However, in different university rankings, based on different ranking indicators and weight, have steadily ranked in the world for many years due to their institutional practices, characteristics, visible achievements and superior performances in the commonly included top higher education institutions (HEIs). Following this notion and by considering the world common ranking systems (ARWU, QS and THEs) in this paper the variation of the NFWCUs will mean disparities among the list of top-ranked HEIs in the global ranking systems from the last eight years (2010 and 2018) respectively. The NFWCUs contributions entail the role, impact and achievements in educational, social, economic and cultural arena in their countries—the literature on variation and contribution of NFWCUs in particular and universities. Today, there is a global debate on the importance of WCUs as a critical element for a quality generation in skills for a knowledge-based economy. Critical to national competitiveness and sustainable development in today's competitive global economy—NFWCUs described as the core, key and forces to build an inclusive and diverse knowledge society the subject of this paper. Figure 1: then summarises and indicates the conceptual framework of the study and showing the sustainability indicators in the three rankings.

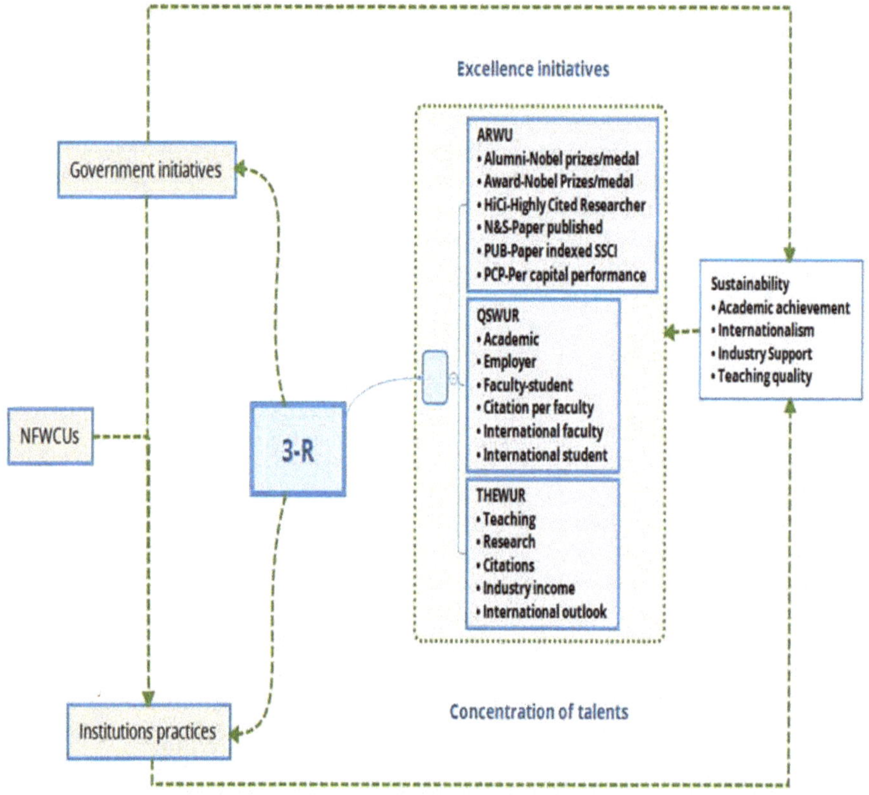

Figure 1. Conceptual Frame Work.

Data Analysis Tool

The data was analyzed using computer program-Statistical Package for the Social Sciences (SPSS v22.0). Then, the regression analysis for overall sustainability and indicators affecting university ranking from ARWU, QS and THE was based on the following regression equation:

$$ARWU = \beta0 + \beta_1 * (alumni) + \beta_2 * (award) + \beta_3 * (HiCi) + \beta_4 * (N\&S) + \beta_5 * (PUB) + \beta_6 * (PCP) + SE$$

$$QS = \beta0 + \beta_1 * (ACADREP) + \beta_2 * (EMPREP) + \beta_3 * (STUDFAC) + \beta_4 * (citations) + \beta_5 * (INTFAC) + \beta_6 * (INTSTUD) + SE$$

$$THE = \beta0 + \beta_1 * (teaching) + \beta_2 * (research) + \beta_3 * (citations) + \beta_4 * (INDINC) + \beta_5 * (INTOUT) + SE$$

Key: $\beta0$ = constant; β_1, β_2, β_3 ... = regression coefficient; SE= standard error

Figure 2 Summarizes and indicates detailing steps and the process of data, raking indicators, characteristics and analysis process of the data, where by data grouping, normative analysis, case study and regression analysis appeared to reach the results of weightings on factors affecting ranking.

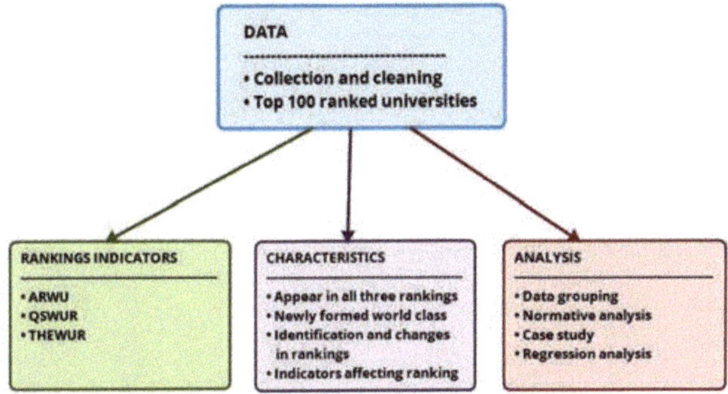

Figure 2. Summary of the process of data collections.

5. Results of the Study

5.1. Sustainability and Indicators of NFWCUs in the Last Eight Years

To address the research question, Tables 2 and 3 summarize the latest findings of the ARWU, QS and THE, comprising data from 2010 and 2018. Taking into consideration the hybrid list of top 100 universities in each ranking, the comparative analysis shows that there were only 47 NFWCUs in 2010. However, the number increased, reaching 57 in 2018. In addition, taking the hybrid list of 2018 NFWCUs, the figure indicates that 44 universities continuously maintained their status among the top 100 in the past eight years but three declined in rank and dropped out of the NFWCU family in 2018: Ecole Normale Supérieure in France and Brown University and the University of Minnesota in the United States. As for 2010 NFWCUs, the trend shows that 13 universities appeared in the hybrid list in 2018 as fresh replacements in the latest group over the past eight years.

As mentioned above, the top 47 universities in 2010 were in seven countries: US, Australia, Canada, France, Germany, Switzerland and UK (Table 2). Generally, US universities dominated the top 47 universities that met the criteria of being in the top 100 in 2010 according to ARWU, QS and THEs. The US universities comprised more than two-thirds, with 37 universities, followed by UK (14.9%), while the rest had less than 10%. Similarly, the US still dominated the top 57 universities in 2018 with 30 (42%) universities (Table 3), followed by the UK (14%) and Australia (9%), with 8 and 5 universities, respectively. Other countries with top 57 universities included Canada (3), Germany (3), China (2), Japan (2), Singapore (2), Belgium (1) and Switzerland (1). Notably, three universities were dislodged from the top 57 in 2018 compared to 2010, two in the US (Brown University and University of Minnesota) and one in France (Ecole Normale Supérieure).

Table 4 shows the most improved 13 universities in 2018 that did not appear in the top 100 in 2010. The US showed the most improvement, with five universities (38.5%), followed by China, Singapore and Australia at 15.4% and finally Belgium and Germany, each with a single university (7.7%). US universities have dominated the top positions globally, hence this study focuses on two Asian and European countries that have had more improvement lately in terms of global ranking. Consequently, 13 NFWCUs in the US, Australia, China, Singapore, Germany and Belgium are the main focus of the study.

Table 2. The combined list of commonly included the top 100 universities in 2010.

Overall Ranking	Institution	Country	Rankings in 2010		
			ARWU	QS	THE
1	Harvard University	US	1	2	1
2	Cambridge University	UK	5	1	6=
3	Massachusetts Institute of Technology	US	4	5	3
4	California Institute of Technology-Caltech	US	6	9	2
5	Stanford University	US	3	13	4
6	Oxford University	UK	10	6	6=
7	Princeton University	US	7	10	5
8	Yale University	US	11	3	10
9	University of Chicago	US	9	8	12
10	Columbia University	US	8	11	18
11	University of California-Berkeley	US	2	28	8
12	Imperial College London	UK	26	7	9
13	Cornell University	US	12	16	14
14	Pennsylvania University	US	15	12	19
15	University College London	UK	21	4	22
16	Johns Hopkins University	US	18=	17	13
17	University of Michigan-Ann Arbor	US	22	15	15=
18	Zurich-ETHZ	Switzerland	23	18	15=
19	University of California, Los Angeles	US	13	35	11
20	University of Tokyo	Japan	20	24	26
21	University of Toronto	Canada	27	29	17
22	Duke University	US	35	14	24
23	Northwestern University	US	29	26	25
24	University of Washington	US	16	55	23
25	University of Wisconsin-Madison	US	48	17	43
26	Kyoto University	Japan	26	25	57
27	University of British Columbia	Canada	36=	44	30=
28	University of California-San Diego	US	14	65	32
29	Carnegie Mellon University	US	58	34	20
30	McGill University	Canada	61	19	35
31	Edinburgh University	UK	54=	22	40
32	University of Illinois	US	25	63	33
33	Australian National University	Australia	59=	20	43
34	University of North Carolina at Chapel	US	41	57	30
35	New York University	US	31	41	60
36	Melbourne University	Australia	62	38	36
37	Washington University, St. Louis	US	30	75=	38
38	Ecole Normale Supérieure, de Paris *	France	71	33	42
39	Brown University *	US	65	39	55
40	King's College London	UK	63=	21	77
41	Manchester University & Umist	UK	30	44	87
42	Bristol University	US	66=	27	68
43	University of Minnesota *	US	28	96	52
44	München University	Germany	51	66	61
45	Heidelberg University	Germany	49	51	83=
46	Boston University	US	77	64	59
47	University of Sydney	Australia	92	37	71

Note. * The universities appeared in the group of NFWCUs in 2010 but disappeared in 2018. Source: Data retrieved on 10–30 September 2018 from THEs (www.timeshighereducation.com/), QS (www.topuniversities.com) and ARWU (www.arwu.org).

Table 3. Combined list of commonly-included top 100 universities in the year 2018.

Overall Ranking	Institution	Countries	Globe Rankings in 2018		
			ARWU	QS	THE
1	Stanford University	US	2	2	3
2	Harvard University	US	1	3	6
3	Massachusetts Institute of Technology-MIT	US	4	1	5
4	Cambridge University	UK	3	5	2
5	Oxford University	UK	7	6	1
6	California Institute of Technology-Caltech	US	9	4	3
7	Princeton University	US	6	13	7
8	University of Chicago	US	10	9	9
9	Zurich-ETHZ	Switzerland	19	10	10
10	Columbia University	US	8	18	14
11	Imperial College London	UK	24	8	8
12	University College London	UK	17	7	16
13	Yale University	US	12	16	12
14	Cornell University	US	12	14	19
15	University of Pennsylvania	US	16	19	10
16	Johns Hopkins University	US	18	17	13
17	University of California-Berkeley	US	5	27	18
18	University of California-Los Angeles	US	11	33	15
19	Duke University	US	26	21=	17
20	University of Michigan-Ann Arbor	US	27	21=	21
21	Northwestern University	US	25	28=	20
22	University of Toronto	Canada	23	31	22
23	University of Edinburgh	UK	32	23=	27=
24	University of California-San Diego	US	15	38=	31
25	Tokyo Kasei University	Japan	22	28=	46
26	Tsinghua University	China	45	25	30
27	University of Washington	US	14	61	25=
28	New York University	US	32	52	27=
29	University of Melbourne	Australia	38	41=	32
30	King's College London	UK	56	23=	36
31	National University of Singapore	Singapore	85	15	22
32	Peking University	China	57	38=	27=
33	University of Manchester	UK	34	34	54=
34	University of Wisconsin-Madison	US	28	55	43
35	University of British Columbia	Canada	43	51	34=
36	Australian National University	Australia	69	20	48
37	McGill University	Canada	70	32	42
38	Kyoto University	Japan	35	36=	74=
39	University of Illinois at Urbana-Champaign	US	41	69	37
40	Technical University of München	Germany	48	64	41
41	University of München	Germany	53	66	34=
42	University of Texas at Austin	US	40	67	49
43	Nanyang Technological University	Singapore	96	11	52
44	University of Heidelberg	Germany	47	68	45
45	Carnegie Mellon University	US	91	47	24
46	University of North Carolina at Chapel	US	30	80	56=
47	University of Queensland	Australia	55	47=	65
48	Washington University in St. Louis	US	20	100	50=
49	University of Sydney	Australia	68	50	61
50	Georgia Institute of Technology	US	79	70	33
51	University of Bristol	UK	74	44	76
52	Catholic University of Leuven	Belgium	86	71=	47
53	Boston University	US	70	81	70=
54	Monash University	Australia	91	60	80=
55	Pennsylvania State University	US	74	93=	77
56	Rice University	US	70	89	86=
57	Ohio State University	US	94	86	70=

Source: Data retrieved on 10–30 September 2018 from THEs (www.timeshighereducation.com/), QS (www.topuniversities.com) and ARWU (www.arwu.org).

Table 4. Newly-formed world-class universities of 2018 from the selected three rankings of ARWU, QS and THE.

			2010			2018	
Institution	Country	ARWU	QS	THE	ARWU	QS	THE
Tsinghua University	China	151–200	54	58	45	25	30
National University of Singapore	Singapore	101–150	31	34	85	15	22
Peking University	China	151–200	47	37	57	38	27
Technical University of München	Germany	56	58	101	48	64	41
Nanyang Tech. University	Singapore	301–400	74	174	96	11	52
Georgia Institute of Technology	USA	101–150	106	27	79	70	33
Catholic University of Leuven	Belgium	101–150	86	119	86	71	47
Monash University	Australia	151–200	61	178	91	60	80
Pennsylvania State University	USA	43	98	109	74	93	77
Rice University	USA	99	115	47	70	89	86
Ohio State University	USA	59	125	66	94	86	70
University of Queensland	Australia	101–150	43	81	55	47	65
University of Texas at Austin	USA	38	67	>200	40	67	49

Source: Data retrieved on September 10–30, 2018 from THEs (www.timeshighereducation.com/), QS (www.topuniversities.com) and ARWU (www.arwu.org).

Identifying clusters and comparing the findings of the ranking systems for 2018, Monash University, Tsinghua University and Nanyang Technological University have shown significant improvement in performance in recent years compared with their previous rankings in 2010. The trend indicates that the Ecole Normale Supérieure, Brown University and the University of Minnesota have shown rapid fluctuation in ranking for the past eight years. Despite the significant improvement of Tsinghua University and Nanyang Technological University, the progress does not reflect a substantial achievement in the ARWU ranking. As indicated in Table 5, Peking University ranked in the top 100 by both THE and QS in all years and had slight progress in ARWU ranking in the past three years. Brown University was included in all versions for six consecutive years, and then steadily dropped from the family of NFWCUs in 2017 and 2018 (101th to 150th). Due to the rapid decline, it was not included in the ARWU ranking for the past two years. Peking University, Tsinghua University and Nanyang Technological University have consistently continued their performance and variation in the rankings, hence transformed to NFWCUs after slightly increasing in ARWU rankings for the last three years. In all versions (2010–2018), the one ranking of QS has had all NFWCUs in the last eight years, reflecting an improvement compared with the other global rankings of ARWU and THE.

The results indicate that among the 5 hypothesized factors affecting THE ranking in the last 8 years included in the model, only 4 were found to have significant influence on THE ranking in 2018. These include teaching (−0.741 *), research (−0.999 **), citations (−0.807 *) and international outlook (−0.293 *) in which all factors are negatively associated with THE rankings (Table 6). The reason for the 2010 is in relation to the substantial progress in the formation of the NFWCUs in the last eight years. All the variable has negative effects on rankings in 2010 except research. All the variables affecting the ranking explain 45% and 95% variation in the model in 2010 and 2018, respectively.

Table 5. Hybrid list of sustainability and fluctuated universities from 2010 to 2018.

Institution	Ranks	2010	2011	2012	2013	2014	2015	2016	2017	2018
Monash University	ARWU	151–200	151–200	101–150	101–150	101–150	101–150	79	78	91
	QS	61	60	61	69	70	70	67	65	60
	THE	178	178	117	99	91	83	73	74	80
Tsinghua University	ARWU	151–200	151–200	151–200	151–200	101–150	101–150	58	48	45
	QS	54	47	48	48	47	47	25	24	25
	THE	58	58	71	52	50	49	47	35	30
Nanyang Technological University	ARWU	301–400	201–300	201–300	201–300	151–200	151–200	101–150	101–150	96
	QS	74	58	47	41	39	39	13	13	11
	THE	174	174	169	86	76	61	55	54	52
Peking University	ARWU	151–200	201–300	151–200	151–200	101–150	101–150	71	71	57
	QS	47	46	44	46	57	57	41	39	38
	THE	37	37	49	46	45	48	42	29	27
Ecole Normale Supérieure de Paris	ARWU	71	69	73	71	67	72	87	69	64
	QS	33	33	34	28	24	24	23	33	43
	THE	42	42	59	59	65	78	54	66	182
Brown University	ARWU	65	65	65	67	74	75	90	101–150	101–150
	QS	39	39	42	47	52	52	49=	49	53
	THE	55	55	49	51	52	54	51	51	50

Source: Data retrieved on 10–30 September 2018 from THEs (www.timeshighereducation.com/), QS (www.topuniversities.com) and ARWU (www.arwu.org).

Table 6. Regression coefficients for factors affecting THE ranking.

Factors Affecting THE Ranking	Coefficient	SER	Coefficient	SER
	←2010→		←2018→	
Constant	318.6	152.8	262.0	37.86
Teaching	−3.115	3.981	−0.741 *	0.305
Research	2.536	3.139	−0.999 **	0.288
Citations	−1.487	1.815	−0.807 *	0.318
Industry Income	−0.792	1.459	−0.072	0.109
International Outlook	−0.570	0.822	−0.293 *	0.106
R Squared	0.45		0.95	

*, ** Significant at $p \leq 0.05$ & $p \leq 0.001$, respectively, THE = Times Higher Education.

Thus, the results of regression coefficient in ARWU, indicate that among the 6 hypothesized factors affecting ARWU ranking in the last 8 years included in the model, only 2 were found to have significant influence on ARWU ranking in 2018. These include Alumni (−0.845 *) and PUB (−0.922 *), in which all factors are negatively associated with ARWU rankings Table 7. All the variables had negative effects on university rankings in 2010 and 2018. The independent variables account for 83% in 2010 and 96% in 2018 variation in ARWU ranking.

Table 7. Regression coefficients for factors affecting ARWU ranking.

Factors Affecting ARWU Ranking	Coefficient	SER	Coefficient	SER
	←2010→		←2018→	
Constant	596.1	191.4	235.6	29.42
Alumni	−0.382	2.172	−0.845 *	0.289
Award	−3.618	3.172	−0.913	0.417
HiCi	−1.191	1.722	−1.395	0.672
N&S	−3.881	3.678	−1.196	0.622
PUB	−4.551	2.453	−0.922 *	0.34
PCP	−4.599	4.517	−0.631	0.367
R Squared	0.83		0.96	

* Significant at $p \leq 0.05$, HiCi = Researchers cited by Thomson Scientific, N&S = Articles published in the Nature and Sciences, PUB = Articles indexed in the SCI and SSCI, PCP= Faculty average score, ARWU = Academic World of World-University.

The results of regression coefficient in QS ranking, indicate that among the 6 hypothesized factors affecting QS ranking in the last 8 years included in the model, 4 factors were found to have a significant influence on QS ranking in 2018. These includes employer reputation (−0.561 *), student-to-staff ratio (−0.357 ***), citation per faculty (−0.565 **) and international faculty ratio (−0.294 **) as reveals in Table 8. It was also observed that two variables that are academic reputation (−2.535 **) and the student-to-staff ratio (−0.755 **), had negative effects on university rankings in 2010. The independent variables accounted for 95% and 99% in 2010 and 2018, respectively on the variation on QS ranking.

Table 8. Regression coefficients for factors affecting QS ranking.

Factors Affecting QS Ranking	Coefficient	SER	Coefficient	SER
	←2010→		←2018→	
Constant	372.1	50.2	232.4	8.86
Academic reputation	−2.535 **	0.477	−0.609	0.267
Employer reputation	0.195	0.305	−0.561 *	0.199
Student-to-faculty ratio	−0.755 **	0.186	−0.367 ***	0.039
Citations per faculty	−0.625	0.268	−0.565 ***	0.081
International faculty ratio	0.098	0.191	−0.294 **	0.060
International student ratio	−0.330	0.142	0.022	0.080
R Squared	0.95		0.99	

*, **, *** Significant at $p \leq 0.05$, $p \leq 0.01$ & $p \leq 0.01$, respectively, QS = *Quacquarelli Symonds*.

The most current challenge for young universities trying to achieve distinguished global recognition is their internationalization. Based on regression analysis on THE several universities are ranked top in 2018 however their international profile which includes factors such as international students and faculty member is still low compared with top US and UK universities.

5.2. Shared Elements for Sustainability of NFWCUs

5.2.1. Institutional Practices and Sustainability of World-Class Universities

Around the world, there is great interest on the part of governments in the capacity and performance of elite research universities within national higher education and global ranking systems [26]. In some ways, the level of interest and initiatives varies and in many countries the motives vary and the measures are different [42,58]. Some countries (notably England and the United States) have a widespread influence due to well-established elite HEIs and research institutes founded several years ago and recently ranked at the top level. Other countries, such as Germany, China and Japan, are focusing on promoting some of their existing universities to become WCU. Some are motivating and appraising the global status of their leading national institutions, while Vietnam, India and Malaysia are focusing on building and designing new institutions at the highest national level to gain a global reputation. Mostly second-world economies such as Canada, Australia, New Zealand and South Africa are seeking to break out from national policies and frameworks of funding in an effort to rise. Reflecting changes in recent years, some efforts have been implemented by both developed and developing countries.

Governments and institutions throughout the world have significantly implemented various policies, huge projects and substantial initiatives to translate the governance, function, missions and performance improvement of universities for global rankings [48]. The literature holds that to achieve success as WCU, the quality of teaching and research needs to be given room to innovate through faculty, programs, curricula and enrollment. Considering several important trends today, US-based research accounts for two-thirds (65%), while Asian countries account for around 8.5%. Research capacity has successfully strengthened HEIs in countries such as China, Taiwan, India, Korea and Singapore. Oxford and Cambridge Universities in the UK and Harvard University in the US demonstrate a long history of superior performance. For young universities to transform to WCU may take a decade if most focus on excellence in teaching, research, publications and internationalization. By contrast, the size of the institution can affect its practices, as most of the early formed WCU are large (≥12,000 students) to extra-large (>30,000 students) in size. It has been reported that the lack of a comparative level of investment and positive initiatives poses a severe threat to institutional practices. Considering Martin Trow's theory on the development stages of higher education, the gross enrollment ratio (GER) is <15% for elite education, 15–50% for mass education and >50% for universal education. The NFWCUs ushered in an era of mass education, as the number of students increased, hence the GER is more than 26.5%. Singapore NFWCUs stand out as an example of a country's that uses internationalization strategy to drive its higher education sector. Thus, by embedding it into core institution missions, expanding participations and align curriculum and institution in order to attract highly talented faculty and students, produce globally impactful research and increase global competitiveness.

5.2.2. World-Class Universities and Government Initiatives

In the ongoing development of newly formed world-class universities (NFWCUs) in most of the desirable countries around the world, as competition for status, several initiatives are implemented to enhance and pursue academic excellence and promote national development. One such effort is huge budgetary allocation of government funding, often in the form of HEI loans and scholarships, projects, mechanisms and reforms [28]. Further, against the background of global competition in science and technology, the pace for governments to develop world-class universities is accelerating. Similarly, world-class universities have undergone fast expansion of development. Governments have supported this growth by introducing reforms in HEIs, such as in governance, taking examples from developed countries such as China. In 2015, the Chinese government released the Developing World-Class Universities and First-Class Disciplines project, known as World-Class 2.0. In 2017, the Ministry of Education announced a list of colleges and universities that replaced Projects 211 and 985, aiming to become a global higher education center.

The literature holds that during the past two decades, previous projects created significant research capacity and contributed to improvements of universities in global ranking systems. With much success, China reached 136 schools in the top 1250 of the US News 2018 rankings, second only to the US, with 221 schools. China currently is also second to the US in some papers and citations in science, although governance and new initiatives are more pragmatic. With the new strategy, the initiatives put great emphasis on top-notch talent education. In short, some studies have found that several countries continue to open up their educational opportunities to other parts of the world.

Recently, some measures have been implemented to improve the quality of NFWCUs. In this context, European and Asian countries has made substantial excellence initiatives and investments in the quality of its higher education institutions. Recently, the Chinese government launched the Belt and Road (B&R) initiatives, which focus on international trade among 65 countries spanned by a common road. The statistics indicate that the B&R project contributes significantly to an increase in the number of international students in the seven countries along the B&R route. Statistics show that in 2003 to 2004, about 25,000 students from B&R counties studied in China. At the end of 2016, the number increased to 200,000, with an average of 22.0% increase per year from 2003 to 2014 from seven top countries: Kazakhstan, Indonesia, Russia, India, Pakistan, Vietnam and Thailand.

Public universities dominate a wide range of performance in the global ranking systems. The federal government in the US has become more comprehensive in a wide range of research and international recognition toward achieving world-class status. Government initiatives and substantial funding affect some of the rankings in American universities and fluctuate compared with those with high-quality support for research, teaching and educational services. Washington University in St. Louis, Missouri, has shown unstable performance, as indicated in Figure 3. The ranking has reshaped the practices of US universities and competition for positions in the ranking has intensified along with competition for resources, especially in the decentralized US system.

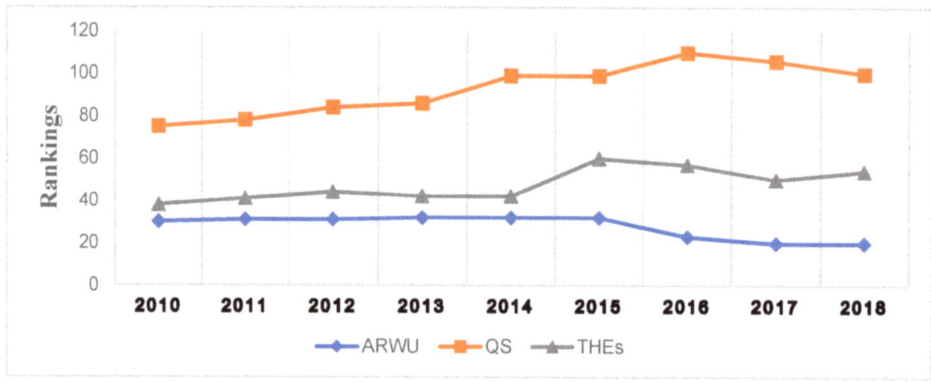

Figure 3. Rankings of Washington University in St. Louis from 2010 to 2018. Source: Data retrieved on 10–30 September 2018 from THEs (www.timeshighereducation.com/), QS (www.topuniversities.com) and ARWU (www.arwu.org).

The results of variation and disparities of university rankings from 2010 to 2018 where by Singapore, China and Belgium have shown a sharp increase and most progress in numbers of NFWCUs in the global list of leading universities as presented in (Table 9). Singapore, China and Belgium stand out as the most improved countries. The main losers are the USA, UK and France with significantly declining in numbers. In ARWU ranking USA and France, lost 7 and 3 universities while compared to QS ranking UK lose 3 and France 2. In the same case, UK and France dropped in THEWUR by 10 and 3 universities respectively. To large extent, it is inevitable that the progress in some countries forces the exit of universities from other countries. It appears that Switzerland, Australia and Germany sustained NFWCUs with high level of funding and excellence initiative.

Table 9. Sustainability per country and evolution in ranking systems between 2010 and 2018.

		2010			2018			Marginal Changes		
S/N	COUNTRY	ARWU	QS	THE	ARWU	QS	THE	ARWU	QS	THE
1	USA	53	31	53	46	31	43	−7	0	-10
2	UK	10	19	14	8	16	12	−2	−3	−2
3	Switzerland	4	3	4	5	4	3	+1	+1	−1
4	Canada	4	4	4	4	4	4	0	0	0
5	Japan	5	6	2	3	5	2	−2	−1	0
6	Australia	4	8	5	6	7	6	+2	−1	+1
7	France	3	2	3	0	0	0	−3	−2	−3
8	Germany	6	4	3	4	4	10	−2	0	+7
9	Singapore	0	0	0	2	2	2	+2	+2	+2
10	China	0	0	0	3	6	2	+3	+6	+2
11	Belgium	0	0	0	2	1	1	+2	+1	+1

THEs (www.timeshighereducation.com/), QS (www.topuniversities.com) and ARWU (www.arwu.org).

5.2.3. Sustainable and Catching-Up Countries

In response to US universities, the success factor for global competition is differentiating the mission between universities. The variety of catching-up countries include Germany, France and Japan. These countries are likely to have disparities in global university rankings (as illustrated in Figure 4) because of the languages they use. In the Japan case, half of the drop is linked to financial crisis and which have prevented additional funding expected for making significant progress on the internationalisation activities [26].

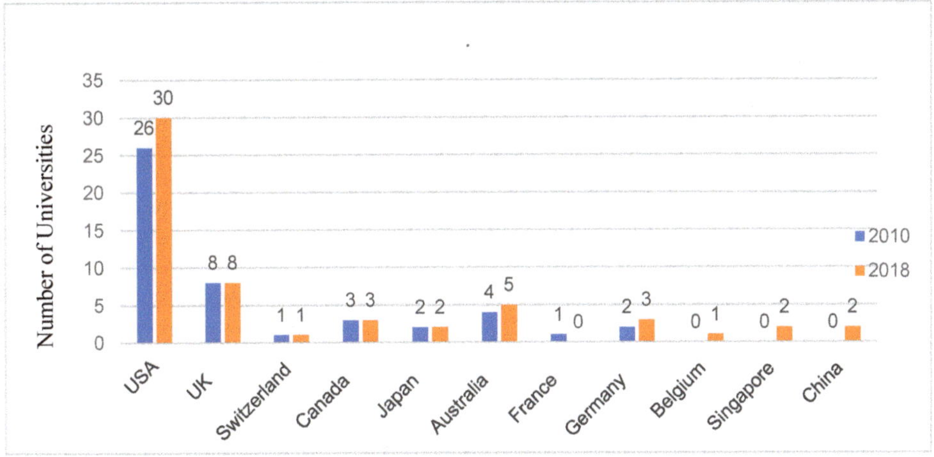

Figure 4. The variation and total number of Newly Formed World Class Universities (NFWCUs) from 2010 to 2018.

Compared with US and UK publications, most of the research published in languages other than English are not counted in the ranking system. Indeed, the French university system differs from the German education system. France has a well-developed research and technology base in their grandes ecoles. French universities are mostly teaching-focused and therefore different from German universities, which combine research and teaching. The majority of its students are recruited directly from French preparatory schools. There is a strong division of labor between universities and grandes ecoles and between research institutes and the higher education sector. Germany adheres to the philosophy and historical idea that all universities and research universities are equal, which mostly emphasizes global competition. Hence, with excellence initiatives, the government abolished a similar university philosophy.

Currently, the Ecole Normale Supérieure is doing better in ranking because France has a long-established system and the government has attempted to merge teaching-focused universities with research institutes. This has shown strong achievement and high status of French universities in global ranking systems.

Japanese university practices and initiatives for NFWCUs and the capacity of their universities feature a wide range of positive outcomes. Although Asian universities are latecomers in the quest for WCU, global ranking demonstrates how fast they are catching up. According to THE, nine of the top 10 universities in Asia ranked among the top 100 universities in the world. THE has a slightly wider range of countries, showing that four of them were among the top 50 in the world. Based on rapid changes, Japan established its WCU status during the 1990s with Research University 11 (RU11). In August 2011, RU11 launched its mission of getting Japan's research universities ready to face fierce global competition. The government and universities in Japan have made efforts such as the Global 30 Project, a new grand design of the higher education system in which universities collaborate with each other as the official policy of fostering WCU in the first decade of the 21st century. To pursue the goal of being comprehensive and maintaining the breadth of capabilities, universities have engaged in dynamic resources and revitalizing partnerships.

5.2.4. Change in System in All Rankings

Table 6 indicates the variations and disparities of countries and university rankings from 2010 and 2018. The results show changes in a system wherein several universities replaced elite universities in ranking. The successful universities are in countries such as China, Singapore and Belgium. Over the past eight years, these countries have stimulated their domestic universities to achieve world-class status. The findings show that to date, most universities in Asia are increasingly experiencing immense pressure to compete internationally. The growing interest in the sustainability of universities in the global league has become the norm. Tables 4 and 9 also shows further details of the system change in 2010 and 2018. As revealed by the three international ranking systems, the US and the UK showed falling rankings by ARWU (7) and THE (10) and ARWU (2), QS (3) and THE (2), respectively. Similarly, France fell in both rankings and Germany showed a relatively advantageous position in THE (7). China, Singapore and Belgium had leading status and demonstrated a fast increase in the emergence in a change of system and global ranking exercises. This also serves as evidence that there have been several improvements in global ranking and many NFWCUs are likely to improve further. In Singapore, two universities scored slightly better. This reveals that they are yet to be among the top 100 world-class universities. The NTU and NUS presents a model and marked the beginning of new chapter as Singapore's global universities.

6. Discussion of the Findings

Considering the determinants of success and achievements for NFWCUs in the last eight years, the following are scientifically based frameworks for sustainability and potential complementary critical indicators leading to universities being among the top 100 global rankings by the three selected systems, ARWU, QS and THE.

Comprehensive and high concentration of talent. In an attempt to propose manageable lessons learned from the results of top NFWCUs from 2010 to 2018, it is found that the dominant feature of all leading universities ranked at the top is the essential attribute of faculty and students (Tables 6–8). The superior results of NFWCUs and changes in a number of countries are highly sought by graduates, leading-edge research and technology transfer. It is widely acknowledged that NFWCUs are distinguished by the presence of outstanding faculty, a critical mass of academically qualified students and the most researchers and qualified professors. The concentration of talent is the key outlook and sustainability indicator for NFWCUs.

It is clear that the foremost determinant of excellence in top-ranked universities is the presence of a critical mass of top students and outstanding faculty Table 8. Top-ranked universities can select and attract the most qualified students, professors and researchers not only from their own country but also internationally. Indeed, it must be pointed out that the concentration of talent is the essential factor that represents the ability and privilege of NFWCUs to select the best students and staff. Principally, this can be observed at elite universities such as Harvard University and the Massachusetts Institute of Technology (MIT), which are among the most selective universities in the US and Tsinghua University, China's top institution, which admits the 50 best students from each province every year. Faculty quality is considered as one of the most important components in defining academic excellence and has an enormous impact on the quality of the program offered. Ascertaining this important condition is extremely important and a good example can be seen in the champions of NFWCUs such as the National University of Singapore (NUS) and Tsinghua University. According to Salmi Jamil, the proportion of international students at Stanford University is 21% and at Harvard University is 19%, hence the proportion of international faculty and sizable foreign academic staff. The proportion of international academic staff at Cambridge University is approximately 36%.

Indeed, this factor has always been seen in both early formed and NFWCUs such as Oxford University in the UK, Tsinghua University in China and the National University of Singapore. For instance, in the United States, the most selective universities, such as Yale University, measure SAT scores in their recruitment of undergraduate students. As shown in (Table 8), enrollment at Stanford University (US) was 17,354; University of Cambridge (UK), 18,977; KU Leuven (Belgium), 56,351; University of Tokyo (Japan), 28,253; and Tsinghua University (China) 47,762. The huge enrollment size is undoubtedly a major factor in the growth rate gap in the top global ranking. At Stanford and Cambridge Universities, overall enrollment is fewer than 30,000 students. In line with global rankings and their desire for a good reputation, NFWCUs are also favored by the strong international dimension of foreign academic staff and students.

Sound financial return base and resources. The lesson learned from the National University of Singapore (NUS) is the university's success in the top ranking of NFWCUs for 2018 in response to substantial endowment funding. Comparing to elite universities such as Oxford and Cambridge, NUS has a steady source of financing through substantial fundraising. Also, the NFWCUs of Singapore managed to use about 775 million to build up an effective, sizeable portfolio. Comparatively, the US has always managed to cover the top 100 world-class universities in all rankings due to large endowments of about $40,000 every year.

Research-based rankings. Rankings of NFWCUs are also based on government research funding. Therefore, the level of expenditure in one way affects the rankings and respective catch-up countries. Universities in the UK and Switzerland are the best competitors with the highest variations due to abundant resources and good funding.

As illustrated in (Table 5), despite being ranked by all three systems in 2010, the Ecole Normale Supérieure dropped from the ranking despite France's status of having the strongest economy in the world. This is in line with German universities, few of which are placed in the top 100, despite its strong economy in the world.

The discussion and the lesson learned regarding Figure 4 is that China and Singapore have shown the most improvement in the ranking system in the last eight years (2010–2018). Considering the NFWCUs of 2018, Tsinghua, National University of Singapore, Monash University and Rice University showed relative progress toward being among the family of world-class universities for a short time compared to previous rankings. When evaluating against changes in scores in both top 100 university rankings, the following discussion offers a perspective for the variation among countries.

Inducing excellence-initiatives: Following the same ideology in constructing world-class universities and enhancing global sustainability and competitiveness, China has been undertaking programs such as the 211 Project and 985 Scheme, which were carried out earlier, compared to other countries such as India. The objectives were measurable and the construction results were obvious. As far as the advantages of cooperation and innovation are concerned, China still has an opportunity to catch up with the UK and the US. The following sustainable strategies for globalization should be used to retain its leading position in the world: modernize the higher education system, attract active research faculty, encourage dissemination of research output, nurture competitive graduate departments and research groups by inviting capable research scholars to join them and integrate resources with full advantages for efficient implementation [39,40]. Thus, Peking University and Tsinghua University are demonstrating better global rankings because they started internationalization earlier than other Asian universities [39]. World-Class 2.0 was released in 2015 with a target goal of developing a number of world-class universities and first-class disciplines by 2020, as well as having higher education and disciplines among the best in the world by 2030 and leading in the number and capacity of WCU and disciplines among the world's best, hence becoming a higher-education powerhouse by 2050. Tsinghua University aims to be a WCU by 2020 and one of the world's best universities by 2050, with greater precision as part of the Chinese dream.

Developing global students: Meanwhile, in China since the 1990s, the mass movement of overseas students from many developed countries has increased, driven by both financial and marketing efforts in countries such as US, UK, Canada and Australia [59]. Internationalization of higher education enables students to exchange and flow in all directions, not only from west to east but from less developed to more developed countries. China is now hosting 7% of the 3.3 million international students worldwide, just behind the US, the UK and France [4,60]. In previous years, the data show that China was the leading nation for sending students to study abroad in those developed countries and the top ranked universities. According to 2017 statistics, China has become the nation that is attracting and receiving more international students from different countries of the world [61–63]. As of the end of 2017, China was the most popular destination for international students in Asia.

Ultimately, nearly two-thirds of all foreign students in the country (65%) come from markets targeted by China's One Belt, One Road initiative. The literature shows that the top 10 source countries were South Korea, Thailand, Pakistan, US, India, Russia, Japan, Indonesia, Kazakhstan and Laos.

China currently has a significant number of international students seeking a degree, increasing from 184,799 to 209,966 in 2016, while those seeking PhDs increased from 14,367 to 18,051 (www.csis.org). *Programming, curriculum and environment:* The Singaporean government has strategically identified the leading global universities and invited them to set up branch campuses in the city-state, actively promoting collaborations between world-renowned academics and local scholars [64]. According to the policy target, the ministry was well placed to gain an estimated \$2.2 trillion of the world education market. Indeed, the Singaporean government has been successful in convincing world-renowned institutions to establish overseas campuses or offer programs in collaboration with local institutions, thus helping Singapore's universities become globally competitive [61,65,66]. At the same time, the sustainability of the highly ranked top university in Asian countries was based primarily on broader goals, including:

- *Governance and organizational structure*
- *Curriculum and teaching*
- *Learning resources and environment*
- *Learner support services*
- *Professional staff development*
- *Capacity and human resources management*
- *Organizational culture and values*

Identifying these critical factors is associated with the sustainability of higher education institutions. According to Shin [67], in their effort to develop NFWCUs, policy makers and institutional leaders pay attention to research productivity, research funding and international faculty and students but other factors such as international alliance, language, geographic location, international climate on campus and economic development are also important for building WCUs.

Experience from Singapore's Newly Formed World-Class Universities

Building NFWCUs is a national strategy implemented and guided by top policy makers. As early as the 1990s, the Singaporean government was proposing the idea of building a world-class university from its elite universities. In its overall development of higher education, Singapore has had a remarkable achievement in the goal of being among the top research universities. This is particularly true when analyzing the milestone at two of the country's oldest universities, the National University of Singapore (NUS; 1905) and Nanyang Technological University (NTU; 1981). In term of rankings, Table 4 demonstrates that NUS moved up from 101 to 150 on the ARWU, 31st on the QS and 34th on THE in 2009–2010 and to 85th on the ARWU, 15th on the QS and 22nd on THE in 2017–2018, very impressive gains among the top 100 in all three ranking systems. The NUS is free to bring top researchers, including foreign faculty. This involves professors from all over the world, who are paid based on the global market and provided with performance incentives to stimulate competition.

Comprehensive internationalization: The NTU and NUS successes are reflective of Singapore's broader agenda of building Singapore as and education and knowledge hub "global war of talent". Thus, being primarily rely on core institution missions, programs, curriculum, and the large number of international faculty and students. NUS is around 60% foreign faculty, while NTU is around 70%.The overall QS data reveal that NTU and NUS have high faculty/student scores at 95% and 91.8%, respectively and are more advanced in international faculty (www.topuniversities.com).

According to studies by many researchers [66], global ranking is arbitrary depending upon which criteria are applied. However, [67] showed how much ranking could change by applying different measures such as the collaboration index, the total or per capita measure of publications and citations. As many academics have said, a world-class university is recognized by global ranking. The best conceptual approach to define WCU is to identify the characteristics of the newly formed WCU. Global university ranking systems are currently a measure of world-class universities and there are many theoretical and methodological issues involved in the ranking process [66].

In this context [11,68,69], there are four models for NFWCUs: (1) Upgrade the institution's existing strategic alternative. Several studies found that China has followed this since the early 1980s. Upgrading existing strategies is less expensive but it is a challenge to reform and transform. (2) Merge institutions' existing strategic alternatives. Several studies found that advanced mergers among institutions could be applied to transform a university into a world-class one. These studies found that France, Denmark and China adopted this strategic alternative recently [70,71].

In France, at institutions such as grandes ecoles, the government is investigating the possibility of merging as a regional foundation. It was found in Reference [72] that the Danish government has encouraged setting up world-class universities through grants and innovation funds. In China, some mergers have taken place; for example, Beijing Medical University was combined with Beijing University in 2000 and Zhejiang University was the outcome of merging five different universities. (3) Create new institution strategic alternatives. In Reference [73], it was found that it is most costly to create WCU from scratch. This strategy can be done faster and more successfully than upgrading [42]. (4) Use collaborative strategic alternatives. In Reference [27], a mixed option of upgrading and merging existing universities was noted. This can be done at both the national and international level at the same time. This measurable strategy has recently been implemented in India and Japan.

Given the increased importance of science-based innovation in new technologies, university–private sector partnerships are expected to assume a prominent economic role in national development, as they did in Singapore. Centering on NUS, Singapore established the national science and technology system, made up of diverse public–private research networks, which has contributed to a shift toward a knowledge-based economy. Currently, NUS accounts for about 52% of scientific publications in Singapore and contributes 50% to 70% of the total research and development output.

Going beyond ranking systems, there are many opportunities and connections for the future. Considering the increased interest in global rankings, young learners gain an opportunity to think deeply and work collaboratively across cultural boundaries and differences. With the distinctive roles and missions of world-class universities, young learners are encouraged to take advantage and reassess the value of education. Hence, it is important to find pathways to become globally competitive and adopt local challenges to address employability, social mobility and a high-quality graduate environment. Apart from ranking, looking at the status of any university demonstrates the importance of adding learning opportunities, providing new insights to deepen the collective understanding of the dynamic world of global higher education and research.

7. Conclusions and Sustainable Policy Implications

This study employed an understanding of sustainability indicators of NFWCUs from 2010 to 2018. This section offers our conclusion of the findings.

Given the recent literature, sustainability for NFWCUs has recently dominated public discourse, mainly regarding their catching up and development. This study investigated the variations and contributions of NFWCUs among countries. It should be noted that NFWCUs represent a concept that manifests in university strength, performance and standards. Global ranking systems are necessary indicators for measuring strengths, sustainability and standards. The study considers the universities included in the top 100 from 2010 to 2018 in three rankings of different methodologies and characteristics. To understand the status and variation among countries, integrating the three rankings helps to minimize the impact of the methodology used, sustainable changes and position gain over time. On the basis of top-ranked world-class universities, it should be recognized that in the last eight years since the THE and QS split into two independent programs, the commonly ranked top 100 universities in 2018 by ARWU, QS and THE were denoted by the top 57, compared with 2010, which was 47 universities. The paper synthesizes the difference between 2010 and 2018, showing that

44 universities maintained their position in all ranking systems among the top 100 without deviation and weight discrepancies and decline of ranking.

Three lower-ranked NFWCUs in the hybrid list of 2010 did not appear in the top 100 universities in 2018, which are covered by some catch-up universities. The Ecole Normale Supérieure, Brown University and the University of Minnesota dropped out of the NFWCU list in 2018, as their rankings fell from the top 100 in THE and ARWU.On the basis of the result of analysis, even though all the three ranking systems have different hypothesized factors, all the variables affecting rankings variations is more observed for 2018 study period.

Likewise, due to outstanding progress made by Tsinghua University, Nanyang Technological University and Peking University, the rankings of these NFWCUs have shown a tremendous sharp increase and homogeneity. These three universities indicate that variations in NFWCUs still have a long way to go to catch up with other world-class universities. A direct implication of the regression analysis also serves as a reminder that to achieve a standard of academic excellence, a university has a chance to become world-class if it continues to increase the quality and significance of research and is highly internationalized with a wide range of subject coverage. Further analysis suggested that in the future, elite universities can lose their ranking if they have no high-level, creative, talent turning out high-quality original research results and making scientific and technological progress.

This study brings the following variable results and conclusions. By systematic comparison, the US and UK dominated the variation of NFWCUs in 2010 and 2018 consecutively.

Climbing up the hierarchy of NFWCUs and since the start of global ranking more than a half-decade ago, elite universities have consistently ranked in the top 100 in all three ranking systems. Monash University represents a particular case and Tsinghua University, National University of Singapore and Peking University have shown overall improvement in their global reputation. Indeed, elements for success at those universities include a regulatory framework governing public universities, structural policies and flexible pathways that enable fluidity of student movements and structural decisions and exceptional innovations such as ongoing government excellence initiatives in Asia and Europe.

Moreover, a more analytical perspective in NFWCUs is that Harvard, Oxford and Cambridge Universities continue to battle year by year for their positions in the ranking systems and these are highly internationalized, comprehensive universities with a wide range of subject coverage. Although most NFWCUs are large in size, academic research quality is a very important element of success. Also, except for private universities, most NFWCUs are relatively large, with an average number of students ranging from 22,000 to 35,000 and correlated faculty ranging from 2400 to 3500. Likewise, most NFWCUs have sufficient technical and administrative personnel to support teaching and research as a critical path to bring the institution to international prominence. Despite the strong constraints of NFWCUs, what distinguishes the top 100 universities in the last eight years from the rest is the enormous amount of available funding. Although some of the top 100 universities are hundreds of years old, they have a concentration of talented academics and students, significant budgets and strategic visions, which are effective approaches to achieve high ranking.

The results are relevant for the strategic planning of universities to improve their reputation. Following empirical data and literature reviews, this paper theorizes that there is a need for this type of study to be undertaken in order to inform society on the gaps in ranking and for NFWCUs to catch up across wide global variation. This involves excellence initiatives, WCU projects, government funding, reforms, policies and strategies to find the correct measurable approaches and key factors to upgrade and transform HEIs into WCU. This study could serve to stimulate new plans and actions toward striving for adequate provision of WCU development. The data could potentially bring forward evidence on institutional practices and sustainable approaches that work well in transforming to WCU status.

Ultimately, for many years, universities in the United States and Europe have been the dominant and best of the WCU, particularly NFWCUs in 2010 and 2018. Despite the remarkable rankings of US and UK, China and Singapore appear to have demonstrated their advantageous positions in the NFWCUs rankings. In the academic year 2017–2018, two universities in China and Singapore were ranked in the ARWU, QS and THEs. The study brings to light the increasing emphasis on the effectiveness and efficiency of government-supported research for more approaches to internationalized initiatives. Hence university expansion is one of the excellent approaches to improve performance in the global ranking systems.

In terms of future implications, our study has endeavored to identify some general patterns in performance among the arguably more research-oriented universities based on the criteria used by ARWU, QS and THE. To improve NFWCUs interpretation factors, it might be better to judge global competitive order and shared indicators of global top best universities by integration of dominant rankings methodology. However, it is debatable whether US and UK universities dominate all global ranking systems. A closer analysis shows that US universities vary and, in some instances, have lost status in the middle and lower end of the ranking systems over recent years and have been replaced by universities from catching-up countries. Our analysis draws attention to how university rankings have generated global variation and sustainability of institutional reputation and the relationship between status and the emerging proliferation of ranking systems. Our paper offers critical contributions to the current literature on success strategies for NFWCUs and their implications for future universities and their quality, not only to create a landscape of new possibilities for reputation but also to reshape sustainability and institutional behavior in the pursuit of enhanced performance. Therefore, considering the sustainability indicators of NFWCUs, many universities benefit from these rankings as an indication of superior output, thus educational and research progress. Universities need to increase their knowledge generation and knowledge transfer and enlarge their focus on research. Providing a more international learning environment, exchange students and researchers with other leading universities in the world and internationalization of NFWCUs policy ant national and university level. In many instances, rankings help to maintain and build institutional position and reputation; rankings are used by students to shortlist their university choices and by universities, stakeholders and policy makers for direct decision-making in higher education, supporting the predominant and leading disciplines and increasing and improving the number and level of foreign teachers, students and international cooperation.

Finally, although the quest for sustainability of NFWCUs remains a challenge around the world, Singaporean and Chinese universities should be applauded for their efforts. Along with various reforms and improvements, in mainland China and Singapore the governments have concentrated their resources on helping their universities become global competitors. Moreover, the Chinese government has focused its grants on a limited number of universities to transform the country's elite universities into world-class universities. In particular, according to policy targets, the Singaporean government has set up campuses for leading universities and has actively promoted active collaboration between world-renowned and local scholars to establish overseas campuses. The empirical results confirm that elite universities are continually leading competitors due to their distinguishing factors: a comprehensive internationalization strategies and activities which driven by; a stable concentration of talented academics and students, a significant budget, research productivity, global and national partnerships and outstanding educational programs. We therefore suggest that it is necessary to take effective major action for NFWCUs in higher education such as maintaining the stability of investments and research productivity, clear mechanism for adjusting the allocation of funds among universities and implementing related strategic initiatives to enhance their international reputation to cope with leading global universities in the future.

Author Contributions: This study was conducted under cooperation between all authors. All Authors had contributed to the intellectual content of this entire research study; L.Z. designated the ideas of this study, supervised the overall directions and workflow of the study and findings, reviewed the final analysis, funding acquisition; G.J.M. conceptualized the study, literature review, drafted methodology, collected and analyzed the data and preparation of the final manuscript and C.M.A. participated in gathered the document database from

ARWU, QS and THE, formal analysis, and discussed the final results, and all the authors have read, discussed and approved the final version of the manuscript.

Funding: The work describe in this paper was funded by the Jiangsu Social Science Fund for the financed research project, on the paths of governance for "Newly formed" world-class Universities (Project ID number: 17JYA003).

Acknowledgments: We would like to acknowledge and express our sincere gratitude to the two anonymous reviewers for their constructive comments and suggestions given for the earlier versions that have significantly improved the quality of this manuscript. We also profoundly thanks Chinese Scholarship Council (CSC), International student's office, and academic staff from the College of Public Administration, Nanjing Agricultural University for the overall comprehensive support to complete this study. We further acknowledge editor's overall guidance and support throughout the process of publishing this manuscript.

Conflicts of Interest: The authors declare that there were no potential conflicts of interest reported in this research study.

References

1. Forster, N. Why are there so few world-class universities in the Middle East and North Africa? *J. Furth. Higher Educ.* **2018**, *42*, 1025–1039. [CrossRef]

2. Altbach, P.G.; Balan, G. *Transforming Research Universities in Asia and Latin America World Class Worldwide*; Johns Hopkins University Press: Baltimore, MD, USA, 2007; Available online: http://muse.jhu.edu/journals/tam/summary/v065/65.3.covert.html (accessed on 15 January 2019).

3. Bing-Lin, G.U. Focus issues in building the world-class university. *Tsinghuajeduc* **2007**, *1*. Available online: http://en.cnki.com.cn/Article_en/CJFDTOTAL-QHDJ200701002.htm (accessed on 17 December 2018).

4. Altbach, P.G. The past, present and future of the research university. *Econ. Polit.* **2011**, *46*, 65–73.

5. Altbach, P.G.; Salmi, J. (Eds.) *The Road to Academic Excellence: The Making of World Class Research Universities*; The World Bank: Washington, DC, USA, 2011; pp. 11–32.

6. Salmi, J. *Global Rankings and the Geopolitics of Higher Education: Understanding the Influence and Impact of Rankings on Higher Education, Policy and Society*; Taylor & Francis: Abingdon, UK, 2016; pp. 216–243.

7. Pizarro Milian, R.; Rizk, J. Do university rankings matter? A qualitative exploration of institutional selection at three southern Ontario universities. *J. Furth. High. Educ.* **2018**, *42*, 1143–1155. [CrossRef]

8. Li, J.; Li, J. *Quest for World-Class Teacher Education?* Springer: Berlin, Germany, 2016. [CrossRef]

9. Kim, D.; Song, Q.; Liu, J.; Liu, Q.; Grimm, A. Building world class universities in China: Exploring faculty's perceptions, interpretations of and struggles with global forces in higher education. *Comp. J. Comp. Int. Educ.* **2018**, *48*, 92–109. [CrossRef]

10. Liu, N.C.; Wang, Q.; Cheng, Y. *Paths to a World-Class University*; Sense Publishers: Leiden, The Netherlands, 2011; pp. 10–14.

11. Salmi, J. Excellence strategies and the creation of world-class universities. In *Matching Visibility and Performance*; Springer: Berlin, Germany, 2016; pp. 15–48. Available online: https://doi.org/10.1007/978-94-6300-773-3_2 (accessed on 17 September 2018).

12. Soh, K. Nearing world-class: Singapore's two universities in QSWUR 2015/16. *J. High. Educ. Policy Manag.* **2016**, *38*, 78–87. [CrossRef]

13. Luque-Martínez, T.; Faraoni, N. Meta-ranking to position world universities. *Stud. High. Educ.* **2019**, 1–15. [CrossRef]

14. Cheng, Y.; Wang, Q.; Liu, N.C. How world-class universities affect global higher education. In *How World-Class Universities Affect Global Higher Education*; SensePublishers: Rotterdam, The Netherlands, 2014; pp. 1–10. Available online: https://link.springer.com/chapter/10.1007/978-94-6209-821-3_1 (accessed on 17 September 2018).

15. Jang, D.H.; Ryu, K.; Yi, P.; Craig, D.A. The hurdles to being world class: Narrative analysis of the world-class university project in korea. *High. Educ. Policy* **2016**, *29*, 234–253. Available online: https://link.springer.com/article/10.1057/hep.2015.23 (accessed on 20 August 2018). [CrossRef]

16. Hazelkorn, E. *Rankings and the Reshaping of Higher Education: The Battle for World-Class Excellence*; Springer: Berlin, Germany, 2015; Available online: https://books.google.com/books?hl=en&lr=&id=rMW_BwAAQBAJ&oi=fnd&pg=PP1&dq=Rankings+and+the+Reshaping+of+Higher+Education:+The+Battle+for+World-Class+Excellence%3B+S&ots=6nPS9S4YMj&sig=cbmViT6L_pAkpSKxBV_OWP8a_o4#v=onepage&q=Rankings%20and%20the%20Reshaping%20of%20Higher%20Education%3A%20The%20Battle%20for%20World-Class%20Excellence%3B%20S&f=false (accessed on 2 January 2019). [CrossRef]

17. Tilak, J.B. Global rankings, world-class universities and dilemma in higher education policy in India. *High. Educ. Future* **2016**, *3*, 126–143. [CrossRef]

18. Zhang, Y.; Zhang, Q.; Qizhen, D.U. World university rankings index system comparison and its enlightenments for building the world-class universities and the world-class disciplines of China. *J. China Univ. Pet.* **2017**. Available online: http://www.en.cnki.com.cn/Article_en/CJFDTOTAL-SYSK201702016.htm (accessed on 17 December 2018).

19. Hazelkorn, E. Reshaping the world order of higher education: The role and impact of rankings on national and global systems. *Policy Rev. High. Educ.* **2018**, *2*, 4–31. Available online: https://doi.org/10.1080/23322969.2018.1424562 (accessed on 20 January 2019). [CrossRef]

20. Yuan, B.T.; Pan, Y.L. Internationalization and the construction of world class universities:A case study of Tsinghua University. *J. High. Educ.* **2009**. Available online: http://en.cnki.com.cn/Article_en/CJFDTOTAL-HIGH200909006.htm (accessed on 10 November 2018).

21. Liu, L.L. Optimization of university teaching faculty and relevant measures—Based on an analysis of world-class universities' experience. *J. Southeast Univ.* **2010**, *6*, 15–24. Available online: http://www.en.cnki.com.cn/Article_en/CJFDTOTAL-DNDS201006028.htm (accessed on 20 April 2019).

22. Douglass, J.A. *The New Flagship University Changing the Paradigm from Global Ranking to National Relevancy*; Springer: Berlin, Germany, 2016.

23. Danson, M.; Halkier, H.; Cameron, G. *Governance, Institutional Change and Regional Development*; Routledge: London, UK, 2018.

24. Salmi, J. Directions in Development—Human Development. In *The Challenge of Establishing World Class Universities*; The World Bank: Washington, DC, USA, 2009; p. 115.

25. Ahmed, H.O.K. Strategic approach for developing world-class universities in Egypt. *J. Educ. Pract.* **2015**, *6*, 125–145.

26. Salmi, J. Excellence initiatives to create world-class universities: Do they work? *High. Educ. Eval. Dev.* **2016**, *10*, 1–29. Available online: http://www.heeact.edu.tw/public/Attachment/73218241446.pdf (accessed on 17 October 2018).

27. Salmi, J. *The Challenge of Establishing World Class Universities*; The World Bank: Washington, DC, USA, 2009.

28. Cui, Y. Times higher education "World university rankings 2010–2011": A new world-class Universities evaluation system. *High. Educ. Eval. Dev.* **2011**. Available online: http://en.cnki.com.cn/Article_en/CJFDTotal-JTGY201103004.htm (accessed on 17 October 2018).

29. Liu, L.; Liu, Z. The variation of universally acknowledged world-class universities (UAWCUs) between 2010 and 2015: An empirical study by the ranks of THEs, QS and ARWU. *High. Educ. Stud.* **2016**, *6*, 54–69. [CrossRef]

30. Liu, X. How government policy is implemented in the private university: A case study from china. *Front. Educ. China* **2018**, *13*, 426–447. Available online: https://doi.org/10.1007/s11516-018-0020-2 (accessed on 2 January 2019). [CrossRef]

31. Sheil, T. Moving beyond university rankings: Developing a world class university system in Australia. *Aust. Univ. Rev.* **2010**, *52*, 69.

32. Liu, Y. On the path choice of building double world-class university in China. *Mod. Educ. Sci.* **2016**. Available online: http://www.en.cnki.com.cn/Article_en/CJFDTOTAL-JLJK201610001.htm (accessed on 2 January 2019).

33. Taylor, P.; Braddock, R. International university ranking systems and the idea of university excellence. *J. High. Educ. Policy Manag.* **2007**, *29*, 245–260. [CrossRef]

34. Wang, Q.; Cheng, Y.; Liu, N.C. *Building World-Class Universities*; Springer: Berlin, Germany, 2013; pp. 1–10.

35. Yang, X.; You, Y. How the world-class university project affects scientific productivity? Evidence from a survey of faculty members in China. *High. Educ.Policy* **2017**, *4*, 1–23. [CrossRef]

36. Kauppi, N. The global ranking game: narrowing academic excellence through numerical objectification. *Stud. Higher Educ.* **2018**, *43*, 1750–1762. [CrossRef]

37. Li, M.; Shankar, S.; Tang, K.K. Catching up with Harvard: Results from regression analysis of world universities league tables. *Camb. J. Educ.* **2011**, *41*, 121–137. [CrossRef]

38. Soh, K. The seven deadly sins of world university ranking: A summary from several papers. *J. High. Educ. Policy Manag.* **2017**, *39*, 104–115. [CrossRef]

39. Hazelkorn, E. World-class universities or world class systems: Rankings and higher education policy choices. In *Rankings and Accountability in Higher Education: Uses and Misuses*; UNESCO: Paris, France, 2013.

40. Liu, C.B. *Speeding up the Pace of Constructing World-class University from the National Strategic View. Research on Education*; Tsinghua University: Beijing, China, 2005.

41. Shin, J.C. *The World-Class University: Concept and Policy Initiatives*; Springer: Berlin/Heidelberg, Germany, 2013; Available online: https://doi.org/10.1007/978-94-007-4975-7_2 (accessed on 10 September 2018).

42. Wang, Y. Comprehensive internationalization and world-class university. *Int. Comp. Educ.* **2018**, *7*. Available online: http://en.cnki.com.cn/Article_en/CJFDTotal-BJJY201807001.htm (accessed on 2 January 2019).

43. Xavier, C.A.; Alsagoff, L. Constructing "world-class" as "global": a case study of the National University of Singapore. *Educ. Res. Policy Pract.* **2013**, *12*, 225–238. [CrossRef]

44. Rhoads, R.A.; Li, S.; Ilano, L. The Global Quest to Build World-Class Universities: Toward a Social Justice Agenda. *New Dir. High. Educ.* **2014**, *2014*, 27–39. [CrossRef]

45. Mok, K.H.; Hallinger, P. The quest for world class status and university responses in Asia's World Cities: An introduction by the guest EDITORS. *J. High. Educ. Policy Manag.* **2013**, *35*, 230–237. [CrossRef]

46. Tayeb, O. *Roadmap to Become a World-Class University*; Springer: Berlin/Heidelberg, Germany, 2016.

47. Ngok, K.; Guo, W. The quest for world class universities in China: Critical reflections. *Policy Futures Educ.* **2008**, *6*, 545–557. [CrossRef]

48. Bejinaru, R.; Prelipcean, G. Successful strategies to be learnt from world-class universities. In Proceedings of the International Conference on Business Excellence, Bucharest, Romania, 30–31 March 2017; Volume 11, pp. 350–358.

49. Bratianu, C.; Pinzaru, F. Challenges for the university intellectual capital in the knowledge economy. *Manag. Dyn. Knowl. Econ.* **2015**, *3*, 609.

50. Weerts, D.J.; Freed, G.H.; Morphew, C.C. Organizational identity in higher education: Conceptual and empirical perspectives. In *Higher Education: Handbook of Theory and Research*; Springer: Berlin/Heidelberg, Germany, 2014; pp. 229–278.

51. Yukl, G.A. *Leadership in Organizations*; Pearson Education: Chennai, India, 2013.

52. Wang, Y.J. Shared governance: UC Berkeley's road of becoming a world-class university. *Comp. Educ. Rev.* **2011**. Available online: http://en.cnki.com.cn/Article_en/CJFDTotal-BJJY201101002.htm (accessed on 2 June 2018).

53. Liu, N.C.; Cheng, Y. The academic ranking of world universities. *High. Educ. Eur.* **2005**, *30*, 127–136. [CrossRef]

54. Li, M.; Shankar, S.; Tang, K.K. Why does the usa dominate university league tables? *Stud. High.Educ.* **2011**, *36*, 923–937. [CrossRef]

55. Sowter, B. The times higher education supplement and quacquarelli symonds (THES–QS) World university rankings: New developments in ranking methodology. *High. Educ. Eur.* **2008**, *33*, 345–347. [CrossRef]

56. Çakır, M.P.; Acartürk, C.; Alaşehir, O.; Çilingir, C. A comparative analysis of global and national university ranking systems. *Scientometrics* **2015**, *103*, 813–848. [CrossRef]

57. Hazelkorn, E. Reflections on a decade of global rankings: What we've learned and outstanding issues. *Eur. J. Educ.* **2014**, *49*, 12–28. [CrossRef]

58. Bo, Y.U.; Yang, X. University positioning issue during the construction process of world-class university and world-class discipline. *Heilongjiang Res. High. Educ.* **2017**. Available online: http://en.cnki.com.cn/Article_en/CJFDTotal-HLJG201702005.htm (accessed on 4 December 2018).

59. Huang, C.; Yang, Y.J.; Jiang, H. Strategy organization to create a world-class university. *J. High. Educ.* **2011**, *32*. Available online: http://en.cnki.com.cn/Article_en/CJFDTOTAL-HIGH201108007.htm (accessed on 4 September 2018).

60. Chen, J.; Tian, Z.Z. Internationalization strategy and practice of world class universities in Japan—Taking the university of Tokyo as an example. *High. Educ. Sci.* **2017**. Available online: http://www.en.cnki.com.cn/Article_en/CJFDTotal-GDLK201704010.htm (accessed on 17 December 2018).
61. Mok, K.H.; Chan, D.K.K. Challenges of Transnational Higher Education in China. 2012, pp. 113–133. Available online: https://works.bepress.com/mokkh/104/ (accessed on 10 June 2018).
62. Ma, J.; Zhao, K. International student education in China: Characteristics, challenges and future trends. *High. Educ.* **2018**, 1–17. [CrossRef]
63. Ma, W.; Yue, Y. Internationalization for quality in Chinese research universities: Student perspectives. *High. Educ.* **2015**, *70*, 217–234. [CrossRef]
64. Chen, K. *International Student Mobility in Higher Education in China: Tensions between Policies and Practices*; The University of Western Ontario: London, ON, Canada, 2016.
65. Rhoads, R.A.; Shi, X.; Chang, Y. *China's Rising Research Universities: A New Era of Global Ambition*; Johns Hopkins University Press: Baltimore, MD, USA, 2014.
66. Sabzalieva, E. The policy challenges of creating a world-class university outside the global 'core'. *Eur. J. High. Educ.* **2017**, *7*, 424–439. [CrossRef]
67. Shin, J.C.; Kehm, B.M. *The World-Class University in Different Systems and Contexts*; Springer: Berlin/Heidelberg, Germany, 2013; pp. 1–13.
68. Salmi, J.; Liu, N.C. *Paths to A World-Class University: Lessons from Practices and Experiences*; Sense Publishers: Leiden, The Netherlands, 2011.
69. Salmi, J. Daring to Soar: A Strategy for developing world-class universities in Chile. *Pensam. Educ.* **2013**, *50*, 130–146.
70. Morphew, C.C.; Fumasoli, T.; Stensaker, B. Changing missions? How the strategic plans of research-intensive universities in Northern Europe and North America balance competing identities. *Stud. High. Educ.* **2018**, *43*, 1074–1088. [CrossRef]
71. Aula, H.M.; Tienari, J. Becoming "world-class"? Reputation-building in a university merger. *Crit. Perspect. Int. Bus.* **2011**, *7*, 7–29. [CrossRef]
72. Shattock, M. The 'world class' university and international ranking systems: what are the policy implications for governments and institutions? *Policy Rev. High. Educ.* **2017**, *1*, 4–21. [CrossRef]
73. Deem, R.; Mok, K.H.; Lucas, L. Transforming higher education in whose image? Exploring the concept of the 'world-class' university in Europe and Asia. *High. Educ. Policy* **2008**, *21*, 83–97. Available online: https://link.springer.com/article/10.1057/palgrave.hep.8300179 (accessed on 10 June 2018). [CrossRef]

Article

Sustainability Strategies in Portuguese Higher Education Institutions: Commitments and Practices from Internal Insights

Carla Farinha [1,*], Sandra Caeiro [1] and Ulisses Azeiteiro [2]

[1] Department of Sciences and Technology (DCET), Universidade Aberta, 1269-001 Lisbon, Portugal and CENSE, Centre for Environmental and Sustainability Research, NOVA University of Lisbon, 2829-516 Lisbon, Portugal; scaeiro@uab.pt

[2] Department of Biology and CESAM, Centre for Environmental and Marine Studies, University of Aveiro, 3810-193 Aveiro, Portugal; ulisses@ua.pt

* Correspondence: carlasofia.farinha@gmail.com

Received: 18 April 2019; Accepted: 1 June 2019; Published: 11 June 2019

Abstract: The Copernicus Declaration of 1994, which was understood as a commitment to sustainable development (SD) by top management in higher education, was signed by many universities. This signature worked as an important driver for these institutions to put different dimensions of SD principles into practice. In Portugal, a Southern European country, six of the fourteen universities belonging to the Portuguese University Rectors Council signed the declaration, but no attempt has been made to evaluate how these public universities integrated education for sustainable development at policy and strategy levels. This paper presents the results of a study aimed at identifying to what extent the integration of sustainability in the fourteen universities was achieved, through their own strategic and activity plans and activity and sustainability reports. A detailed content analysis was conducted on these plans and reports within the period from 2005 to 2014 (the time frame of the United Nations Decade of Education for Sustainable Development), to identify the main commitments and practices. Notwithstanding a lack of national integrated strategies or policies related to education for SD, the results show that the movement made progress at the university level, with good examples and initiatives at several universities. This paper highlights the importance of analyzing the content of plans and reports from higher education institutions (HEIs) when intending to assess and define a country profile for the implementation of sustainability in the educational sector. In addition, this research, conducted in Portugal, may be helpful to understand and value how SD is being applied in the policies and strategies of other European HEIs, as well as to share and encourage best practices and ways of improvement.

Keywords: commitments; education for sustainable development; Portuguese; practices; sustainability reports; universities

---

1. Introduction

For the decade from 2005 to 2014, much research has focused on how sustainable development (SD) was incorporated in universities, especially because higher education institutions (HEIs) signed declarations, charters, and initiatives (DCIs) to demonstrate their top management's commitment to sustainability in their system [1–3].

By the end of the above-mentioned decade, more than 1000 universities had ratified DCIs, so HEIs were engaged in fostering transformative SD [2]. Until now, there is a scarcity of investigation looking at the extent to which planning for SD can help HEIs to assess their performance and to determine whether the aims of their strategies and practices have been met [3].

In Portugal, earlier research showed that embedding sustainability (the "top-down" approach) is insufficiently developed in Portuguese governmental institutions at university level [4,5].

In addition, the debate concerning HEIs' role towards SD has recently begun [6,7] and the few events organized so far were mostly dedicated to the environmental perspective [8]. Moreover, SD policies are key factors for a university's successful engagement concerning sustainability matters and indicate how active they are in this field [8]. One of the levels of sustainability integration in higher education (HE) is at the institution level within the macro HE public policy system [9]. Nonetheless, no attempt has been made to assess how Portuguese public HEIs are integrating education for sustainable development (ESD) at policy and strategy levels, and how the documental analysis of HEI plans, reports, and strategies can be a useful approach to evaluate SD integration in universities. The research question is to what extent ESD has been integrated in the Portuguese public HEIs' policies within the United Nations Decade of Education for Sustainable Development (UN DESD) 2005–2014, and consequently to provide insights about their (best) practices.

The purpose of this study, conducted within the timeframe of the United Nations Decade of Education for Sustainable Development (UN DESD) 2005–2014, is to evaluate the extent to which ESD has been integrated in Portuguese public HEIs through the treatment and analysis of the universities' (i) strategic activity plans (PEs), strategic plans and development plans (PDEs), and activity and operational plans (PAs); (ii) activity reports (RAs), strategic activity reports, sustainability reports (RSs), and annual financial reports (RCs); as well as (iii) responsibility and assessment frameworks (QUARs) (QUAR ("Quadros de avaliação e responsabilização") illuminate the universities' mission, their strategic and operational goals, their key performance indicators and aims, as well as the financial and human resources available to facilitate moving towards targets and the achievement and effectiveness of such targets).

These plans relate to what HEIs are planning to accomplish in the short or medium term, depending if it is an annual or a quinquennial program, and the reports relate to what has been achieved from within the plan or beyond the plan.

1.1. Universities' Commitments to Implement ESD

In October 1990, the Taillores Declaration was signed by 30 universities worldwide. This early declaration recognized the fundamental role that universities should have in the future concerning the implementation and dissemination of sustainability:

> *Universities have a major role in the education, research, policy formation, and information exchange necessary to make these goals possible. Thus, university leaders must initiate and support mobilization of internal and external resources so that their institutions respond to this urgent challenge* [10].

Later, the 1992 Conference of European Rectors at the United Nations Conference on Environment and Development (UNCED), which took place in Rio de Janeiro, made an urgent appeal for the involvement of universities in SD and for an inclusive strategy for building a sustainable future which is equitable for all. In Europe, this declaration was signed by more than 320 HEIs in 38 countries [11].

In 1994, the Copernicus program developed its own strategy on the ten action principles to preserve the environment and promote SD, which was signed by 196 universities [12]. The universities' role was defined as follows:

> *It is consequently their [universities] duty to propagate environmental literacy and to promote the practice of environmental ethics in society, in accordance with the principles set out in the Magna Carta of European Universities and subsequent university declarations, and along the lines of the UNCED* [Rio Conference in 1992] *recommendations for environment and development education* [12].

In May 2005, at the European Higher Education Ministerial Conference held in Bergen, Norway, there was a strong reference to SD for the first time. It was said, when describing the Bologna Process, that "our contribution to achieving education for all should be based on the principle of sustainable development and be in accordance with the ongoing international work on developing guidelines for quality provision of cross-border higher education" [11].

At the United Nations Rio + 20 conference in 2012, the commitment of Higher Education Sustainable Initiatives (HESI) was announced, including teaching sustainable development concepts, encouraging research on SD, making campuses more sustainable, and involving the community in all these actions, committing institutions to concrete results and actions [13].

Additionally, the UNESCO World Conference on ESD, held in Aichi-Nagoya (Japan) in 2014, adopted a declaration and a call for urgent action to further strengthen and scale up ESD, where HEIs have a special role [14], namely in transforming societies and in key aspects of citizenship.

In the post-2015 DESD agenda, these characteristics were emphasized and linked to the establishment and achievement of the sustainable development goals (SDGs) defined by the United Nations in 2015 [15]. In fact, the seventeen SDGs were set placing education at the heart of the promotion of SD [16], proposing a HE field that is greatly influenced by the global sustainability agenda as well as by the management education requirements [17].

From a worldwide survey linked to the seven dimensions of the recognized university system [2], it was concluded that there is a strong relationship between SD commitment, integration, and the signing of DCIs, showing that there are two HEI clusters:

"the ones at the forefront, which show high commitment, have signed a declaration or belong to a charter, and have engaged in implementing SD; and those HEIs, which are lagging in commitment, implementation, and declaration signing" [2].

1.2. A Worldwide Integration of ESD in Universities' Strategies and Policies

HEIs can implement ESD in several dimensions in order to be as holistic as possible. The more common dimensions are: (1) Institutional framework (i.e., the HEIs' commitment); (2) campus operations; (3) education: courses on SD, programs on SD, transdisciplinary curricular reviews, including "educate-the-educators" programs (which promote competencies in EDS to enable an integrated approach of knowledge, procedures, attitudes, and values in teaching through multidisciplinary and transdisciplinary teams [18]); (4) research; (5) outreach and collaboration; (6) SD through on-campus experiences, working groups, policies for students and staff, among other practices; and (7) assessment and reporting [2,19].

Universities worldwide are experiencing an increasing trend towards responding to the need for sustainability and various knowledge gaps [20], as well as collaborating and contributing to the generation of sustainability values, attitudes, and behaviors within future regenerative societies [21]. Regarding some European countries, access to quality education is so critical for development [22] that the European Parliament has continuously called for the allocation of its budget to investment in this sector [23]. Universities can use low-carbon campuses as living laboratories in shaping the leaders of future sustainability thought. Many HEIs are already involved in mainstreaming the environment and sustainability into their curricula, training, research, and community engagement activities [24].

From the results of surveying a sample of universities from Germany, Greece, United Kingdom (UK), United States of America (USA), South Africa, Brazil, and Portugal [8], it was reported that there is a widely-held belief that SD policies are essential for HEIs to successfully engage in matters related to sustainability and that such policies show how active they are in this field. Therefore, a university must be considered active and have formal policies on SD as a pre-condition for successful sustainability efforts [25].

Considering HEIs' degree of commitment to and institutional trust in sustainability in USA, it was noted [25] that universities are uniquely positioned as knowledge disseminators, behavior consolidators, and idea innovators towards a resilient and impartial society, as they offer a superior learning environment and campus lifestyle experience to initiate a more holistic understanding and contemplation around sustainability.

Therefore, HEIs have embedded sustainability initiatives into their core activities, curriculum, research, community, and operational, to respond to the worldwide transformation towards a sustainable future [26].

1.3. An Implementation Research Gap in Portuguese Public Universities

Despite international studies on ESD in European universities ,which provide best practices and examples [27–29], this area represents a gap in higher education research in some countries (e.g., Czech Republic, Poland, Spain) [30–32] and the insufficient number of studies in Portugal concerning strategic environmental assessment was emphasized [33].

These detailed, national-scale studies can contribute to a better evaluation of HEIs' levels of effort and success in contributing towards encouraging worldwide sustainable development and the role of academia in meeting this purpose [17].

In 2007, which falls within the decade 2005–2014, the Portuguese Government passed the Decree-Law 242/2007, which transposed the Directive 2001/42/EC, promoted the effective institutional autonomy of universities [34], and facilitated environmental assessments regarding the effects of certain plans and programs [32].

In comparison to other European countries, Portugal was far behind in externally-oriented activities aimed at building capacity within local communities to promote SD, and Portuguese HEIs were classified as "laggards" and/or "late majority" in integrating SD in education, in research on sustainability, and in inclusive development in universities, in particular when compared with other Southern European countries [6].

Despite having signed Declarations and/or Charters, Portuguese public HEIs may or may not have implemented SD, while others that did not sign any commitment have engaged in implementing sustainability.

Regardless of previous research, it is important to comprehend how Portuguese public universities are applying ESD at policy and strategy levels (between 2005 and 2014), since no attempt has been made to evaluate their commitments and practices in a systematic and detailed way.

2. Methodology

2.1. University's Sample of Universities

Considering the UN DESD 2005–2014, the University Higher Education Institutions (UHEI) sample was based on the effective members of the Portuguese University Rectors Council (CRUP) during the analysis period (2005–2014), which correspond to all public universities. These HEIs comprised: UAc—University of the Azores [35], UMinho—University of Minho [36–38], UAb—Universidade Aberta [39,40], UP—University of Porto [41,42], UAlg—University of Algarve, UTAD—University of Trás os Montes e Alto Douro [43], UÉ—University of Évora [44,45], UBI—University of Beira Interior [46–49], UC—University of Coimbra [50,51], UTL—Technical University of Lisbon, UL—University of Lisbon, ULisboa – Universidade de Lisboa [52], UNL—NOVA University of Lisbon [53], UA—University of Aveiro [54], and UMa—University of Madeira.

In July 2013, two large public universities, UTL and UL merged to increase their scale, attract a larger volume of students, capitalize on the prestige of their faculties, and help them to achieve a greater leadership role in the European context. ULisboa "brings together various areas of knowledge and has a privileged position for facilitating the contemporary evolution of science, technology, arts and humanities [52]"

These public HEIs, together with ISCTE-IUL—University Institute of Lisbon [55] and UCP—Universidade Católica Portuguesa [56], represent the core of the Portuguese national higher education system [57].

The creation in 1979 of CRUP—Portuguese University Rectors Council, a Portuguese university associative structure, constituted a major step in the decentralization of the Ministry of Science,

Technology, and Higher Education (MCTES) responsibilities for Higher Education [58]. One of its major working areas is guaranteeing universities' coordination and their representativeness, while ensuring their autonomy [57] (see Appendix A, Figure A1).

Despite the researchers' efforts, it was not possible to obtain supplementary documentation from all the universities that belong to CRUP.

The final UHEI sample turned out to be 14 public universities and some had similar characteristics such as geographical location, number of students, and campus area (see Table 1).

Confidentiality was ensured by allocating an alphanumeric identification to each public university (HEI_01 to HEI_14) so that the names of the respective institutions did not appear in the publication findings and results.

Table 1. Characteristics of Portuguese public universities.

Public UHEIs		Number of Students	Campus Area (m^2)	Emissions of CO_2 Eq. (ton) * 1000 ton CO_2 Eq.	
Acronymous	Founded	Academic Year			
			400.000 m^2	Value	Reference year
UMinho	1973	19.500 (f)	Green space aprox. 40% within 3 Polos: Gualtar, the largest Polo (Braga), Azurém and Couros Polos (both in Guimarães). Areas are unavailable	16	2015
UP	1911	29.796 (c)	Consisting of 3 main Polos spreading out all over the city of Porto: Centro (the largest), Asprela and Campo Alegre. The polos areas are unavailable	2.849	2011
UBI	1979	7.262 (f)	4 Polos whose areas are unavailable	NA	NA
UNL	1973	19.867 (c)	30.000 m^2 is aprox. the area of FCT/UNL (Caparica Campus) which is one out of 9 Faculties of UNL, in Monte da Caparica (Almada)	NA	NA
UTAD	1986	6.609 (d)	3 Polos whose area is unavailable	NA	NA
UC	1290	21.390 (c)	3 Polos whose area is unavailable	NA	NA
ISCTE	1972	9.234 (c)	2 buildings and 1 autonomous ala	NA	NA
UA	1973	14.280 (c)	921.500 m^2 With its 3 campi, UA has its main Campus (Santiago), others in Águeda and Oliveira de Azeméis	NA	NA
ULisboa	2013	48.47 (b)	8 campuses make up Ulisboa which are: Ajuda, Alameda, Chiado, Cidade Universitária, Jamor, Loures, Quelhas, Tagus Park	NA	NA
UTL	1911	25.574 (a)	—	NA	NA
UL	1911	22.143 (a)	—	NA	NA
UAb	1988	8.590 (b)	2 sites, Rua da Escola Politécnica and Rua Braancamp in Lisbon	NA	NA
UAlg	1979	9.708 (f)	63.084 m^2 as UAlg has 4 campus: Penha (centre of the city of Faro), Gambelas, Saúde and Portimão	NA	NA
UÉ	1979	8.970 (f)	UÉ has 9 sites, one is outside the city (Mitra), other is the gimnosdesportiv pavillion; others are buildings	NA	NA
UMa	1988	3.389 (f)	The university has only 1 campus	NA	NA

Legend: Information not available (NA), square meters (m^2), carbon dioxide emissions (CO_2), Eq. (Equivalent), tonnes (ton); Each year corresponds to academic year; academic year. (a) 2012/2013 (b) 2014/2015; (c) 2015/2016; (d) 2016/2017; (e) 2017/2018; (f) unknown. Source: CRUP, 2018.

2.2. Data Collection and Time Frame

This study used a qualitative approach [59] and a detailed content analysis method. Institutional documents were analyzed to:

(a)   Find out how each public HEIs integrated sustainability, whether under any DCI or not;
(b)   Discover the commitment of each public university to SD;
(c)   Provide insights about (best) practices in implementing ESD at public universities.

The following types of documents corresponding to the period 2005 to 2014 (i.e., a 10-year period; see Table 2) for each HEI, were:

- Plans (PAs, PDEs, and PEs),
- Reports (RAs, Strategic Activity, RCs, and RSs), and
- QUARs.

The data were collected between 1 January 2015 and 30 June 2016, through public university websites, email contacts, and some UHEIs' documentation centers, mainly due to their willingness to participate in this study. After the data collection period, no further documentation was considered despite its availability on websites.

Eventually, universities might publish this type of documentation, but it was not available for the researchers during the time frame of the collection period despite their efforts.

Overall, 168 documents from the 14 public universities were gathered for treatment and analysis.

2.3. Documental Approach of Public Universities' Sustainability Integration

HEI_01, HEI_02, HEI_03, HEI_04, HEI_05, HEI_06, and HEI_07 contributed 85% of all the collected documents (see Appendix A, Figure A2). Even though seven universities provided the vast majority of the institutional document sample, the aim was to find out how each public HEI implemented sustainability and their commitment to SD, and to provide insights about best practices.

The year 2011, which was the year in which Portugal came under the international financial assistance program, corresponded to the highest number of documents gathered. This may be explained by the increased need to support financial reports with long-term planning.

Considering the first half of the UN decade 2005–2014, corresponding to the period from 2005 to 2009, concerning document type, PAs, PDEs, PEs, RAs, Strategic Activity Reports, and RSs represented 83% of all the documents.

In the second half of the period 2010–2014 there was not much difference (80%). RS accounted for 10% and 4% of the collected documents in the first and second half of the decade 2005–2014, respectively, and were published either by HEI_01 or HEI_03. From the second half of the DESD, around 33% and 43% were RAs/Strategic Activity Reports and PAs/PDEs/PEs, respectively. There seems to have been more activity planning than reporting, which might not be so true if RSs and RCs were combined.

The scenario was quite different when analyzing the documentation obtained in the period 2005 to 2009, as it seems there was more reporting and less planning. Adding RC (5%) and RS (10%) accounted for almost 66% of reporting activity altogether (see Figure 1).

From 2005 to 2014, almost 80% of the collected documentation was related to activity planning or reporting (see Appendix A, Figure A2). Despite the few sustainability reports published by the public HEIs (only two did so, UMinho and UP), they are of utmost importance for the content analysis concerning sustainability implementation because they were published during the UN Decade.

Table 2. Number of documents by typology in each Portuguese public university.

Documents	HEI_01	HEI_02	HEI_03	HEI_04	HEI_05	HEI_06	HEI_07	HEI_08	HEI_09	HEI_10	HEI_11	HEI_12	HEI_13	HEI_14	sum
Activity reports (RAs)/Strategic Activity Reports	10	9	8	10	6	7	1	-	2	2	4	3	1	-	63
Activity and operational plans (PAs)	3	7	5	9	4	5	5	6	1	1	-	-	-	-	46
Strategic plans and development plans (PDE)/Strategic activity plans (PEs)	2	4	3	2	3	2	-	2	1	1	-	-	-	1	21
Responsibility and assessment frameworks (QUARs)	5	6	-	-	-	-	4	-	-	-	-	-	-	-	15
Annual financial reports (RCs)	7	-	2	-	3	1	-	-	-	-	-	-	1	-	14
Sustainability reports (RSs)	3	-	6	-	-	-	-	-	-	-	-	-	-	-	9
sum	30	26	24	21	16	15	10	8	4	4	4	3	2	1	168
	30	56	80	101	117	132	142	150	154	158	162	165	167	168	

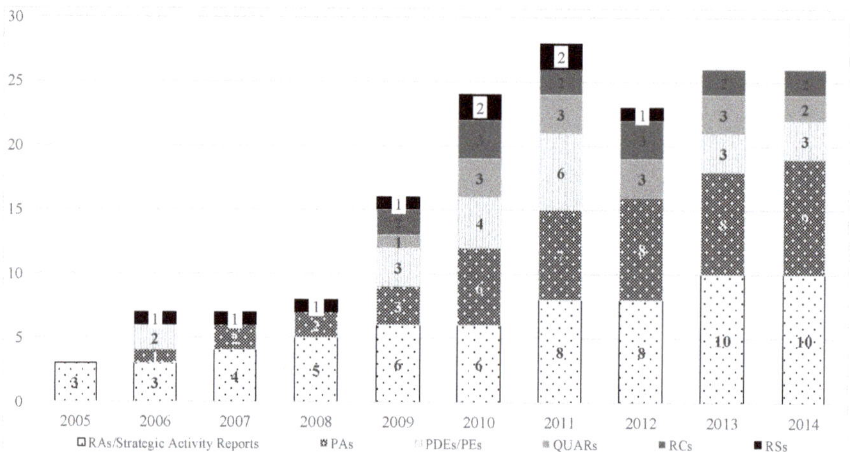

Figure 1. Distribution of document type through the UN decade 2005–2014 in Portuguese public universities.

2.4. Documental Sample Data Treatment and Analysis

The data treatment and analysis were divided in a four-step approach:

1. When collecting documents, few universities possess documents such as RC and QUAR. Since this is the case, this constitutes a drawback in the study to (better) assess policies and strategies at university level, so it was the first cut in the treatment phase. From an overall sample of 168 documents it was reduced to 139 (the "major documents") (see Table 3). From here, the data treatment was made.

Table 3. Four-step approach in data treatment and analysis.

Steps	Document Type							Processs
	RAs/Strategic Activity Reports	PAs	PDEs/PEs	QUARs	RCs	RSs	HEIs´sum	
Step 0	63	46	21	15	14	9	168	
	RAs/Strategic Activity Reports	PAs	PDEs/PEs			RSs		Data collection
Step 1	63	46	21			9	139	Data treatment
Step 2	63	46	21			9	139	Data treatment & analysis in a coding system
Step 3	63	46	21			9	139	Content analysis in a systematic review
Step 4	63	46	21			9	139	Content analysis with defined nodes (HEIs and Dimensions) [2] and subcategories (themes)

Note: The documents that were treated and analyzed from step 1 onwards neither include QUARs nor RSs.

The documents were selected, taking into account neither type nor university origin, to be treated and analyzed considering the highest frequency of keywords (see Table 4) in the defined coding system obtained in the content analysis of a previous study [4]. The following results were, in descending order, "Integration or intervention or implementation" (the main reference found), followed by "Environmental Education" (these two were the main references), then "University Higher Education or University" and "Sustainability (ies) or sustainable (s)".

2. The content was then analyzed in a systematic review, where a node corresponds to a public UHEI and each subcategory to a type of document. This coding technique was used to analyze

the documents. As coding is a process to generate categories, the analysis started by using descriptive coding, where words and sentences from document transcripts were labeled using relevant words or phrases [60].

3. Other nodes were built hereinafter as "Dimensions" relating to the recognized university system [2]:

- Institutional framework (Dimension #1);
- Campus operations (Dimension #2);
- Education (Dimension #3);
- Research (Dimension #4);
- Outreach and collaboration (Dimension #5);
- SD through on-campus experiences (Dimension #6); and
- Assessment and reporting (Dimension #7).

Table 4. The highest frequency of keywords.

1. DESD—Decade for Education for Sustainable Development	8. Development
2. Environmental Education	9. Transdisciplinary
3. Sustainable Development	10. Holistic
4. Science for Sustainability	11. Integration
5. Environmental Management	12 Higher Education/Universities
6. Sustainability/Sustainable	13. Curricula/Curricular Plan/Curricular Plan Programme
7. Environment/Environmental	14. Campus
15. Education for Sustainable Development *	

* We added this keyword as it was found to be important in many of the documents analyzed.

The themes where ESD has been implemented in HEIs were organized in dimensions and corresponded to subcategories. Each subcategory was called a sustainability implementation action (SIA) within the content analysis methodology [59]. In the end, the coding system was rearranged again based on the number of codified references, and the sustainability implementation actions (SIA) renamed, which were obtained after the treatment and analysis of the major documents.

The process consisted of organizing the disclosed data into distinct categories and/or new nodes, through a classification.

Every time a document was treated and analyzed; the code was modified to reflect the correct adjustments. This was; therefore, a collaborative process based on diversified readings before treating and analyzing the available documentation—139 documents from the 2005 to 2014 period—from which at least three adjustments were made to some of the items (a suggested procedure [61]).

The dimensions of the recognized university system [2] were used, as well as the themes associated with each aspect as a proxy of integration sustainability in each HEI. This was a cataloguing method in which an organized codebook was produced.

Lastly, all data contributed to the definition of a country profile for the implementation of sustainability in the HE sector.

For the qualitative content analysis, NVIVO (version 11) software (QSR International Pty Ltd, Victoria, Australia) was used [62].

3. Results

3.1. The Sustainability Implementation Actions in Portuguese Public HEIs

Overall, considering the seven dimensions [2], 66 themes were found as sustainability implementation actions (see Figure 3).

All Portuguese public universities seemed to have been implementing sustainability and more than 50% of actions were not exclusive to a single UHEI (see Table 1). Among the seven dimensions, "campus operations," "outreach and collaboration," and "SD through on-campus experiences" represented almost two thirds of the total sustainability implementation actions (see Table 5 and Figures 2 and 3). It; thus, seems that these were the main dimensions by which the Portuguese UHEIs implemented sustainability through strategies and policies.

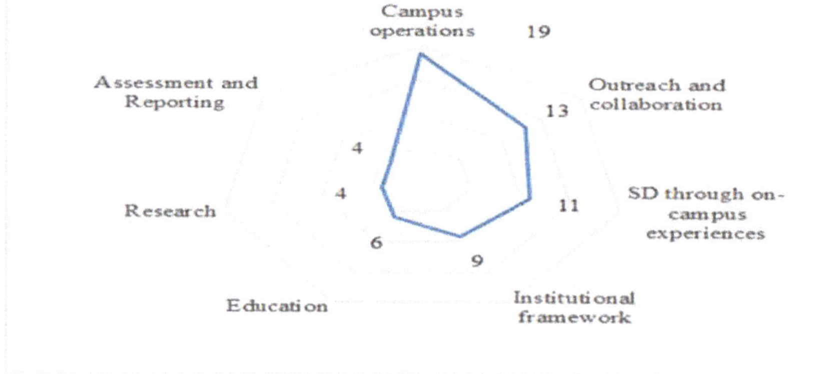

Figure 2. Number of sustainability implementation actions by each dimension.

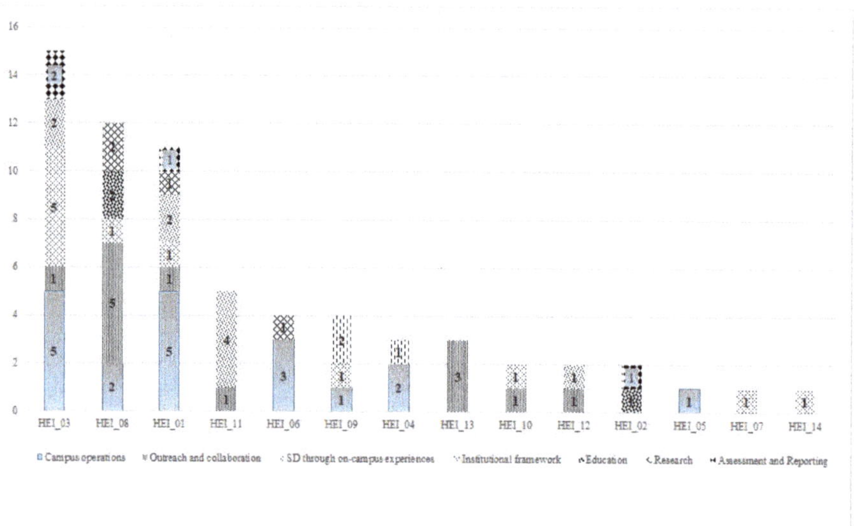

Figure 3. Number of sustainability implementation actions in each public university in each dimension.

Table 5. Number of sustainability implementation actions by dimension and university.

Sustainability Implementation Actions by Dimension	University														Sum *
	HEI_03	HEI_08	HEI_01	HEI_11	HEI_06	HEI_09	HEI_04	HEI_13	HEI_10	HEI_12	HEI_02	HEI_05	HEI_07	HEI_14	
Campus operations	5	2	5	0	3	1	2	0	0	0	0	1	0	0	**19**
Policies and activities to reduce paper consumption such as e-communications, or double-sided copying	1	-	1	-	1	-	-	-	-	-	-	-	-	-	3
Energy efficient equipment	1	-	1	-	-	-	1	-	-	-	-	-	-	-	3
Plans to improve energy efficiency	-	-	1	-	1	-	1	-	-	-	-	1	-	-	4
Sustainable landscaping	2	-	-	-	-	-	-	-	-	-	-	-	-	-	2
Renewable energy usage	1	1	-	-	-	-	-	-	-	-	-	-	-	-	2
Plans and efforts to reduce GHG emissions	-	-	-	-	-	1	-	-	-	-	-	-	-	-	1
Sustainable food & Diet practices	-	1	-	-	-	-	-	-	-	-	-	-	-	-	1
Biodiversity	-	-	-	-	1	-	-	-	-	-	-	-	-	-	1
Green purchasing from environmentally and socially responsible companies	-	-	1	-	-	-	-	-	-	-	-	-	-	-	1
Plans to improve management waste (waste bins to separate and recycle waste (recycling solid waste))	-	-	1	-	-	-	-	-	-	-	-	-	-	-	1
Outreach and collaboration	1	5	1	1	0	0	0	3	1	1	0	0	0	0	**13**
SD partnerships with other society stakeholders	-	4	-	-	-	-	-	2	-	-	-	-	-	-	6
Academic staff involved in voluntary advisory activities in SD	1	1	1	-	-	-	-	-	1	-	-	-	-	-	4
Joint degrees with other universities	-	-	-	1	-	-	-	-	-	1	-	-	-	-	2
Part of interdisciplinary SD expert networks	-	-	-	-	-	-	-	1	-	-	-	-	-	-	1
SD through on-campus experiences	5	1	1	0	0	1	0	0	1	0	0	0	1	1	**11**
Policies that promote SD for all students and staff	1	1	1	-	-	-	-	-	1	-	-	-	-	1	5
SD efforts are visible throughout the campus	1	-	-	-	-	-	-	-	-	-	-	-	1	-	2
SD working group with members from different departments	1	-	-	-	-	-	-	-	-	-	-	-	-	-	1
Sustainable practices for students	1	-	-	-	-	-	-	-	-	-	-	-	-	-	1
Student participation in SD activities	1	-	-	-	-	-	-	-	-	-	-	-	-	-	1
SD awareness raising in the campus	-	-	-	-	-	1	-	-	-	-	-	-	-	-	1
Institutional framework	2	0	2	4	0	0	0	0	0	1	0	0	0	0	**9**
Signature of a Declaration, Charter or Initiative (DCI) within SD, ESD or sustainability during UN DESD 2005-2014	1	-	1	2	-	-	-	-	-	1	-	-	-	-	5
Existence of policy for implementing SD in University	-	-	1	1	-	-	-	-	-	-	-	-	-	-	2
Inclusion of SD in the vision and mission, goals and objectives of the University	1	-	-	-	-	-	-	-	-	-	-	-	-	-	1
Existence of a Strategic Plan for implementing sustainability in University	-	-	-	1	-	-	-	-	-	-	-	-	-	-	1
Education	0	2	0	0	0	2	1	0	0	0	1	0	0	0	**6**
Courses on SD, programmes on SD	-	2	-	-	-	-	1	-	-	-	1	-	-	-	4
Teaching across (fostering the link between) the natural sciences and social sciences faculties	-	-	-	-	-	2	-	-	-	-	-	-	-	-	2

Table 5. *Cont.*

Sustainability Implementation Actions by Dimension	University														Sum *
	HEI_03	HEI_08	HEI_01	HEI_11	HEI_06	HEI_09	HEI_04	HEI_13	HEI_10	HEI_12	HEI_02	HEI_05	HEI_07	HEI_14	Sum *
Research	0	2	1	0	1	0	0	0	0	0	0	0	0	0	4
Providing fund-raising for SD Research	-	1	-	-	-	-	-	-	-	-	-	-	-	-	1
Existence of Patents in the field of SD	-	-	-	-	1	-	-	-	-	-	-	-	-	-	1
Creation of SD new knowledge and technologies	-	-	1	-	-	-	-	-	-	-	-	-	-	-	1
Existence of na SD Institute or Research Centre	-	1	-	-	-	-	-	-	-	-	-	-	-	-	1
Assessment and Reporting	2	0	1	0	0	0	0	0	0	0	1	0	0	0	4
Sustainability reports	1	-	1	-	-	-	-	-	-	-	-	-	-	-	2
Assessment of SD issues as SD integration instruments and tools within their University	1	-	-	-	-	-	-	-	-	-	1	-	-	-	2
Sum *	15	12	11	5	4	4	3	3	2	2	2	1	1	1	66

* Sum of Sustainability implementation actions; Legend: Green House Gases (GHG).

Considering the number of sustainability implementation actions (see Table 5) throughout the HEIs, the top three were:

- SD partnerships with other society stakeholders (#6), which are linked to "outreach and collaboration";
- Policies that promote SD for students and staff (#5), which are linked to "SD through on-campus experiences"; and
- Signature of DCIs within SD, ESD, or sustainability during United Nations (UN) DESD 2005–2014 (#5), which is linked to "institutional framework".

Taking into consideration the treated and analyzed documents, universities' actions relating to ESD seemed to have been taken in "isolation" and were not integrated in a whole institution approach. Each HEI acted according to a tank of actions—"think tank" (see Figure 4).

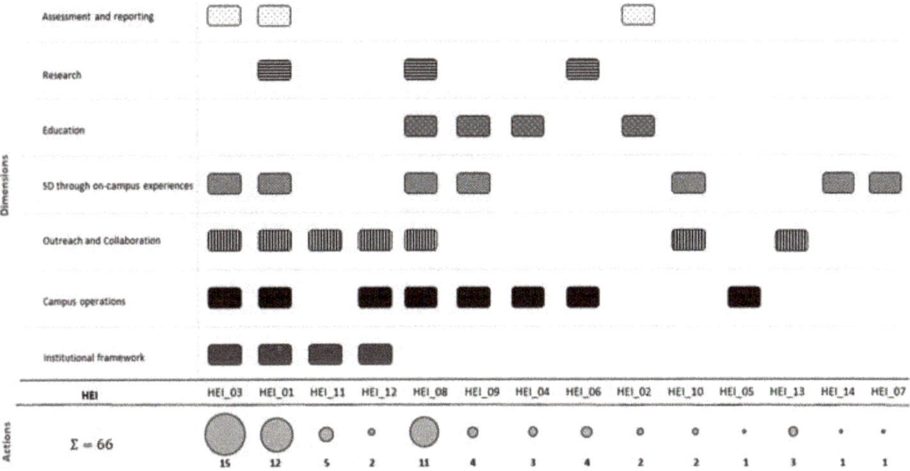

Figure 4. Think tank of initiatives in Portuguese public universities.

Each university may have taken one, or more than one, path to integrate their strategies and policies on sustainability, but it seems any integration did not keep up with the simultaneous pace of action.

The findings in Portuguese public HEIs also suggested that identical SD integration during the DESD 2005–2014 could have occurred for different reasons:

- Some universities had some similar characteristics such as geographical location, number of students, and campus area (see Table 1). There were no available data concerning campus areas of 10 HEIs.
- Some universities established partnerships in their best interest as a win–win strategy, concerning mainly education and research dimensions.

3.2. DCIs and the Commitments of the Portuguese Public Universities

As of 1 February 2018, 502 institutions had signed the Taillores Declaration. However, in 1990, the NOVA University of Lisbon (UNL) was listed as the only Portuguese signatory HEI, according to the Association of University Leaders for a Sustainable Future [10].

The findings indicate that UNL was deeply involved in the outreach and collaboration and institutional framework Dimensions through the following sustainability implementation actions:

(1) Joint degrees with other universities, and (2) the existence of policy and a strategic plan for implementing SD in the University.

Besides having signed the Taillores Declaration, UNL belonged to the Copernicus Charter in 1994. According to these documents' principles, sustainability should be incorporated in a university's faculties, departments, and other entities. The signature by UNL of both the Declaration and the Charter signaled an official commitment to SD by this university.

Nevertheless, other Portuguese HEIs also signed the Copernicus Charter, such as UTL, UP, UMinho, UL, and UCP. The results concerning UMinho and UP will be shown in Section 3.3.

The results indicate that like UNL, UTL was involved in the outreach and collaboration Dimension through the creation of joint degrees with other universities.

The overall results indicate that UNL and UTL (which, after the merger with UL, resulted in ULisboa), representing almost 30% of all HEIs' students, were both involved in the creation of a joint degree as mentioned. Nonetheless, it cannot be assured through any DCIs that this fact is due to their commitment to SD.

3.3. Commitment to SD of Universities with Sustainability Reports (RS) and DCI

UMinho and UP were the only two out of the six Portuguese Copernicus Charter signatories that developed the "assessment and reporting" through sustainability reports. RSs enable organizations to take into consideration the impact of a wide range of sustainability issues, allowing them to be more transparent about the risks and opportunities [63].

Owing to UMinho's strong cultural activity, this HEI uses the Global Report Initiative (GRI) as guidelines for sustainability reporting (2010 and 2011) and improved its methodology in 2012/2013 [36] (pp. 113–114) by including a new (cultural) dimension [37].

According to the RS from 2011 [36] (pp. 113–114), globally UMinho is on its way to sustainability considering economic, environmental, and social indicators, namely due to its direct and indirect impact in the local economy. As an example, the production of dangerous solid waste had been reduced by 2.5 ton from 2009 to 2011 and the 2015 emissions of CO_2 equivalent (ton) $\times$ 1000 ton. CO_2 equivalent were 16 in a campus area of 40 ha (see also Table 1).

Nevertheless, environmental performance should be improved to reinforce UMinho's commitment to sustainability, according to the University Rector (see Table 6). From the analysis of the documents, the sustainability implementation actions of UMinho were mainly based (almost 50% of the total number of UMinho's initiatives) on the "campus operations" Dimension, either through (1) plans to improve energy efficiency; (2) energy efficient equipment; (3) policies and activities to reduce paper consumption; (4) plans to improve the management of waste; or (5) green purchasing from environmentally and socially responsible companies. There were also actions based on "institutional framework" through the existence of policies for implementing SD in the university.

The National Strategy for Ecological Public Purchases by Resolution of the Council of Ministers (i.e., a government decision) was found to be used by UMinho concerning green purchasing as well as the Energetic Efficiency Program in Public Administration (Eco.AP) regarding energy efficiency.

There are some best practices in this university seen in the Institute of Science and Innovation for Bio-Sustainability (IB-S) and Landscape Laboratory.

The first Portuguese HEI that used GRI guidelines was the Engineering Faculty of University of Porto (FEUP) in 2006, and from 2008 onwards; however, the RS are only related to the faculty and not the whole university. The GRI model was used to assess, monitor, and report sustainability with a focus on the academic community, operations, teaching, and impact on society, which seems to have some similarities with the Sustainability Assessment Questionnaire (SAQ).

Table 6. Sustainability reports: UP and UMinho.

TAILLORES DECLARATION (1990)	COPERNICUS CHARTER (1994)	Periods Years 2006	2007	Available Methodology	# Indicators	Reference Years	Sustainability Reports Periods Years 2008	2009	2010	2011	2012/2013	Available Methodology	# Indicators	Reference Years
-	UP	▓	▓	Global Reporting Initiative (GRI) Guidelines. Dimensions: (1) economic; (2) social; (3) environmental	24 (2006) to 31 (2007)	2006–2007	▓	▓	▓	▓		Model developed to assess, monitor and report sustainability in Universities. Dimensions and categories: (1) Academic community; (2) Operations; (3) teaching; (4) Impact on the society	47 (2008; 2009) to 44 (2010) and to 42 (2011)	2008–2011
-	UMinho	-	-	-	-	-	-	-	-	▓	▓	Global Reporting Initiative (GRI) Guidelines. Dimensions: (1) economic; (2) social; (3) environmental (4) Cultural (NEW in 2012/2013)	24 (2010) to 26 (2011) to 62 (2012/2013)	2007–2013

Note: The use of the methodology or model by each HEI on their commence year when publishing their RSs; in case of UP its evolution.

It should be noted that FEUP is concerned with all Dimensions and not only environmental ones [41].

These sustainability implementations actions by the University of Porto seem to have been based on many different Dimensions. Concerning the "campus operations" Dimension, actions seem to occur through (1) sustainable landscaping; (2) policies and activities to reduce paper consumption, such as e-communications or double-sided copying; (3) renewable energy usage, through the implementation of photoelectric performance systems; and (4) energy-efficient equipment.

There were also actions relating to "SD through on-campus experiences," through (1) policies that promote SD for all students and staff; (2) sustainable practices for students; (3) a SD working group with members from different departments; (4) SD efforts that are visible throughout the campus; and (5) student participation in SD activities, such as collaboration in multiple social solidarity projects.

Concerning the "assessment and reporting" Dimension, UP seemed to have implemented sustainability through (1) RS, and (2) the assessment of SD issues using SD integration instruments and tools within the University through the total management system (SGT); the implementation of consumption monitoring routines (namely, student participation in SD activities through collaboration in multiple social solidarity projects, and the disclosure of RS); and some best practices (namely the optimization of equipment and system schedules through the centralized technical management system (SGTC) and the "paper calculator" software developed by the "Environmental Paper Network" [42] and G.A.S.PORTO - Oporto Social Action Group).

There seems to have been be special care taken regarding the publication of RS by UP/FEUP between 2008 and 2011 and the integration of instruments and tools to assess SD issues.

Regarding the "outreach and collaboration" Dimension, the action related to the involvement of academic staff in voluntary advisory activities in SD seemed to be one of the initiatives.

The UP's "institutional framework" demonstrates a commitment to the inclusion of SD in the vision, mission, goals, and objectives of the University.

The extent to which UMinho and UP were able to integrate sustainability into their strategies or policies can be found through the actions organized in themes. From there, not only did these HEIs seem to have implemented sustainability internally through campus activities and on-campus experiences, but they also did it through outreach and collaboration (external routes). Both HEIs were committed to SD within their institutional framework and deeply involved in the assessment and reporting Dimensions.

3.4. Commitment to SD of Universities without DCIs or RS

There were universities that had not signed any DCI or published any RS but were committed to SD and implemented sustainability actions.

Many HEIs used the Energetic Efficiency Program in Public Administration (Eco.AP) regarding energy efficiency in the "campus operations" Dimensions (which was the case of HEI_04, HEI_05, HEI_06, and HEI_08; see Table 5 and Figure 5).

The implementation of "SD through on-campus experiences" was found in many of the studied universities, as well as other sustainability implementation actions, such as policies that promote SD for all students and staff; in these areas, SD efforts were visible throughout the campus and some best practices were found (e.g., "knowledge sharing" and a "cultural training program").

Regarding "outreach and collaboration," the actions found were: (1) SD partnerships with other society stakeholders (HEI_08 and HEI_13), and (2) academic staff involved in voluntary advisory activities in SD (e.g., HEI_08).

One of the universities played a role in the environmental area with the creation of a sustainable campus that resulted from a partnership with GALP Energia (a Portuguese energy company) and others. Another initiative by this university involved the creation of synergies between sports and health, involving a stadium in the promotion of common projects with schools (best practice). Moreover, another university had a role in the promotion of sports and adapted sports, like canoeing, sailing, and adapted sailing, as well as in the creation of research centers and/or associated laboratories (hosting researchers from other universities).

Concerning the "education" Dimension, some HEIs created study programs (e.g., Masters—Sustainable Energy, Environment and Sustainability, PhD—Sustainable Energy Systems, which was financed by the Massachusetts Institute of Technology (MIT) program in 2007 [64], Global Change (Climate Change and Sustainable Development Policies), Social Sustainability and Development) in areas such as energy, global change, sustainability, environment and sustainability, social sustainability and development, or a combination of these terms.

In one university, the gathering of professors from different faculties, departments and research and development (R&D) units was a path to promote interdisciplinary collaboration in teaching and development. This leverages talent and financial resources and creates awareness on sustainability issues, namely in the areas of energy and SD.

At one of the studied universities, the commencement of a doctoral program in the academic year 2010/2011, which is an interdepartmental program between two departments, is a good example of a university offering education with a transdisciplinary focus. The sustainability implementation action was evidenced by course syllabuses of courses or programs on SD.

In the "research" Dimension, one university showed the existence of patents in the field of SD.

Towards a country profile for Portugal for the implementation of sustainability in Higher Education, on the basis of their likelihood in the "think tank" (Figure 4), the sustainability implementation actions were classified according to the quartiles (see Figure 5) for the overall number, for the dimensions of campus operations, outreach and collaboration, and SD through on-campus experiences (the top three).

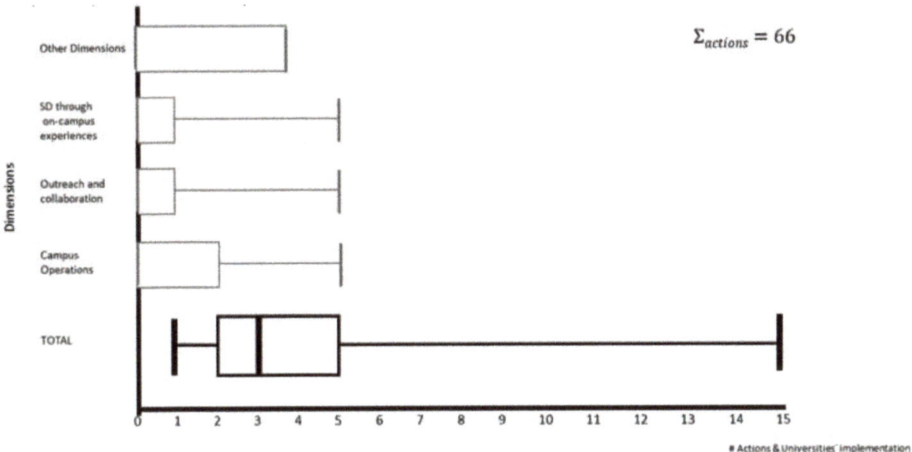

Figure 5. Box plot for the top three sustainability implementation actions in Portuguese public universities.

- Group I corresponds to the first quartile (one to two actions overall) including six universities
- Group II corresponds to the second quartile (three actions overall) including two universities
- Group III corresponds to the third quartile (four to five actions overall) including three universities
- Group IV corresponds to the fourth quartile (more than five actions overall) including three universities

Figure 5 is a box plot. Considering the top three Dimensions, the first, second, and third quartiles overlapped. This means that 75% of the Universities have taken this path to implement one or two sustainability actions. For the Dimensions education and research combined, four actions were taken in 75% of the universities.

Universities seemed to have integrated SD through multiple and simultaneous actions at their own rhythm and pace.

These findings showed no apparent relationship with the number of students or campus area because the results followed all of the steps explained in Section 2.

4. Discussion

Many European universities have integrated SD into their academic systems. There are also important connections between commitment, integration, and the signing of a DCI [2], relating to the leverage of values, attitudes, and behaviors within present and future regenerative societies [21].

The results presented in this paper show that if a university signs a declaration or a charter it seems to lead to a commitment to SD, no matter how narrow it may be, partly through the implementation of several sustainability actions. This was the case of at least four universities in Portugal (UP, UMinho, UNL, and UTL). However, sustainability implementation was present in all the other studied universities.

During the DESD 2005–2014, the results show that Portuguese public universities implemented sustainability through diverse and multiple actions, mostly by (i) establishing partnerships with other society stakeholders; (ii) implementing policies that promote SD for all students and staff; (iii) signing DCIs within SD, ESD, or sustainability during the UN decade; and also (iv) by promoting best practices.

Aleixo et al. (2018) and Arroyo et al. (2017) [65,66] refer not only to the importance of putting into practice universities' transformative role in SD by including sustainability in an institution's agenda, strategies, and best practices to promote said agenda, but also by the institution remaining engaged in the field despite facing the usual implementation problems, varying from restricted resources to lack of trained staff [3], deficient organizational structure, inertia, and resistance [66].

Based on the evidence of sustainability implementation actions, concrete proof for whether universities were committed to SD, a four-group classification was built to measure how far the policies and strategies were integrated. It showed that despite some universities having done more than others regarding the dimensions [2], all of them were engaged in SD implementation at their own pace. This is in line with published literature about Portuguese HEIs [65] that recommend a further development of sustainability initiatives for several Portuguese universities.

More than 50% of the actions in Portuguese public universities were not exclusive to a single university. Additionally, the "campus operations," "outreach and collaboration," and "SD through on-campus experiences" Dimensions represented about two thirds of the total sustainability implementation actions. Therefore, the way by which ESD has been integrated in Portuguese public universities within the United Nations Decade of Education for Sustainable Development 2005–2014 seems to have been a bottom-up approach. A university must have policies on SD which are in line with [18] when mentioning them as a pre-condition for successful sustainability efforts.

Sustainability reports are a suitable tool for universities concerning SD incorporation, but this is not a common practice [1]. RSs are a tool increasingly used by accreditation bodies, governments, and students [67]. This seems to correspond to the presented findings, as UP and UMinho were the only two universities that produce RSs.

RSs have a large potential for the process of sustainability development integration in HE, namely for organizational change, stakeholder engagement processes in RS, link between RS and general sustainability management, and relationships between existing reporting indicators, tools, and management standards [68]. Thus, the development of RSs at universities in Portugal should be widely encouraged. Aleixo et al. (2018) [66] mention that UMinho is in a SD implementation phase, due to university sustainability reports, and so this university seems to be an early adopter.

From this study's findings, best practices regarding green campus procedures were found in many of the studied universities. Indeed, campus operations are among the more commonly applied

ESD domains in universities ([9,66,69]. At this point, it should be said that the data used for this characterization can be underestimated and differences between institutions may be attributed to cataloguing methods, lack of documentation, or a less systematic search where the terms (e.g., "green campus procedures") were not formally stated.

Regarding the "outreach and collaboration" Dimension, namely "partnerships with other civil stakeholders (e.g., Non-Governmental Organizations (NGOs), municipality, regional government, etc.)," many best practices were found in Portuguese public universities (e.g., UBI and UA) which seems to be not quite in line with [70], who reported that Portugal was far behind in externally-oriented activities aimed at building capacity within local communities to promote SD.

Implementation actions relating to the "education" and "research" Dimensions were not intensely found, which is in accordance with [6] that classified Portuguese universities as "laggards" and/or "late majority".

There may be significant advancements in the operational dimensions of a university, in curricular and educational transformation as well as in research and outreach activities [71,72], but in most cases, sustainability has not yet become an integral part of the university system [73].

Notwithstanding its improvement in recent years, the requested paradigm change from un-sustainability to sustainability in university systems is not yet fully identifiable [74].

Even so, Portuguese universities show good examples of sustainability interdisciplinary curricula, particularly at the post-graduate level. The breadth and interconnectedness required for implementing the SDGs make it evident that experts from different subjects and sectors must work together to deliver the goals [16], as well as that future research should concentrate on the challenge of measuring and assessing the differing conceptualizations of "sustainability" within what the curricular offers [68,75].

Many universities are already involved in sustainability through the curricula, training, research, and community engagement activities [24]. This difference may be attributed either to the localization of public universities and/or the lack of documentation from some universities.

Communication is a core function of higher education [9]. In terms of ESD coordination and communication at the national level, it should be mentioned that there is an existing gap arising from the lack of ESD at governmental policy and strategy levels either by the Portuguese Government or the Ministry of Public Universities [4,5].

Nevertheless, there has been effective coordination between universities regarding national and international programs like Eco.AP and the MIT 2007 Program.

A detailed and deep content analysis of several documents, namely the strategic and activity plans, showed that, during the UN DESD 2005–2014, Portuguese public universities implemented sustainability actions in many different ways and Dimensions when compared with earlier studies.

Nevertheless, the initiatives found in each university were not integrated within a whole-school approach [19]. A whole-university approach for embedding sustainability in the university is fundamental for a transformation in learning and education for sustainability with interdisciplinary collaboration between academics. This is critical for promoting the needed transformation in students to become agents of a sustainable future [9,25].

Usually in these types of studies, where a profile of a region is drawn, data are gathered only by questionnaire or interview survey [75]. This systematic analysis of gathered documental data was the basis for the characterization of a country profile for Portugal for ESD implementation in universities and allowed a detail analysis usually not possible through surveys, in which response rates are often low.

Based on the searched and identified actions, a "think tank" (a tank of actions) may be widened, and a cooperation network—SharingSustainability4U—established with a list of best practices and areas for sustainability improvement, irrespective of the university's dimensions. Single universities may support and benefit from being a node in a university network for sustainability [76]. Collaboration and support among universities are key success factors as universities have not implemented sustainability at the same pace, to the same extent, and in the same Dimension(s). The Portuguese University Rectors Council can have a key role as mediator or even coordinator of this network.

5. Conclusions, Future Lines of Research, and Limitations of the Study

5.1. Main Conclusions

During the United Nations Decade of Education for Sustainable Development 2005–2014, Portuguese public universities integrated sustainability into university policies and strategies mainly through "campus operations," "outreach and collaboration," and "SD through on-campus experiences" Dimensions. Universities implemented sustainability through actions, many of which were not exclusive to only one university. One hundred and thirty-nine documents from fourteen universities were treated and analyzed to provide a better understanding of the progress regarding ESD implementation in Portuguese public universities and to find the main commitments and practices. The step-by-step treatment and systematic analysis of those documents helped to understand and value the possible sustainability implementation actions and university results on the strategies and policies of the public universities.

From this research, some important conclusions may be drawn:

1. As the largest number of codified references in public universities' documents were about integration and environmental education, it might seem that universities were not sufficiently engaged in SD during UN DESD 2005–2014, compared to the terms sustainable or sustainability, which had few references. Nevertheless, at this point some sustainability implementation actions in public universities were found in the documentation. However, outcomes show that the movement has made progress at the university level, with good examples and initiatives in several Portuguese universities, notwithstanding the insufficiency of national combined strategies or policies related to ESD;
2. UN DESD 2005–2014 was not found to be, in itself, a common motivation for implementing university sustainability, as it is not one of the most well-found codified references in universities' documents. Nevertheless, the results show that Portuguese public universities implemented sustainability through different and multiple actions whether under any DCI or not;
3. Universities' actions related to ESD seemed to have been taken in "isolation" and were not integrated according to a whole-institution approach;
4. The implementation of ESD at public universities provides insights about (best) practices regarding green campus procedures, which were found in many of the studied universities;
5. This study contributed to a country profile for the implementation of sustainability in the HE sector, highlighting the importance of analyzing the content of strategic and activity plans of HEIs. The information gathered by this systematic documental analysis is more thorough than that obtained through questionnaire surveys, a tool usually used in this kind of study.

The aims regarding the institutional document analysis from internal insights were accomplished.

5.2. Limitations of the Study and Future Research

This study had some methodological limitations. For the relevant period (2005–2014), most universities published all the documentation necessary for treatment and analysis. Nevertheless, in relation to some universities, and despite best efforts to obtain further documentation either through websites or direct contact with staff and documentation centers, it was confirmed that only a limited number of documents were actually published or made available.

In order to overcome this hindrance and to complete and/or deepen the analysis, if possible, an investigation will be pursued through interviews with the persons in charge of sustainability integration in each university to assess what has been done to implement ESD during DESD and what is being done at the present to propose strategies and policies for sustainability improvements and to share them among all universities. It is expected that a more complete country profile for ESD implementation will emerge.

Based on the country profile developed in this research, each Portuguese university could share with all stakeholders (teaching staff, students, and community) all the initiatives and (best) practices in order to increase knowledge of the work that has been done, namely in terms of partnerships, fundraising, and other actions implementing sustainability. A platform—SharingSustainability4U—sharing sustainability initiatives based on this partnership idea is suggested.

This may be widened to a European or even to a worldwide platform, as universities are not all at the same stage concerning ESD. In the near future, this platform could be a worldwide reference for all universities to share and communicate activities, projects, and results concerning their ESD implementation. From there, the policies and strategies of multiple universities may be designed towards the implementation of ESD.

Author Contributions: C.F. had the idea for the paper. All the authors read, revised, and approved the final manuscript.

Funding: This research was funded by Portuguese Science and Technology Foundation (FCT) for its funding to the Centre for Environmental and Sustainability Research (CENSE), NOVA School of Science and Technology, NOVA University of Lisbon, grant number (UID/AMB/04085/2019).

Acknowledgments: We are grateful to Madalena Carvalho from documentation Centre in Universidade Aberta, which facilitated the contacts with other Public HEIs´ librarians that beforehand were not so fruitful in our numerous attempts.
Thanks also to the all the librarians for sharing their valuable but necessarily different experiences regarding research.

Conflicts of Interest: The authors declare no conflicts of interest.

Appendix A

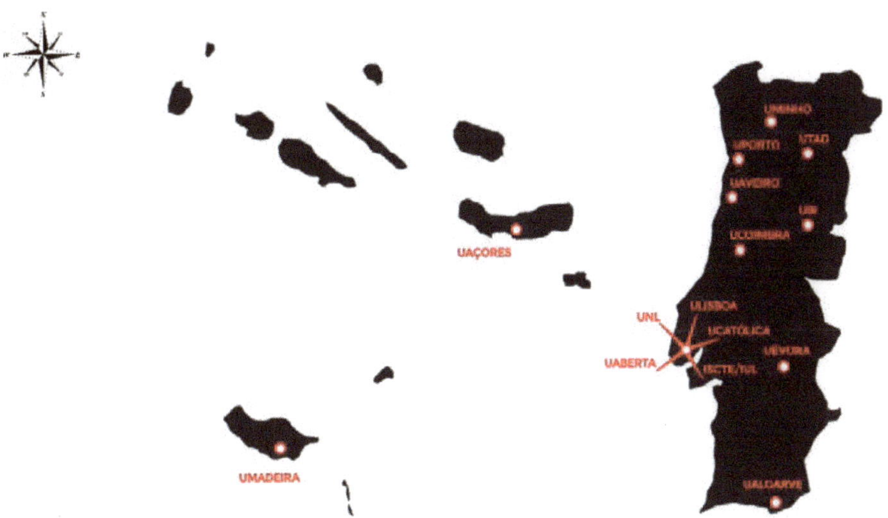

Figure A1. Distribution map of Portuguese public universities. Scale: 1:65,000 km; source: CRUP, 2018. Remark: The acronyms are not the ones by which the HEIs are generally known.

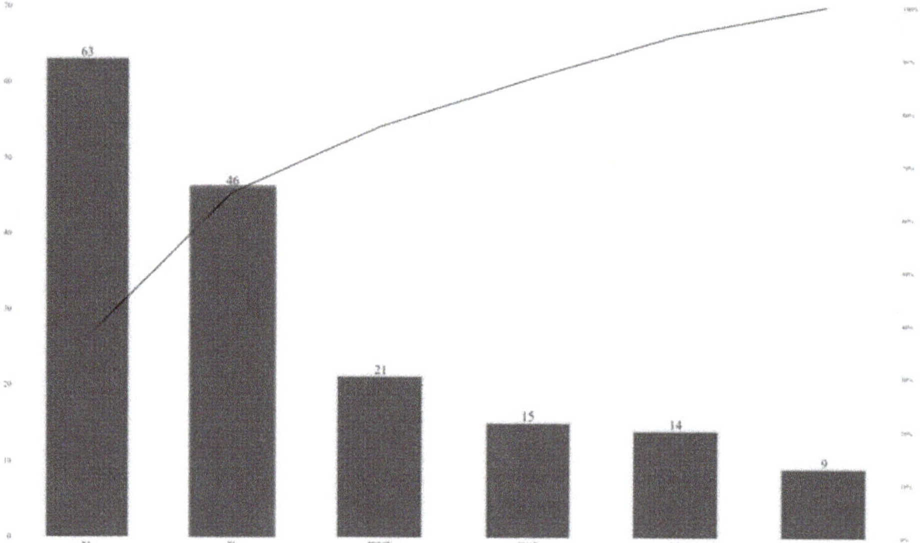

Figure A2. Type of document by each public university regarding the Pareto graph.

References

1. Alonso-Almeida, M.M.; Marimon, F.; Casani, F.; Rodriguez-Pomeda, J. Diffusion of Sustainability reporting in universities: Current situation and future perspectives. *J. Clean. Prod.* **2014**, *106*, 144–154. [CrossRef]
2. Lozano, R.; Ceulemans, K.; Alonso-Almeida, M.; Huisingh, D.; Lozano, F.; Waas, T.; Lambrechts, W.; Lukman, R.; Hugé, J. A review of commitment and implementation of sustainable development in higher education: results from worldwide survey. *J. Clean. Prod.* **2015**, *108*, 1–18. [CrossRef]
3. Leal Filho, W.; Azeiteiro, U.; Alves, F.; Pace, P.; Mifsud, M.; Brandli, L.; Caeiro, S.; Disterheft, A. Reinvigorating the sustainable development research agenda: The role of the sustainable development goals (SDG). *Int. J. Sustain. Dev. World Ecol.* **2018**, *25*, 131–142. [CrossRef]
4. Farinha, C.; Azeiteiro, U.M.; Caeiro, S. Education for sustainable development through policies and strategies in the public portuguese higher education institutions. In *Handbook of Theory and Practice of Sustainable Development in Higher Education*; Leal Filho, W., Azeiteiro, U., Alves, F., Molthan-Hill, P., Eds.; In the Series World Sustainability Development Series; Springer: Berlin, Germany, 2017; pp. 275–290. [CrossRef]
5. Farinha, C.; Azeiteiro, U.; Caeiro, S. Education for sustainable development in Portuguese universities: The key actors' opinions. *Int. J. Sustain. High. Educ.* **2018**, *19*, 912–941. [CrossRef]
6. Aleixo, A.M.; Azeiteiro, U.M.; Leal, S. Towards Sustainability through Higher Education: Sustainable Development Incorporation into Portuguese Higher Education Institutions. In *Challenges in Higher Education for Sustainabiliy*; Davim, J., Leal Filho, W., Eds.; Management and Industrial Engineering; Springer: Cham, Swizerland, 2016; pp. 159–187, ISBN 978-3-319-23705-3. [CrossRef]
7. Aleixo, A.M.; Azeiteiro, U.M; Leal, S. UN Decade of Education for Sustainable Development: Perceptions of Higher Education Institution's Stakeholders. In *Handbook of Theory and Practise of Sustainable Development in Higher Education*; Leal Filho, W., Azeiteiro, U., Alves, F., Molthan-Hill, P., Eds.; In the series World Sustainability Series; Springer: Berlin, Germany, 2017; pp. 417–428. [CrossRef]
8. Leal Filho, W.; Brandli, L.L.; Becker, D.; Skanavis, C.; Kounani, A.; Sardi, C.; Papaioannidou, D.; Paco, A.; Azeiteiro, U.M.; Raath, S.; et al. Sustainable development policies as indicators and pre-conditions for sustainability efforts at universities: Fact or fiction? *Int. J. Sustain. High. Educ.* **2017**, *19*, 1–33. [CrossRef]
9. Kapitutcinova, D.; AtKisson, A.; Perdue, J.; Will, M. Towards integrated sustainability in higher education: Mapping the use of the Accelerator toolset in all dimensions of university practice. *J. Clean. Prod.* **2018**, *172*, 4367–4382. [CrossRef]

10. Association of University Leaders for a Sustainable Future (AULSF). Talloires Declaration Signatories List. Available online: http://ulsf.org/96-2/#Portugal (accessed on 3 September 2017).

11. COPERNICUS-CAMPUS. Copernicus Guidelines for Sustainable Development in the European Higher Education Area: How to Incorporate the Principles of Sustainable Development into the Bologna Process. Available online: http://www.ehea.info/media.ehea.info/file/COPERNICUS_Olderburg_2006/92/6/COPERNICUSGuidelines_587926.pdf (accessed on 3 September 2017).

12. CRE (Conference of European Rectors). The University Charter for Sustainable Development [Carta Copernicus]. Available online: https://www.iau-hesd.net/sites/default/files/documents/copernicus.pdf (accessed on 3 September 2017).

13. United Nations. Higher Education Sustainable Initiative. Sustainable Development Knowledge Platform. Available online: https://sustainabledevelopment.un.org/sdinaction/hesi (accessed on 8 September 2017).

14. UNESCO. Aichi-Nagoya Declaration on Education for Sustainable Development. Available online: https://sustainabledevelopment.un.org/content/documents/5859Aichi-Nagoya_Declaration_EN.pdf (accessed on 10 July 2018).

15. McCowan, T. Universities and the post-2015 developments agenda: an analytical framework. *High. Educ.* **2016**, *72*, 505–523. [CrossRef]

16. Anna-Diab, F.; Molinari, C. Interdisciplinarity: Practical approach to advancing education for sustainability and for the Sustainable Development Goals. *Int. J. Manag. Educ.* **2017**, *15*, 73–83. [CrossRef]

17. Cicmil, S.; Gough, G.; Hills, S. Insights into responsible education for sustainable: The case of UWE, Bristol. *Int. J. Manag. Educ.* **2017**, *15*, 293–305. [CrossRef]

18. Fuertes-Camacho, M.; Graell-Martin, M.; Fuentes-Loss, M.; Balaguer-Fábregas, M. Integrating Sustainability into Higher Education Curricula through the Project Method, a Global Learning Strategy. *Sustainability* **2019**, *11*, 767. [CrossRef]

19. UNESCO. United Nations Decade of Education for Sustainable Development (2005–2014). Exploring Sustainable Development: A Multiple-Perspective Approach. Education for Sustainable Development in Action. Learning & Training Tools, N. 3. Available online: https://unesdoc.unesco.org/ark:/48223/pf0000215431 (accessed on 10 September 2017).

20. Soini, K.; Jurgilevich, A.; Pietikainen, J.; Korhonen-Kurki, K. Universities responding to the call for sustainability: A typology of sustainability centres. *J. Clean. Prod.* **2018**, *170*, 1423–1432. [CrossRef]

21. Sonetti, G.; Brown, M.; Naboni, E. About the Triggering of UN Sustainable Development Goals and Regenerative Sustainability in Higher Education. *Sustainability* **2019**, *11*, 254. [CrossRef]

22. Salvia, A.; Leal Filho, W.; Brandli, L.; Griebeler, J. Assessing research trends related to Sustainable Development Goals: Local and global issues. *J. Clean. Prod.* **2019**, *208*, 841–849. [CrossRef]

23. European Union. Quality Education Is Essential for Development. Available online: http://www.europarl.europa.eu/sides/getDoc.do?pubRef=-//EP//TEXT+TA+P8-TA-2018-0075+0+DOC+XML+V0//EN (accessed on 30 January 2019).

24. Pradhan, M.; Mariam, A. The Global Universities Partnership on Environment and Sustainability (GUPES): Networking of Higher Educational Institutions in Facilitating Implementation of the UN Decade of Education for Sustainable Development 2005–2014. *J. Educ. Sustain. Dev.* **2014**, *8*, 171–175. [CrossRef]

25. Leal Filho, W.; Raath, S.; Lazzarini, B.; Vargas, V.R.; Souza, L.; Anholon, R.; Quelhas, O.L.G.; Haddad, R.; Klavins, M.; Orlovic, L. The role of transformation in learning and education for sustainability. *J. Clean. Prod.* **2018**, *199*, 286–295. [CrossRef]

26. Casarejos, F.; Gustavson, L.; Frota, M. Higher Education Institutions in the United States: Commitment and coherency to sustainability vis-à-vis dimensions of the institutional environment. *J. Clean. Prod.* **2017**, *159*, 74–84. [CrossRef]

27. Goni, F.; Chofreh, A.; Mukhtar, M.; Sahran, S.; Shukor, S.; Klemes, J. Strategic alignment between sustainability and information systems: A case analysis in Malaysian public Higher Education Institutions. *J. Clean. Prod.* **2017**, *168*, 263–270. [CrossRef]

28. Lotz-Sisitka, H. Stories of transformation. *J. Sustain. High. Educ.* **2006**, *N.1*, 8–10. [CrossRef]

29. Glover, A.; Peters, C. A whole sector approach: Education for Sustainable Development ang Global Citizenship in Wales. In *Sustainability Assessment Tools in Higher Education Institutions: Mapping Trends and Good Practices around the World*; Caeiro, S., Ed.; Springer International: Cham, Switzerland, 2013; pp. 205–222.

30. Koscielniak, C. A consideration of the changing focus on the sustainable development in higher education in Poland. *J. Clean. Prod.* **2014**, *62*, 114–119. [CrossRef]

31. Dlouhá, J.; Glavic, P.; Barton, A. Higher Education in central European countries and critical factors for sustainability transition. *J. Clean. Prod.* **2017**, *151*, 670–684. [CrossRef]

32. Jorge, M.; Madueño, J.; Cejas, M.; Peña, F. An approach to the implementation of sustainability practices in Spanish universities. *J. Clean. Prod.* **2014**. [CrossRef]

33. Ramos, T.; Montaño, M.; Joanaz de Melo, J.; Souza, M.; Carvalho de Lemos, C.; Domingues, A.; Polido, A. Strategic Environmental Assessment in higher education: Portuguese and Brazilian cases. *J. Clean. Prod.* **2015**. [CrossRef]

34. Heitor, M.; Horta, H. Reforming higher education in Portugal in times of uncertainty: The importance of illities, as non-functional requirements. *Technol. Forecast. Soc. Chang.* **2015**, *113*, 146–156. [CrossRef]

35. Universidade dos Açores (UAc). História. Available online: http://novoportal.uac.pt/pt-pt/historia (accessed on 27 May 2018).

36. Universidade do Minho (UMinho). Sustainability Report 2011. Available online: www.uminho.pt (accessed on 19 December 2015).

37. Universidade do Minho (UMinho). Sustainability Report 2012–2013. Available online: www.uminho.pt (accessed on 19 December 2015).

38. Universidade do Minho (UMinho). Homepage. 2015. Available online: www.uminho.pt (accessed on 20 January 2019).

39. Universidade Aberta (UAb). Activity Report 2013. Available online: www.uab.pt (accessed on 15 December 2015).

40. Universidade Aberta (UAb). Homepage. Available online: http://portal.uab.pt/en/auab/ (accessed on 19 December 2017).

41. Universidade do Porto (UP). Sustainability Report 2010. Available online: www.up.pt (accessed on 19 December 2015).

42. Universidade do Porto (UP). Environmental Paper Network. Available online: www.papercalculator.org (accessed on 19 December 2015).

43. Universidade de Trás os Montes and Alto Douro (UTAD). Homepage. Available online: https://www.utad.pt/en/university/ (accessed on 23 April 2018).

44. Universidade de Évora (UÉ). Homepage. Available online: www.ue.pt (accessed on 23 January 2019).

45. Universidade de Évora (UÉ). General Description of University. Available online: http://www.en.mobilidade.uevora.pt/University/General-description-of-university (accessed on 17 April 2018).

46. Universidade da Beira Interior (UBI). Activity and Account Plan 2011. Available online: www.ubi.pt (accessed on 11 October 2015).

47. Universidade da Beira Interior (UBI). Activity Plan 2013. Available online: www.ubi.pt (accessed on 29 October 2015).

48. Universidade da Beira Interior (UBI). Strategical Plan 2010–2020. 2016. Available online: www.ubi.pt (accessed on 15 May 2016).

49. Universidade da Beira Interior (UBI) History. Available online: www.ubi.pt/en/page/history (accessed on 22 December 2017).

50. Universidade de Coimbra (UC). Homepage. Available online: http://www.uc.pt/en/international (accessed on 16 May 2018).

51. EfS [Energy for Sustainability Initiative]. *Contribution for Institutional Formation of University of Coimbra in Interdisciplinary Energy Domain [Contribuição para a Formação Institucional da Universidade de Coimbra no Domínio Interdisciplinar da Energia]*; Universidade de Coimbra: Coimbra, Portugal, 2011.

52. ULisboa. *Action Plan 2014–2017: Proposal of the Rector to the General Council*; ULisboa: Lisbon, Portugal, 2014.

53. Universidade Nova de Lisboa (UNL). Home/Nova/History. Available online: http://www.unl.pt/en/nova/history (accessed on 19 May 2018).

54. Universidade de Aveiro (UA). Home Page. Available online: http://www.ua.pt (accessed on 30 May 2018).

55. ISCTE-IUL. Home Page. Available online: ISCTE-IUL:https://www.iscte-iul.pt/ (accessed on 27 May 2017).

56. Universidade Católica Portuguesa (UCP). Institutional Introduction. Available online: http://www.ucp.pt/site/custom/template/ucptplportalpag.asp?sspageID=5&lang=2 (accessed on 27 May 2017).

57. CRUP. Quem Somos? Available online: www.crup.pt (accessed on 27 May 2017).

58. CRUP. Estatuto Jurídico do Conselho de Reitores das Universidades Portuguesas. Available online: http://www.crup.pt/crup/sitecrup/wp-content/uploads/2016/10/Estatutos_Juridicos.pdf (accessed on 27 May 2017).

59. Bardin, L. *Content Analysis [Análise de Conteúdo]*; Edições 70: Lisbon, Portugal, 1977.

60. Bryman. *Social Research Methods (Volume 4)*; University Press: Oxford, UK, 2012.

61. Saunders, M.; Lewis, P.; Thornhill, A. *Research Methods for Business Students*, 5th ed.; Prentice Education: Harlow, UK, 2009.

62. QRS International. NVIVO10 and NVIVO 11. Available online: https://www.qsrinternational.com/ (accessed on 8 October 2016).

63. Global Report Initiative (GRI). Information. Available online: https://www.globalreporting.org/information/sustainability-reporting/Pages/default.aspx (accessed on 19 November 2017).

64. MIT Portugal. Available online: www.mitportugal.org (accessed on 17 April 2017).

65. Aleixo, A.M.; Leal, S.; Azeiteiro, U.M. Conceptualizations of Sustainability in Portuguese Higher Education: Roles, Barriers and Challenges toward Sustainability. *J. Clean. Prod.* **2018**, *172*, 1664–1673. [CrossRef]

66. Arroyo, P. A new taxonomy for examining the multi-role of campus sustainability assessments in organizational change. *J. Clean. Prod.* **2017**, *140*, 1763–1774. [CrossRef]

67. Stough, T.; Ceulemans, K.; Lambrechts, W.; Cappunyns, V. Assessing sustainability in higher education curricula: A critical reflection on validity issues. *J. Clean. Prod.* **2018**, *172*, 4456–4466. [CrossRef]

68. Ceulemans, K.; Molderez, I.; Liedekerke, L.V. Sustainability reporting in higher education: A comprehensive review of the recent literature and paths for further research. *J. Clean. Prod.* **2015**, *106*, 127–143. [CrossRef]

69. Findler, F.; Schonherr, N.; Lozano, R.; Stacherl, B. Assessing the Impacts of Higher Education Institutions on Sustainable Development: An Analysis of Tools and Indicators. *Sustainability* **2019**, *11*, 59. [CrossRef]

70. Shiel, C.; Leal Filho, W.; Paço, A.; Brandli, L. Evaluating the engagement in universities in capacity building for sustainable development in local communities. *Eval. Program Plann.* **2016**, *54*, 123–134. [CrossRef]

71. Global University Network for Innovation (GUNI). *Higher Education in the World 4. Higher Education's Commitment to Sustainability: From Understanding to Action*; Palgrave Macmillan: Hampshire, UK, 2011.

72. Lozano, R.; Lozano, F.; Mulder, K.; Huisingh, D.; Waas, T. Advancing higher education for sustainable development: international insights and critical reflections. *J. Clean. Prod.* **2013**, *48*, 3–9. [CrossRef]

73. Disterheft, A.; Caeiro, S.; Azeiteiro, U.; Leal Filho, W. Sustainability science and education for sustainable development in universities: A Way for Transition. In *Sustainability Assessment Tools in Higher Education Institutions*; Caeiro S., Filho W., Jabbour C., Azeiteiro U., Eds.; Springer: Cham, Swizerland, 2013; pp. 3–27, ISBN 978-3-319-02374-8.[CrossRef]

74. Zutsh, A.; Creed, A. Declaring Taillores: Profile of sustainability communications in Australia signatory universities. *J. Clean. Prod.* **2018**, 687–698. [CrossRef]

75. Aleixo, A.M.; Azeiteiro, U.; Leal, S. The implementation of sustainability practices in Portuguese higher education institutions. *Int. J. Sustain. High. Educ.* **2017**, *19*, 146–178. [CrossRef]

76. Kahhe, J.; Risch, K.; Wanke, A. Strategic networking for sustainability: lessons learned from two case studies in higher education. *Sustainability* **2018**, *10*, 4646. [CrossRef]

Article

Reform of Chinese Universities in the Context of Sustainable Development: Teacher Evaluation and Improvement Based on Hybrid Multiple Criteria Decision-Making Model

Sung-Shun Weng [1], Yang Liu [1,*] and Yen-Ching Chuang [2]

[1] Department of Information and Finance Management, National Taipei University of Technology, Taipei 10608, Taiwan; wengss@ntut.edu.tw
[2] Department of Industrial Engineering and Management, National Taipei University of Technology, 1, Sec. 3, Zhongxiao E. Rd., Taipei 10608, Taiwan; yenching.chuang@gmail.com
*   Correspondence: t106749404@ntut.edu.tw; Tel.: +86-138-1191-7538

Received: 13 July 2019; Accepted: 29 September 2019; Published: 2 October 2019

Abstract: China is pushing universities to implement reforms in order to achieve the sustainable development goals, but with the development level of teachers becoming the key restricting factor. In this sense, teacher evaluation and improvement act as positive factors for China to achieve the 2030 sustainable development goals. Previous studies on teacher evaluation have usually assumed that the relationship between the evaluation criteria is independent, with the weights of each standard derived from this assumption. However, this assumption is often not in line with the actual situation. Decisions based on these studies are likely to waste resources and may negatively impact the efficiency and effectiveness of teachers' sustainable development. This study developed an integrated model for the evaluation and improvement of teachers based on the official teacher evaluation criteria of China's International Scholarly Exchange Curriculum (ISEC) programme and a multiple criteria decision-making methodology. First, a decision-making trial and a laboratory-based analytical network process were used to establish an influential network-relation diagram (INRD) and influential weights under ISEC standards. Next, an important performance analysis was used to integrate the weight and performance of each standard to produce a worst-performance criterion set for each university teacher. Finally, the worst performance set used an INRD to derive an improvement strategy with a cause–effect relationship for each teacher. This study chose a Chinese university that has implemented teaching reform for our case study. The results show that our developed model can assist decision-makers to improve their current evaluations of teachers and to provide a cause–effect improvement strategy for education reform committees and higher education institutions.

Keywords: sustainable development; International Scholarly Exchange Curriculum (ISEC) standards; university teacher evaluation and improvement; multiple criteria decision-making (MCDM); decision-making trial and evaluation laboratory (DEMATEL); DEMATEL-based analytical network process (DANP); importance-performance analysis (IPA)

---

1. Introduction

Education is crucial to sustainable development. The action plan entitled 'Transforming our World: The 2030 Agenda for Sustainable Development' jointly concluded among 193 countries was released at the World Summit on Sustainable Development in September 2015 [1]. Since then, the goal of sustainable development has become a new goal of global development. One of the goals in the agenda is about ensuring inclusive and equitable quality education and lifelong learning opportunities for all [2,3]. The Talloires Declaration announced in 1990 pioneered the inclusion of sustainable development in higher education, highlighting the role of higher educational institutions in promoting global sustainable development [3]. The higher educational institutions subsequently began to shoulder the task of training human resources with the vision of sustainable development. Following the identification of the Global Education Roadmap 2030 at the World Education Forum 2015, the Global Action Program on Education for Sustainable Development was launched. Sustainable development and education for sustainable development thus gained stronger momentum to be promoted worldwide [4]. According to UNESCO, education for sustainable development can guarantee the future of the economy, the environment, and society. To achieve the goal, universities are undergoing teaching and research reforms [5]. China published the National Plan on Implementation of the 2030 Agenda for Sustainable Development in September 2016. The Plan conveyed China's resolution to advance the said 2030 Agenda [6]. A sub-plan was proposed to fulfil the educational development goal, further facilitating the reform of China's higher educational institutions from the perspective of sustainable development. In 2019, China's Ministry of Education (MOE) came up with the plan to develop nearly 10,000 national top and 10,000 provincial top majors for the undergraduates, thus boosting the reform of teaching activities in higher education institutions [7].

Teachers play a fundamental role in the reforms [3,8]. While higher education institutions are carrying out reforms in response to the goal of sustainable development, sustainability-based teaching activities pose new challenges to teachers at universities. The Teaching Staff Development Plan is a feasible approach to accelerating the integration of education for sustainable development [5,9]. For China's universities, particularly the regional ones (i.e., universities established by governments under the provincial level), the quality of teachers has become a key obstacle to reform. Therefore, in 2018, the Action Program to Rejuvenate Education for the Teaching Staff 2018–2022 was released by five ministries in China, including the MOE [10]. In this context, regional universities began to invest heavily in the training and re-education of the teaching staff, in the hope that they would stand out in the competition that takes sustainable development as its goal.

However, a key defect with China's higher education is the lack of, and uneven distribution of sufficient quality education resources [11,12]. At the same time, existing education resources have not been fully leveraged yet [13,14]. Among such resources, there is an important issue of university teacher development wherein massive amounts of invested resources are inefficiently used. Unless addressed properly, this issue would hinder universities in China from realizing the goal of education for sustainable development. If teacher evaluations can identify the core reasons why and where they perform poorly before resources are invested in specific areas, the information can be used to effectively improve the resource investment and use rate in Chinese universities. To solve this problem, a subsidiary department of the Chinese MOE developed the International Scholarly Exchange Curriculum (ISEC) program based on the directive of "globalizing education to deepen reforms in higher education". University teacher evaluation and improvement is part of the ISEC program, where the assessment criteria of university teachers are also known as the ISEC standard in this study.

Teacher evaluation as a management tool of education aims to facilitate the growth of the teaching staff [15–18] and are within the scope of multiple criteria decision-making (MCDM). For example, Ghosh [19] combined the analytic hierarchy process (AHP) and the technique for order of preference by similarity to ideal solution (TOPSIS) in order to evaluate faculty performance in engineering education. Pavani et al. [20] developed an expert-based group model for evaluating teacher performance using fuzzy AHP and TOPSIS. Xu et al. [21] evaluated teaching performance on a smart campus. Wang et al. [22] proposed a hybrid model for classroom teaching performance based on TOPSIS and the triangle fuzzy number. These studies provided valuable contributions to teacher evaluations. However, the relationships among the criteria in these models are independent and do not reflect real conditions. To address this issue, the suggestions for teacher improvement provided by these MCDM models are often directed toward the improvement of poor performance, but the factors that lead to poor performance may not always be addressed by these improvement strategies. As the

Chinese proverb goes, "treat the head when the head aches, treat the foot when the foot hurts". Thus, the improvement of the teaching capacity is not satisfactory, resulting in resource wastage and scarcity.

To address this issue, the study combined the MCDM with the teacher evaluation standards adopted by ISEC in order to create a new evaluation model. First, the evaluation model is the use of China's ISEC standards. These standards in China constitute the practical application of ISEC teachers to choose and improve problems. Based on the ISEC standards, the decision-making trial and evaluation laboratory (DEMATEL)-based analytic network process (ANP; together DANP) was then used to establish an influential network-relation diagram (INRD), and subsequently to obtain the influential weights within ISEC standards. The derived INRD has been proven to be an effective tool to explore the cause and effect relations in many papers [23–27]. Second, the importance-performance analysis (IPA) was used to combine the influential weights and performance of attributes, and to capture a set of criteria for identifying where each teacher performs worst. Finally, decision-makers can use the INRD to focus on the causation of poor performance to determine the actual factors related to each teacher's performance, and to prioritise a direction for improvement. The integrated model of DEMATEL and IPA methods has been successfully applied to different studies [28–32]. The model focuses on analyzing the factors influencing poor teacher performance according to certain criteria with the intention of using fewer but more focused resources to produce effective improvements. This method provides a new mechanism for the sustainable development of university teachers based on evaluations and also supplements the inadequacies of the existing studies to a certain degree.

In this study, empirical data from 15 domain experts from the ISEC management institute were applied to demonstrate our proposed model. The results show that "Professional ethics and literacy (C_1)" is the primary influential standard and "Teacher ability and development (C_2)" has the highest influential weight. Teachers A and B both performed poorly in "Teaching performance (C_{31})" and "Research cooperation (C_{32})". In other words, teachers A and B must improve their performance in these two standards. According to traditional performance improvement strategies, the ISEC management institute should invest resources in encouraging teachers to publish their teaching results and promote research cooperation among teachers. In practice, teacher A and teacher B enjoy the same training and development resources. However, the essential causes of their under-performance vary, which, accordingly, requires different training resources. This phenomenon reveals the issue of the untargeted allocation of teaching staff development resources in the reform of China's universities.

The rest of this paper is structured as follows. Section 2 provides a brief introduction of current teacher evaluations and outlines the literature documenting research methods. Section 3 introduces the DANP and IPA methods used in our new model. Section 4 details the implementation of ISEC topics in this model. Section 5 discusses the results and features of the model, and Section 6 summarizes our contributions and directions for future research.

2. Review of University Teacher Evaluation Models

The previous university teacher evaluation models can be roughly divided into three research stages: (1) the selection of appropriate criteria in the evaluation model, (2) building the decision-making model using the MCDM methodology, and (3) building the decision-making model using statistical or data analysis methodologies.

2.1. Selection of Appropriate Criteria in the Evaluation Model

This stage of research is focused on the selection of subjective and objective indicators in an evaluation model. Contradictions and conflicts exist between indicators, such as encouraging teacher vision and personal development versus increasing wages, and between objective evaluation results and critical feedback [33–35]. For example, when Mills and Hyles [36] evaluated university teachers at Oklahoma State University's Stillwater College (Stillwater, OK, USA), the results provided limited feedback and drawbacks included scattered goals, unclear standards, and inconsistencies in management perspectives. Based on this, they used interviews and surveys to understand opinions on teacher evaluations, and their development process established a consolidated solution with a hierarchical architecture that integrated goals, directions, and procedures for annual performance evaluations. For this reason, some studies began to explore the establishment of more reasonable evaluation indicators and weights. For example, Desselle et al. [37] studied university teacher evaluation systems at Duquesne University's Mylan School of Pharmacy (Stillwater, OK, USA). They used a modified Delphi procedure to confirm 29 teaching activities and 44 academic actions, including their weights, in an evaluation standard. Filipe et al. [38] provided guidelines for avoiding conflict between processes and goals by developing a message management system to evaluate teaching activity—however, they encountered difficulties similar to those of previous studies due to differing interpretations of how to implement and assess teaching. Subjective indicators are an inevitable aspect of the assessment process, creating ambiguity that hinders the transparency and fairness of teaching performance assessments. However, some studies have shifted from subjective indicators that rely on the evaluator (e.g., enthusiasm for one's work) to relatively objective indicators (e.g., the number of reference papers). Although this can prevent fuzziness due to subjectivity, it results in fewer facets of evaluation that cannot fully reflect a teacher's true performance level [39–41]. Based on these lessons, others have proposed to combine decision analysis models with objective and subjective indicators or mathematical programming models, but they often lack real-time scoring systems based on theoretical rationality or have weight settings that vary depending on the evaluator's subjectivity. This neglects the principle of value trade-off and results in a total evaluation score that may have no substance or value. Using a methodology for systematic decision modelling to establish the factors and their weights in an evaluation system is a key problem that remains unsolved [42–44].

2.2. Building Decision-Making Models Using MCDM Methodology

The MCDM is specifically applied to solve evaluative decision-making problems with multiple criteria, i.e., evaluation, selection, and improvement problems. Some researchers have used MCDM methods to build decision models. For example, Ghosh [29] used AHP and TOPSIS to review teacher evaluations in engineering schools. Filipe et al. [38] developed a multi-criteria information system to review teaching practices. Hein et al. [45] used tools such as consensus theory, information entropy, and TOPSIS to construct a multi-criteria decision analysis method to evaluate 56 university professors. However, people's representations of their opinions with regard to objects or events, in reality, contain fuzziness. Therefore, some scholars have developed various fuzzy-based MCDM models. For example, Chen et al. [46] proposed a framework for teaching evaluation based on a combination of fuzzy AHP and fuzzy comprehensive evaluation methods. Chang and Wang [47] proposed a type of multi-criteria decision-making model oriented toward teachers in an attempt to solve the issues of pervasive subjectivity, imprecision, and fuzziness within the faculty. Dey Mondal and Ghosh [48] used AHP, fuzzy AHP, complex proportional assessment of alternatives with grey relations (COPRAS), and TOPSIS in combination with game theory and compromise planning methods in MCDM to evaluate the performance level of teachers. These models provide decision-makers with a simple and easy-to-use method for evaluating and selecting university teachers. However, in these models, decision-making is based on independent relationships among criteria that cannot provide decision-makers with suitable systemic improvement strategies for all university teachers.

2.3. Building Decision-Making Models Using Statistical or Data Analysis Methodologies

The last research stage has involved overcoming the fuzzy defects of these MCDM models that use statistical or data analysis methods to construct decision-making models. For example, Nikolaidis and Dimitriadis [49] established a framework based on statistical quality control to use student feedback to a maximum degree. Lyde et al. [50] used a multisource method for evaluation (MME) and improved constraints such as the timing of reflections, accountability from year to year, and mentoring in order to construct a more comprehensive formative teaching assessment tool. Bi [51] evaluated five years of teaching at a management school of a university by creating a mean and standard deviation diagram based on statistical process control theory. Xu et al. [31] used principal component analysis (PCA) to calculate and identify six primary components and then used AHP to calculate the weight of each hierarchy before using grey correlation to improve the TOPSIS target decision analysis algorithm to avoid errors in decision-making due to subjective factors. These models offer a perspective on data behaviour as a basis for decision-making, but they rely on massive amounts of data and are unable to provide decision-makers with causal influence relationships affecting teacher performance.

2.4. Research Gaps in Their Decision-Making Models

Past decision-making models presented different contributions to teacher evaluation, selection, and improvement problems. However, these MCDM models have a major defect in that the relationship between criteria in the evaluation model is independent. Therefore, current MCDM models cannot help decision-makers to obtain a guide for the performance improvement of each teacher. To fill the research gap, this study developed a novel MCDM model that uses the DANP method to construct the INRD and influential weights for criteria and the IPA method to search the worst performances of criteria for each teacher. The detailed modelling process and its corresponding method are described in Section 3. The comparison of the three categories of decision-making models is shown in Table 1.

Table 1. Comparison with different decision-making models.

Category	Characteristic	Limitations or Current Defects
Selection of appropriate criteria in the evaluation model	The research is focused on the selection of subjective and objective indicators in an evaluation model.	• The modelling process of the scoring system is not considered the relationship between indicators. • The weight setting of the indicator depends on the evaluator's subjectivity.
Building the decision-making model using MCDM methodology	The research used MCDM methods to construct various decision-making models for multi-criteria evaluation, selection, and improvement problems.	• The decision-making depends on a group of experts' domain knowledge. • The current MCDM models are based on independent relationships among criteria that cannot provide decision-makers with suitable systemic improvement strategies for each teacher.
Building the decision-making model using statistical or data analysis methodologies	The research used statistical or data analysis methods to construct decision-making models for overcoming the fuzzy defects of original MCDM models.	• The models are unable to provide decision-makers with causal influence relationships affecting teacher performance.

3. Our Proposed Hybrid DANP-IPA Model

This study developed an integrated hybrid MCDM model that combines the DANP method and IPA analysis. The former can be used to derive the influential network-relationship diagram (INRD) and influential weights that can help decision-makers to understand the cause-effect direction based on a systemic perspective. The latter can help decision-makers to easily capture the worst performance of each teacher in all attributes. Finally, the worst performance attributes of each teacher can be based on the INRM to develop a series of the most appropriate improvement strategies. The modelling flow diagram and corresponding methods of this hybrid DANP-IPA model are depicted in Figure 1.

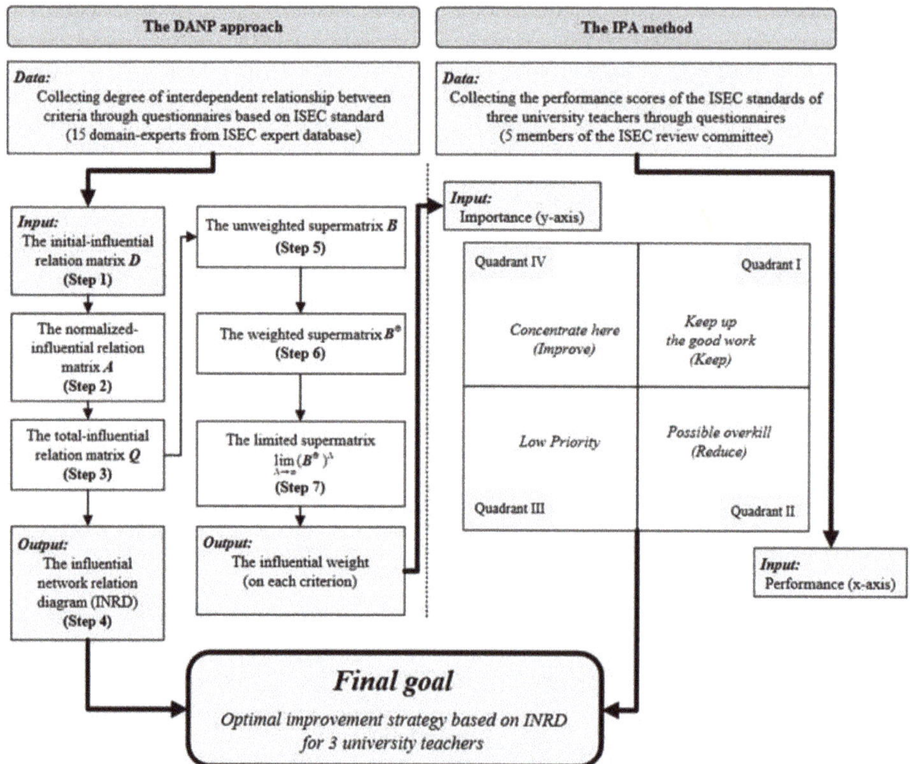

Figure 1. Modelling flow diagram of the decision-making trial and evaluation laboratory (DEMATEL)-based analytic network process (DANP)-importance–performance analysis (IPA) model.

3.1. DANP Method

The DANP method was developed by Lee et al. [52] by combining the DEMATEL technique [53] and the ANP method [54]. The DANP method retains interdependent relationships among criteria and further derives an influential network-relation diagram (INRD) and influential weights for all criteria. The INRD established by the DANP method can help to form decision-making equations for various systemic plans to improve alternative/objective performance after evaluation [55]. Based on this advantage, the method has been applied in many areas, such as public open space development [56], creative communities [57], quality of life [58], supplier management [59], airline performance [60], green buildings [61], and international airports [62]. The detailed steps in the DANP calculation are as follows.

Step 1: Build an initial influence–relationship matrix.

For an evaluation criteria model, respondents assess the degree of influence between criteria using a pairwise comparison based on a five-point Likert scale (ranging from 0 = "no influence" to 4 = "extremely high influence"). Then, the influential matrices of all respondents are integrated into a matrix by averaging to produce the initial influence relationship matrix D. Matrix D represents the actual experience within the group of all respondents:

$$D = \left[\left(\sum\nolimits_{\Theta=1}^{n} c_{ij}^{\Theta} \right) / \alpha \right]_{n \times n} \tag{1}$$

where c_{ij}^{Θ} is the result of the respondent Θ, indicating the degree of influence between criteria i and j; α is the total number of respondents, and n is the total number of criteria.

Step 2: Derive a normalized influence–relationship matrix.

The initial–influence relationship matrix D derives a normalized influence-relationship matrix A using Equations (2) and (3), in which all diagonal terms are 0 and the maximum sum of a row or column is 1:

$$\rho = \max_{i,j}\left[\max_i \sum\nolimits_{j=1}^{n} d_{ij}, \max_j \sum\nolimits_{i=1}^{n} d_{ij}\right] \tag{2}$$

$$A = C/\rho \tag{3}$$

where ρ is the maximum value of the sum of a row or column.

Step 3: Obtain a total influence–relationship matrix.

Matrix A calculates and adds the influence degree of each iteration through the Markov chain process and produces a total influence relation matrix Q, as shown in Equation (4):

$$Q = A + A^2 + \cdots + A^{\delta} = A(I - A)^{-1}, \text{ when } \lim_{\delta \to \infty} A^{\delta} = [0]_{n \times n} \tag{4}$$

Step 4. Build an influential network–relationship diagram (INRD).

First, the sum of each row and column can obtain vectors u_i and v_i through Equations (5) and (6). Then, $(u_i + v_i)$ is the total strength of influences given and received, or prominence, as shown in Equation (7). Otherwise, $(u_i - v_i)$ is the net influence degree between given and received influences, also called the cause/effect. A positive cause/effect value indicates that factor i affects other factors and belongs to the cause group; if the value is negative, factor i is affected by other factors and belongs to the effect group. Finally, the INRD is established based on the vectors of prominence and cause/effect:

$$u_i = (u_i)_{n \times 1} = \left[\sum\nolimits_{j=1}^{n} q_{ij}\right]_{n \times 1}, \ i \in \{1, 2, \ldots, n\} \tag{5}$$

$$v_i = (v_j)'_{1 \times n} = \left[\sum\nolimits_{i=1}^{n} q_{ij}\right]_{1 \times n}, \ j \in \{1, 2, \ldots, n\} \tag{6}$$

where $\prime$ denotes transposition, u_i indicates the sum of direct and indirect effects of the factor i on the other factors, and v_i indicates the sum of direct and indirect effects factor i received from the other factors.

Step 5: Transfer to an unweighted supermatrix.

First, the total influence–relation matrix Q can be divided into two matrices: the attribute level Q_C and the dimension level Q_D. Second, each row within a dimension in the total influence relation matrix Q_C uses Equations (7)–(9) to obtain the normalized total influence relation matrix Q_C^{ρ}, as shown in

Equation (7), in which $Q_C^{\rho 11}$ is an example to demonstrate the basic concept of normalizing, as shown in Equations (8) and (9).

$$
Q_C^{\rho} = \begin{array}{c}
\\ D_1 \\ \\ \vdots \\ \\ D_i \\ \\ \vdots \\ \\ D_m
\end{array}
\begin{array}{c}
c_{11} \\ c_{12} \\ \vdots \\ c_{1m_1} \\ \vdots \\ c_{i1} \\ c_{i2} \\ \vdots \\ c_{im_i} \\ \vdots \\ c_{m1} \\ c_{m2} \\ \vdots \\ c_{mm_m}
\end{array}
\begin{array}{c}
\begin{array}{ccccc} D_1 & & D_j & & D_m \\ c_{11}...c_{1m_1} & \cdots & c_{j1}...c_{jm_j} & \cdots & c_{m1}...c_{mm_m} \end{array} \\
\left[\begin{array}{ccccc}
Q_c^{\rho 11} & \cdots & Q_c^{\rho 1j} & \cdots & Q_c^{\rho 1m} \\
\vdots & & \vdots & & \vdots \\
Q_c^{\rho i1} & \cdots & Q_c^{\rho ij} & \cdots & Q_c^{\rho im} \\
\vdots & & \vdots & & \vdots \\
Q_c^{\rho m1} & \cdots & Q_c^{\rho mj} & \cdots & Q_c^{\rho mm}
\end{array} \right]_{n\times n | m<n, \sum_{j=1}^{m} m_j = n}
\end{array}
\tag{7}
$$

$$
q_i^{11} = \sum_{j=1}^{m_1} q_{ij}^{11}, \; i = 1, 2, \ldots, m_1
\tag{8}
$$

$$
Q_C^{\rho 11} = \begin{bmatrix}
q_{11}^{11}/q_1^{11} & \cdots & q_{1j}^{11}/q_1^{11} & \cdots & q_{1m_1}^{11}/q_1^{11} \\
\vdots & & \vdots & & \vdots \\
q_{i1}^{11}/q_i^{11} & \cdots & q_{ij}^{11}/q_i^{11} & \cdots & q_{im_1}^{11}/q_i^{11} \\
\vdots & & \vdots & & \vdots \\
q_{m_11}^{11}/q_{m_1}^{11} & \cdots & q_{m_1j}^{11}/q_{m_1}^{11} & \cdots & q_{m_1m_1}^{11}/q_{m_1}^{11}
\end{bmatrix}
= \begin{bmatrix}
q_{11}^{\alpha 11} & \cdots & q_{1j}^{\alpha 11} & \cdots & q_{1m_1}^{\alpha 11} \\
\vdots & & \vdots & & \vdots \\
q_{i1}^{\alpha 11} & \cdots & q_{ij}^{\alpha 11} & \cdots & q_{im_1}^{\alpha 11} \\
\vdots & & \vdots & & \vdots \\
q_{m_11}^{\alpha 11} & \cdots & q_{m_1j}^{\alpha 11} & \cdots & q_{m_1m_1}^{\alpha 11}
\end{bmatrix}
\tag{9}
$$

Lastly, the normalized influence relation matrix Q_c^{ρ} is transposed to obtain the unweighted supermatrix $B = (Q_c^{\rho})'$, as shown in Equation (10):

$$
B = (Q_C^{\rho})' = \begin{array}{c}
\\ D_1 \\ \\ \vdots \\ \\ D_j \\ \\ \vdots \\ \\ D_m
\end{array}
\begin{array}{c}
c_{11} \\ c_{12} \\ \vdots \\ c_{1m_1} \\ \vdots \\ c_{j1} \\ c_{j2} \\ \vdots \\ c_{jm_j} \\ \vdots \\ c_{i1} \\ c_{i2} \\ \vdots \\ c_{mm_m}
\end{array}
\begin{array}{c}
\begin{array}{ccccc} D_1 & & D_i & & D_m \\ c_{11}...c_{1m_1} & \cdots & c_{i1}...c_{im_i} & \cdots & c_{m1}...c_{mm_m} \end{array} \\
\left[\begin{array}{ccccc}
B^{11} & \cdots & B^{i1} & \cdots & B^{m1} \\
\vdots & & \vdots & & \vdots \\
B^{1j} & \cdots & B^{ij} & \cdots & B^{mj} \\
\vdots & & \vdots & & \vdots \\
B^{1m} & \cdots & B^{in} & \cdots & B^{mm}
\end{array} \right]_{n\times n | m<n, \sum_{j=1}^{m} m_j = n}
\end{array}
\tag{10}
$$

Step 6: Obtain a weighted supermatrix.

Each row within the goal in the total influence–relation matrix Q_D uses Equations (11)–(13) to obtain the normalized total influence relation matrix Q_D^ρ, as shown in Equation (11). Matrices Q_C^ρ and Q_D^ρ produce a new matrix through Equation (14), the weighted supermatrix B^Θ:

$$Q_D = \begin{bmatrix} q_{11} & \cdots & q_{1j} & \cdots & q_{1m} \\ \vdots & & \vdots & & \vdots \\ q_{i1} & \cdots & q_{ij} & \cdots & q_{im} \\ \vdots & & \vdots & & \vdots \\ q_{m1} & \cdots & q_{mj} & \cdots & q_{mm} \end{bmatrix}_{m \times m} \tag{11}$$

$$d_i = \sum_{j=1}^{m} q_D^{ij}, i = 1, 2, \ldots, m \text{ and } q_D^{\rho ij} = q_D^{ij}/d_i, j = 1, 2, \ldots, m \tag{12}$$

$$W = (Q_D^\rho)' = \begin{bmatrix} f_D^{11}/d_1 & \cdots & f_D^{1j}/d_1 & \cdots & f_D^{1m}/d_1 \\ \vdots & & \vdots & & \vdots \\ f_D^{i1}/d_i & \cdots & f_D^{ij}/d_i & \cdots & f_D^{im}/d_i \\ \vdots & & \vdots & & \vdots \\ f_D^{m1}/d_m & \cdots & f_D^{mj}/d_m & \cdots & f_D^{mm}/d_m \end{bmatrix}_{m \times m} = \begin{bmatrix} f_D^{\rho 11} & \cdots & f_D^{\rho 1j} & \cdots & f_D^{\alpha 1m} \\ \vdots & & \vdots & & \vdots \\ f_D^{\rho i1} & \cdots & f_D^{\rho ij} & \cdots & f_D^{\rho im} \\ \vdots & & \vdots & & \vdots \\ f_D^{\rho m1} & \cdots & f_D^{\rho mj} & \cdots & f_D^{\rho mm} \end{bmatrix}_{m \times m} \tag{13}$$

$$B^\Theta = W \times B = \begin{bmatrix} q_D^{\rho 11} \times B^{11} & \cdots & q_D^{\rho i1} \times B^{i1} & \cdots & q_D^{\rho m1} \times B^{m1} \\ \vdots & & \vdots & & \vdots \\ q_D^{\rho 1j} \times B^{1j} & \cdots & q_D^{\rho ij} \times B^{ij} & \cdots & q_D^{\rho mj} \times B^{mj} \\ \vdots & & \vdots & & \vdots \\ q_D^{\rho 1m} \times B^{1m} & \cdots & q_D^{\rho im} \times B^{im} & \cdots & q_D^{\rho mm} \times B^{mm} \end{bmatrix} \tag{14}$$

Step 7: Limit the weighted supermatrix and derive the influential weights.

The weighted supermatrix B^Θ convergences the influence degree of each time through the Markov chain process and finally obtains the influential weights for all criteria/dimensions, as shown in Equation (15).

$$\lim_{\Lambda \to \infty} (B^\Theta)^\Lambda \tag{15}$$

3.2. Importance–Performance Analysis Method

Importance–performance analysis (IPA) is a well-known business management method [63] that was first developed by Martilla and James [64] to identify the critical performance criteria of products or services [65,66]. The method is used to create an IPA matrix or priority map using standard performance and importance scores, which can be divided into four quadrants (Q1–Q4), as shown in Figure 2 [31,63,67]:

(1) Q1: Keep up the good work, indicating the main strengths and potential competitive advantages of a product or service.
(2) Q2: Possible overkill, representing that these criteria are a low priority for customers. That is, the organization should reduce resources directed toward these criteria because resources are limited.
(3) Q3: Low priority, representing criteria that are not important to customers and not performing exceptionally well. The organization should not care too much about these attributes.
(4) Q4: Concentrate here, representing the service's primary weaknesses and threats to its competitiveness. For the organization, these criteria have the highest priority in terms of investment.

This approach can help decision-makers to easily understand the performance and importance of criteria. The method is widely used in many different areas, such as supplier management [31], tourism development [66,68], and strategy management [69].

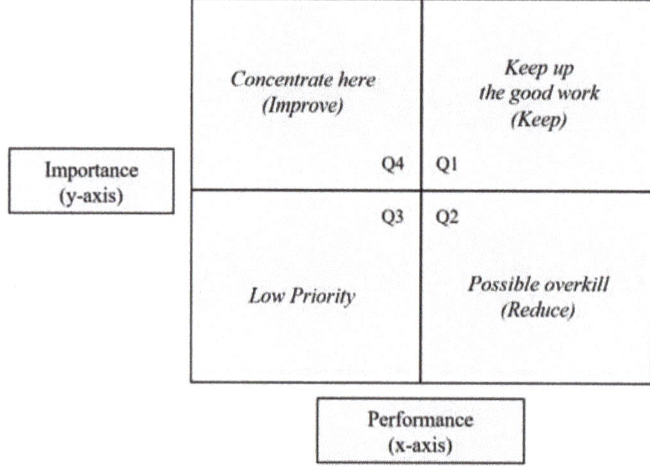

Figure 2. Importance-performance analysis (IPA) map.

4. Empirical Case

In this section, an empirical study using data from the ISEC in China is presented to illustrate the application of the proposed DANP-IPA model for evaluating and improving the performance of university teachers.

4.1. Case Background Problem Description

Compared with developed countries, China lacks higher education resources and their distribution is imbalanced, especially at regional universities, which constitute the majority of universities in China. To solve this problem, a subsidiary department of the Chinese MOE developed the International Scholarly Exchange Curriculum (ISEC) program based on the directive of "globalizing education to deepen reforms in higher education". Universities that participated in the program received assistance with curriculum, teaching, and quality assurance reforms—however, these reforms were based on teachers. For this reason, ISEC sought to establish a sustainable development mechanism in order to cultivate quality teaching teams that would autonomously reform curriculums and teaching and achieve the goal of comprehensive teaching reform. This line of thinking is a departure from China's current method of promoting change from the top downward. The core of this development program involves teachers and the sustainable development of promoting reforms in higher education. The corresponding content includes (1) curriculum systems, (2) support for teacher development systems, and (3) service systems. Currently, the program has been implemented in approximately 30 test universities in nine provinces with more than 1500 teachers listed in ISEC, accumulating practical experience with both successes and failures. These ISEC teachers have met the ISEC inclusion criteria, i.e., age, academic qualifications, foreign study or exchange experience, international curriculum teaching experience, critical thinking, and ISEC mission acceptance. The program prioritizes the development of teachers on the front line as its core. ISEC is the coordinating mechanism and training platform for teacher development, partnering with universities to use global educational resources to support the improvement of teacher abilities. Next, ISEC teachers act as leaders to push teaching reforms throughout the university and faculty. Therefore, establishing a decision-making model with practical value to assist decision-makers with effectively evaluating ISEC teachers and improving their abilities is a critical problem.

With assistance from the ISEC management institute, 15 ISEC domain experts were selected from the ISEC expert database (one American, two Australian, and 12 Chinese). These experts were teachers, administrative staff members, or ISEC teacher representatives (associate professor or above) at renowned universities. The ISEC domain experts represent elite ISEC teachers. As such, this study is based on the ISEC university teacher standard indicator system (Table 2), and it integrated MCDM and IPA tools to establish a mixed multi-criteria decision-making model. This DANP-IPA model can be used to evaluate and improve ISEC standards for teachers.

Table 2. University teachers' evaluation system based on ISEC standards.

Dimension	Criterion	Content
Professional ethics and literacy (C_1)	Professional ethics (C_{11})	According to the Code of Professional Ethics of Teachers in Higher Education formulated by the MOE, teachers are examined for their moral performance in the teaching process. Evaluation of a teacher's professionalism, professional sense of belonging, and physical and mental health
	Professional literacy (C_{12})	
Teacher ability and development (C_2)	Basic teaching skills (C_{21})	Evaluation of future planning ability and level of teaching
	Teaching implementation (C_{22})	Achievement of teaching goals and amount of teaching resources used
	External review (C_{23})	Evaluation of teaching performance by students, peers, and experts
	Professional development (C_{24})	Planning and implementation of personal career development
Teacher performance and contributions (C_3)	Teaching performance (C_{31})	Contributions to teaching performance, awards, research standards, and teaching teams
	Research cooperation (C_{32})	Status of personal and group research results

4.2. INRD and Influential Weight Using the DANP Method

The degree of influence between standards was calculated using a five-point measurement scale for all experts, and then Equation (1) was used to consolidate and obtain an initial influence–relationship matrix (Table 3). In this matrix, Equations (2)–(4) allowed the inference of a total influence–relationship matrix (Table 4). Using Equations (5) and (6), the influence structure of each standard was obtained (Table 5) to draw an INRD (Figure 3).

Table 3. Initial influence–relation matrix.

Criteria	C_{11}	C_{12}	C_{21}	C_{22}	C_{23}	C_{24}	C_{31}	C_{32}
C_{11}	0.00	3.20	2.67	3.00	2.87	2.93	2.73	2.07
C_{12}	2.87	0.00	2.87	3.00	3.00	2.93	2.87	2.40
C_{21}	1.67	2.13	0.00	3.80	3.73	3.13	3.00	2.33
C_{22}	2.07	2.33	3.00	0.00	4.00	3.13	3.53	2.33
C_{23}	1.80	2.33	3.00	3.20	0.00	3.33	3.47	2.40
C_{24}	2.20	2.60	3.07	3.20	3.20	0.00	3.27	2.93
C_{31}	1.67	2.13	2.93	3.00	3.13	3.27	0.00	3.13
C_{32}	1.73	2.27	1.93	2.13	2.33	3.53	2.87	0.00

Table 4. Total influence–relation matrix.

Criteria	C_{11}	C_{12}	C_{21}	C_{22}	C_{23}	C_{24}	C_{31}	C_{32}
C_{11}	0.58	0.81	0.90	0.97	1.01	1.00	0.98	0.81
C_{12}	0.70	0.70	0.92	0.99	1.03	1.02	1.00	0.83
C_{21}	0.66	0.78	0.80	1.01	1.05	1.02	1.01	0.83
C_{22}	0.68	0.81	0.94	0.89	1.08	1.05	1.05	0.85
C_{23}	0.65	0.78	0.91	0.98	0.89	1.02	1.01	0.82
C_{24}	0.69	0.82	0.94	1.01	1.05	0.92	1.04	0.87
C_{31}	0.64	0.76	0.89	0.96	1.00	1.00	0.86	0.84
C_{32}	0.58	0.69	0.77	0.83	0.87	0.91	0.88	0.63

Table 5. Sum of given influence (r_i) and received influence (d_i).

Dimension	r_i	d_i	$r_i + d_i$	$r_i - d_i$	Criterion	r_i	d_i	$r_i + d_i$	$r_i - d_i$
C_1	2.59	2.10	4.69	0.48	C_{11}	7.06	5.18	12.24	1.88
					C_{12}	7.20	6.16	13.36	1.04
C_2	2.64	2.86	5.50	−0.22	C_{21}	7.16	7.08	14.25	0.08
					C_{22}	7.35	7.66	15.00	−0.31
					C_{23}	7.07	7.98	15.05	−0.91
					C_{24}	7.35	7.94	15.29	−0.59
C_3	2.38	2.64	5.02	−0.27	C_{31}	6.96	7.82	14.77	−0.86
					C_{32}	6.17	6.49	12.65	−0.32

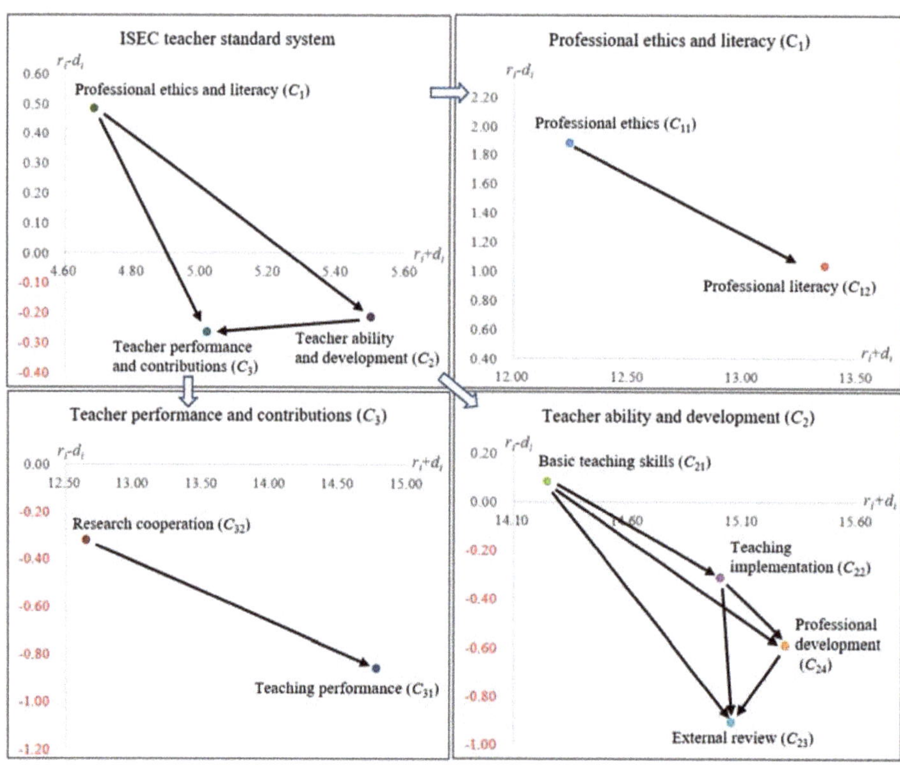

Figure 3. Influential network-relation diagram (INRD) of ISEC teacher standard evaluation system.

Figure 3 shows the entire mutual influence network within the ISEC teacher standard evaluation system, where "Professional ethics and literacy (C_1)" is the primary standard influencing "Teacher ability and development (C_2)" and "Teacher performance and contributions (C_3)"—this shows that "Professional ethics and literacy (C_1)" is the basis of two criteria. That is, a teacher's professional ethics and literacy impact their abilities and future development, which is reflected in performance and teaching contributions. Further analysis showed that "Professional ethics (C_{11})" and "Teaching implementation (C_{22})" are causal groups (i.e., $r_i - d_i > 0$). "Research cooperation (C_{32})" and "Teaching performance (C_{31})" are effect groups (i.e., $r_i - d_i < 0$). Past studies focused on decision-makers investing resources to improve specific standards to correct poor performance. However, this may not address the true cause of the problems as decision-makers may neglect the influence structure of standards being composed of interdependent, not independent, relationships. When decision-makers focus solely on poorly performing standards (effect) and invest massive amounts of resources, they do not recognize that the problem may stem from causal standards. For instance, "Research cooperation (C_{32})" maybe a teacher's poorest performing standard and the decision-maker may hope that the teacher can cooperate with other researchers. They then host workshops to provide cooperative opportunities and invest massive amounts of resources to encourage teachers to work together on research projects. However, teachers may want to focus on lectures and teaching due to their own state of "Professional ethics (C_{11})" and "Professional literacy (C_{12})", which is conservative. They may simply lack an assertive attitude toward learning, leading to poor research ability. In other words, various factors can cause "Research cooperation (C_{32})" to be the area of poorest performance. Here, INRD provides a systemic view that assists decision-makers in understanding the relationship structure influencing the standards to pinpoint each teacher's problems. The aspect of "Basic teaching skills (C_{21})" in "Teacher ability and development" is a basic standard in teaching because it influences other standards such as "Teaching implementation (C_{22})", "Professional development (C_{24})", and "External review (C_{23})".

For the weights of the influencing relationships, the total influence–relationship matrix (Table 4) uses Equations (7)–(10) to establish an unweighted supermatrix (Table 6). Next, Equations (11)–(14) are used in the matrix to establish a weighted supermatrix (Table 7). Finally, Equation (15) is used to achieve a stable, extreme supermatrix. The influence weights for each standard are shown in Table 8.

Table 6. Unweighted supermatrix.

Criteria	C_{11}	C_{12}	C_{21}	C_{22}	C_{23}	C_{24}	C_{31}	C_{32}
C_{11}	0.42	0.50	0.46	0.46	0.46	0.46	0.46	0.45
C_{12}	0.58	0.50	0.54	0.54	0.54	0.54	0.54	0.55
C_{21}	0.23	0.23	0.21	0.24	0.24	0.24	0.23	0.23
C_{22}	0.25	0.25	0.26	0.22	0.26	0.26	0.25	0.25
C_{23}	0.26	0.26	0.27	0.27	0.23	0.27	0.26	0.26
C_{24}	0.26	0.26	0.26	0.26	0.27	0.23	0.26	0.27
C_{31}	0.55	0.55	0.55	0.55	0.55	0.54	0.51	0.58
C_{32}	0.45	0.45	0.45	0.45	0.45	0.46	0.49	0.42

Table 7. Weighted supermatrix.

Criteria	C_{11}	C_{12}	C_{21}	C_{22}	C_{23}	C_{24}	C_{31}	C_{32}
C_{11}	0.11	0.14	0.13	0.13	0.13	0.13	0.13	0.13
C_{12}	0.16	0.13	0.15	0.15	0.15	0.15	0.15	0.15
C_{21}	0.09	0.09	0.08	0.09	0.09	0.09	0.09	0.09
C_{22}	0.10	0.09	0.10	0.08	0.10	0.10	0.09	0.09
C_{23}	0.10	0.10	0.10	0.10	0.09	0.10	0.10	0.10
C_{24}	0.10	0.10	0.10	0.10	0.10	0.09	0.10	0.10
C_{31}	0.19	0.19	0.19	0.20	0.19	0.19	0.17	0.20
C_{32}	0.16	0.16	0.16	0.16	0.16	0.16	0.17	0.14

Table 8. Influence weights for each criterion of ISEC standard.

Dimension	Local Weight	Ranking	Criterion	Local Weight	Ranking	Global Weight	Ranking
C_1	0.277	3	C_{11}	0.458	2	0.127	4
			C_{12}	0.542	1	0.150	3
C_2	0.376	1	C_{21}	0.231	4	0.087	8
			C_{22}	0.249	3	0.094	7
			C_{23}	0.260	1	0.098	5
			C_{24}	0.260	2	0.098	6
C_3	0.347	2	C_{31}	0.545	1	0.189	1
			C_{32}	0.455	2	0.158	2

Table 8 shows the influential weight of each standard within the entire system and the degree of influence for each standard, which is beneficial for subsequent teacher evaluation processes, as the performance in each standard considers the degree of influence. Judging from the results, the dimension of "Teacher ability and development (C_2)" has the highest influential weight and reflects teacher ability and development as the most influential relationship in the evaluation system. In addition to being driven by "Professional ethics and literacy (C_1)", performance is also reflected by "Teacher performance and contributions (C_3)". A teacher's long-term planning and development in each period of "Teacher performance and contributions (C_3)" impacts improvements in future "Teacher ability and development (C_2)", which is why "Teacher ability and development (C_2)" is the most influential dimension. "Teaching performance (C_{31})" and "Research cooperation (C_{32})" are the top two criteria in terms of influential weight because they reflect performance in the other dimensions, so if other dimensions perform poorly, this is reflected in "Teaching performance (C_{31})" and "Teacher ability and development (C_2)".

4.3. University Teacher Evaluation Using the IPA Method

Based on teacher performance and weights, this section outlines our use of the IPA method to analyze and gather the standard in which each teacher performs the poorest. With the assistance of the ISEC management institute, five members of the review committee and three university teachers participated in this study. All five members had experience in evaluating ISEC teachers for more than one semester; they used a 0–10-point scale to evaluate three university teachers. These scores were averaged and consolidated into a performance score (Tables 9–11). Next, the centre values of weights and performance were used as threshold values to separate standards into four groups as follows: Group I—high weights and performance; group II—low weights and high performance; group III—low weights and performance, and group IV—high weights and low performance. Decision-makers must focus on group IV, as standards within the group are categorized as high weight but the performance in these standards is the poorest. The investment of resources should prioritize the standards in this group to effectively improve the performance in the group. The analysis results for the three university teachers are provided in Table 12 and Figures 4–6.

Figures 4–6 show that teacher C performed the best overall. Teachers A and B perform poorest in the standards of "Teaching performance (C_{31})" and "Research cooperation (C_{32})". In other words, teachers A and B must improve their performance in these two standards. Based on traditional performance improvement strategies, the ISEC management institute should invest resources to encourage these teachers to publish their teaching results in order to promote academic cooperation among teachers. However, their problems may not actually stem from these two standards because their performance simply reflects the existence of a problem. To avoid this issue, the cause–effect relationship analysis of INRD (Figure 3) can be used to understand the entire issue and propose appropriate improvement measures for each university professor—this is also the solution that allows both the minimization of resources and the maximization of benefit. This will be discussed in further detail in the next section.

Table 9. The average performance of university teacher A.

Criteria	Member_1	Member_2	Member_3	Member_4	Member_5	Average
C_{11}	8	7	8	7	8	7.6
C_{12}	8	7	8	7	8	7.6
C_{21}	7	8	8	6	7	7.2
C_{22}	7	8	7	8	8	7.6
C_{23}	7	8	7	6	7	7
C_{24}	6	6	6	6	6	6
C_{31}	5	5	5	6	5	5.2
C_{32}	5	5	5	5	5	5

Table 10. The average performance of university teacher B.

Criteria	Member_1	Member_2	Member_3	Member_4	Member_5	Average
C_{11}	8	7	8	7	8	8
C_{12}	8	7	7	7	8	8
C_{21}	9	7	8	8	8	9
C_{22}	8	7	8	8	7	8
C_{23}	8	8	8	9	8	8
C_{24}	6	6	6	6	6	6
C_{31}	8	8	6	7	7	8
C_{32}	6	6	5	6	6	6

Table 11. The average performance of university teacher C.

Criteria	Member_1	Member_2	Member_3	Member_4	Member_5	Average
C_{11}	8	8	8	8	8	8
C_{12}	8	8	8	8	8	8
C_{21}	8	9	8	8	8	8
C_{22}	9	9	9	9	7	9
C_{23}	9	8	9	9	8	9
C_{24}	9	9	9	9	9	9
C_{31}	10	10	10	8	10	10
C_{32}	10	10	10	10	8	10

Table 12. IPA method for university teacher evaluation.

Criterion	Weight	Teacher A		Teacher B		Teacher C	
		Performance	Group	Performance	Group	Performance	Group
C_{11}	0.127	7.6	I	7.6	I	8.0	I
C_{12}	0.150	7.6	I	7.4	IV	8.0	I
C_{21}	0.087	7.2	III	8.0	II	8.2	II
C_{22}	0.094	7.6	II	7.6	II	8.6	II
C_{23}	0.098	7.0	III	8.2	II	8.6	II
C_{24}	0.098	6.0	III	6.0	III	9.0	II
C_{31}	0.189	5.2	IV	7.2	IV	9.6	I
C_{32}	0.158	5.0	IV	5.8	IV	9.6	I

Note: Center value as a threshold value (7.5, 0.125). The values of x and y are derived from the central point between the maximum and minimum of performance and weight, respectively.

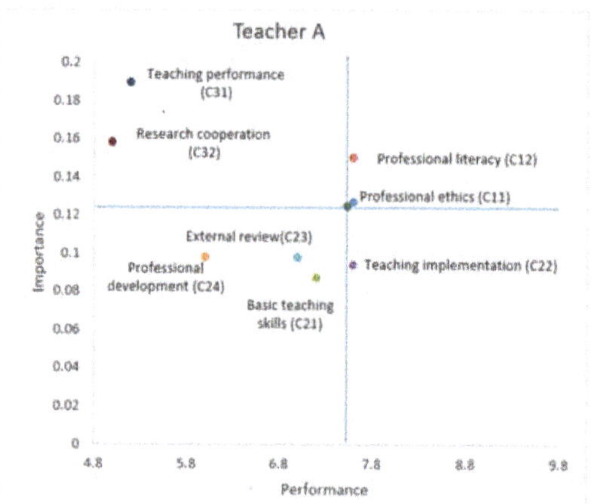

Figure 4. IPA of teacher *A* using the standard evaluation system.

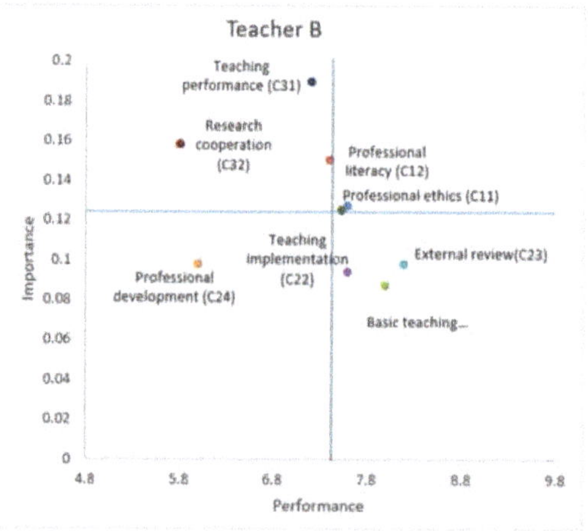

Figure 5. IPA of teacher *B* using the standard evaluation system.

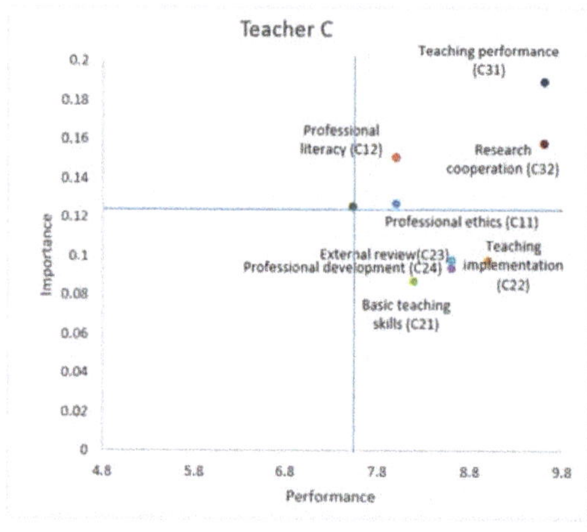

Figure 6. IPA of teacher *C* using the standard evaluation system.

5. Discussion

This section presents the results of the IPA with a cause-and-effect relationship analysis based on the INRD to create plans and strategies for teacher improvement. Finally, the influential weights produced from this study and actual current weights are explored.

5.1. Proposal for Improvement Suggestions Based on the INRD

To understand how INRD is used, this study used the case of teacher B to explain how INRD is used in analyzing and proposing improvement measures. Figure 7 shows that teacher B performed poorly in "Teaching performance (C_{31})" and "Research cooperation (C_{32})". Of the two, "Research cooperation (C_{32})" was associated with the poorest performance. However, the standards that impact "Teacher performance and contributions (C_3)" are "Teacher ability and development (C_2)" and "Professional ethics and literacy (C_1)". The teaching-related standards are "Basic teaching skills (C_{21})", "Teaching implementation (C_{22})", and "External review (C_{23})", which all had good scores (7.6 to 8.2). This was also reflected in the "Teaching performance" score (7.2). Of these, "Professional development (C_{24})" was the poorest performing standard (6.0), which may be due to different factors, such as not listing research cooperation in professional development planning, using the majority of their daily time for teaching or administrative tasks, not having the ability or opportunity to conduct research cooperation, or not having a suitable research budget. All these factors can lead to poor performance in "Research cooperation (C_{32})". Based on the above analysis, this study conducted discussions with ISEC management staff and proposed suggestions for improvements based on the perspectives of "Professional ethics and literacy (C_1)" and "Teacher ability and development (C_2)". The university should reassess teacher B's professional ethics and literacy, research ability, and time spent teaching, and adjust these three aspects, for instance, by improving a teacher's professional acknowledgment, adjusting their lecture and teaching time, arranging for them to learn the skills required for research, and arranging opportunities for research cooperation. The ISEC Management Institute could arrange a series of comprehensive courses to improve research capability and post-curricular meetings for research cooperation.

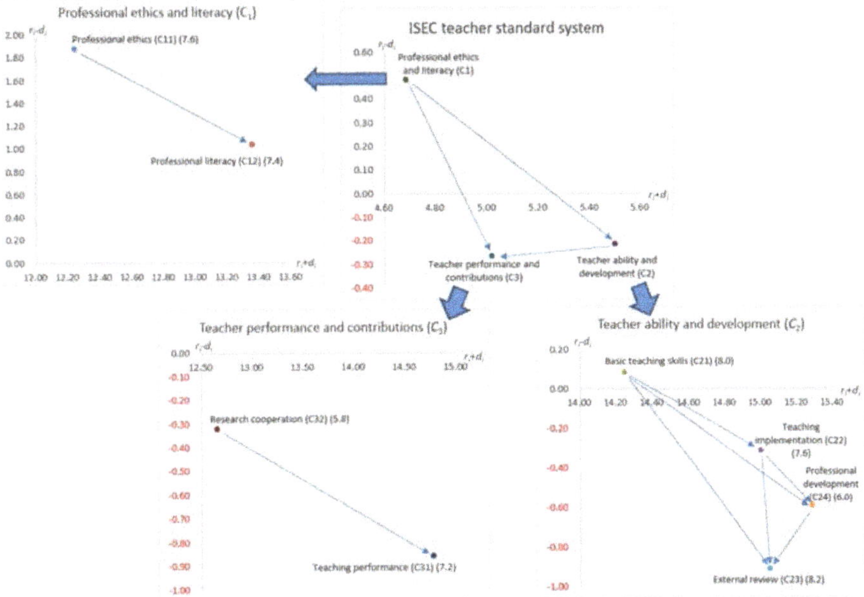

Figure 7. INRD for teacher B.

5.2. Comparison of Weights

Next, this section compared the weights obtained by the analysis of ISEC teacher standards using the DANP and AHP methods to actual weights currently being used, as shown in Table 13. This study discovered the following: (1) actual weights and DANP weights stem from the cumulative practical teaching experience of experts, and the results show that the rankings in both are close. This result shows that the establishment of actual weights contains implicit systemic perspectives. (2) Actual weights do not provide specific values but approximate values. For instance, the standards "Basic teaching skills (C_{21})" and "Professional development (C_{24})" are both weighted 0.2, whereas "Teaching implementation (C_{22})" and "External review (C_{23})" are both weighted 0.3. The same weight values provide the impression that the standards "Basic teaching skills (C_{21})" and "Professional development (C_{24})" are equally important, and "Teaching implementation (C_{22})" and "External review (C_{23})" are equally important. However, the DANP weights show that: "External review (C_{23})" > "Professional development (C_{24})" > "Teaching implementation (C_{22})" > "Basic teaching skills (C_{21})". By comparison, precise weights allow for precise evaluations along with subsequent use and investment of resources for teacher development. From a practical standpoint, this does not merely involve categorizing the importance of standards. (3) Currently, actual weights do not precisely describe factors. Further, past studies have often used the AHP method as weight analysis of criteria in teacher evaluation models [19,20,29,46,48]. In this study, the ranking of dimensions and criteria between the AHP method and the other two methods is a little different, however, AHP weights can provide specific values for each criterion. However, the AHP method assumes that the relationship between the criteria is independent, which is inconsistent with the operation of the real world. On the contrary, the DANP method used to obtain the weights of the standards from the perspective of systemic influence allows these systemic perspectives to identify cause-and-effect relationships using INRD. This assigns greater significance and high explanatory power to standard weights. Based on this feature, the DANP has been successfully applied to different issues [56–62]. Compared to the previous model, the DANP method employed in this study can provide more specific information to help decision-makers to obtain a complete systematic solution.

Table 13. Comparative analysis of weights.

Dimension/Criterion	Real Case		DANP		AHP	
	Local Weight	Ranking	Local Weight	Ranking	Local Weight	Ranking
Professional ethics and literacy (C_1)	0.2	3	0.277	3	0.459	1
Professional ethics (C_{11})	0.4	2	0.458	2	0.519	1
Professional literacy (C_{12})	0.6	1	0.542	1	0.481	2
Teacher ability and development (C_2)	0.5	1	0.376	1	0.324	2
Basic teaching skills (C_{21})	0.2	2	0.231	4	0.237	2
Teaching implementation (C_{22})	0.3	1	0.249	3	0.220	3
External review (C_{23})	0.3	1	0.260	1	0.354	1
Professional development (C_{24})	0.2	2	0.260	2	0.188	4
Teacher performance and contributions (C_3)	0.3	2	0.347	2	0.218	3
Teaching performance (C_{31})	0.7	1	0.545	1	0.816	1
Research cooperation (C_{32})	0.3	2	0.455	2	0.184	2

6. Conclusions and Remarks

ISEC teacher evaluation standards are a key aspect of Chinese education reform. In the past, the ISEC management institute has focused on teachers' poorest performing standards to propose a series of improvement measures. However, these measures did not consider the mutual influence structures amongst the standards. Therefore, the improvements often treat the symptoms but not the source, and resources are not used efficiently for maximum benefit. Standard weights are expressed in integers that cannot effectively differentiate the degree of relative importance among standards and explain the management significance behind each weight. This type of evaluation process does not truly reflect the abilities of each teacher. To solve this issue, this study developed a mixed multi-criteria decision-making (DANP-IPA) model based on ISEC teacher standards.

First, this model provides the INRD and influential weights based on the systemic perspective. INRD assists decision-makers in understanding the influential relationship structure among standards. Influential weights integrate influential perspectives into subsequent processes of teacher evaluation so that improvement measures are based on cause and effect. In the practical case studies used in this research, INRD identified "Professional ethics and literacy (C_1)" as the primary influential standard that impacts "Teacher ability and development (C_2)" and "Teacher performance and contributions (C_3)". In other words, the basic factors that affect teacher ability and performance are the individual's own professional ethics and literacy. "Teacher ability and development (C_2)" was shown to have the highest influential weight because it is driven by "Professional ethics and literacy" and is reflected in "Teacher performance and contributions (C_3)". A teacher's long-term planning and development and "Teacher performance and contributions (C_3)" in each period were found to be influenced by future improvements in "Teacher ability and development (C_2)". This is why "Teacher ability and development (C_2)" was found to have the highest influential ranking of all dimensions. "Teaching performance (C_{31})" and "Research cooperation (C_{32})" were the two highest in terms of influential weight because they reflect the performance of other standards, meaning that if other standards perform poorly, this is eventually reflected in "Teaching performance (C_{31})" and "Research cooperation (C_{32})". Finally, the IPA performance analysis showed that teacher C had the best overall performance—in comparison, teachers A and B both performed poorly in "Teaching performance (C_{31})" and "Research cooperation (C_{32})". Teachers A and B must improve their performance in these two standards. According to traditional performance improvement strategies, the ISEC management institute should invest resources to encourage teachers to publish their teaching results and promote research cooperation among teachers. However, these problems may not actually stem from these two standards because they are simply a reflection of the problem.

Based on the above analysis, this study conducted discussions with ISEC management staff and provided suggestions for improvements based on the perspectives of "Professional ethics and literacy (C_1)" and "Teacher ability and development (C_2)". Specifically, the university should reassess teacher B's professional ethics and literacy, research ability, and time spent teaching and adjust these three aspects, for example, by improving the teacher's professional acknowledgment, adjusting their lecture and teaching time, arranging for them to learn the skills required for research, and arranging opportunities for research cooperation. The ISEC management institute could arrange a series of comprehensive courses to improve research capability and post-curricular meetings for research cooperation. Therefore, this study contributes to this topic as follows: (1) our method can be combined with INRD (Figure 3) to analyze the cause-effect relationship and understand the entire problem to propose the most suitable improvement measures for each university professor. (2) Our solution both minimizes resources and maximizes benefits to improve the efficiency of resource investment in the development of university teachers. Our method will have a catalyzing effect on the continued development and cultivation of teachers.

Although this study provides a scientific decision-making model, the orientation of future research involves the exploration of multiple facets. This study focused on the establishment of the model and the application of ISEC teacher assessment and improvement. In our current research, the limitations are the ISEC evaluation standards and improvement strategies that must depend on domain-experts' knowledge (i.e., influential network-relation diagram and influential weights). The data source requires the expertise of a group of ISEC domain experts with practical experience in the issue area. Therefore, the data sources (i.e., DANP and performance) require support from the ISEC management institute. In addition, the student's perspective in the evaluation process and the subsequent development of the teacher were not included in the scope of this research, which forms another limitation of this study. Based on these limitations, one direction of future studies could involve the consideration of the perspective of students in the evaluation process for university teachers. Another research opportunity could be the integration of research on data exploration and multi-criteria decision-making into the model with the use of big data to analyse and understand the correlations between standards for filtering key standards and their weights—as a way of ultimately proposing improvement measures based on expert knowledge with objective behavioural rules and subjective practical experience. As the Chinese education reform deepens, various problems relating to building decision-making models will require corresponding solutions from future researchers.

Author Contributions: Conceptualization, S.-S.W.; methodology, Y.-C.C.; numeration: Y.-C.C.; investigation, Y.L.; resources, Y.L.; data curation, Y.L.; writing—original article, Y.L.; writing—review and editing, S.-S.W. and Y.-C.C.; supervision, S.-S.W.; project administration, S.-S.W.

Funding: This research received no external funding

Acknowledgments: The authors would like to thank the management department of China's ISEC program for its strong support for this study. We are grateful to the editor of the Special Issue and the two anonymous reviewers for their constructive and helpful comments and suggestions.

Conflicts of Interest: The authors declare no conflict of interest.

References

1. Albareda-Tiana, S.; García-González, E.; Jiménez-Fontana, R.; Solís-Espallargas, C. Implementing Pedagogical Approaches for ESD in Initial Teacher Training at Spanish Universities. *Sustainability* **2019**, *11*, 4927. [CrossRef]
2. The United Nations. Sustainable Development Goals: 17 Goals to Transform our World. Available online: https://sustainabledevelopment.un.org/ (accessed on 15 September 2019).
3. Bürgener, L.; Barth, M. Sustainability competencies in teacher education: Making teacher education count in everyday school practice. *J. Clean. Prod.* **2018**, *174*, 821–826. [CrossRef]
4. UNESCO. *Roadmap for Implementing the Global Action Programme on Education for Sustainable Development*; UNESCO: Paris, France, 2014.
5. Brito, R.; Rodríguez, C.; Aparicio, J. Sustainability in Teaching: An Evaluation of University Teachers and Students. *Sustainability* **2018**, *10*, 439. [CrossRef]
6. The Chinese Government. A National Plan of China for Implementing the 2030 Agenda for Sustainable Development. Available online: http://www.gov.cn/xinwen/2016-10/13/content_5118514.htm (accessed on 15 September 2019).
7. Ministry of Education. Notice from the Ministry of Education on Implementing the "Dual Ten-Thousand Plan" for Establishing First Rate Undergraduate Majors. Available online: http://www.moe.gov.cn/srcsite/A08/s7056/201904/t20190409_377216.html (accessed on 15 September 2019).
8. Varela-Losada, M.; Arias-Correa, A.; Pérez-Rodríguez, U.; Vega-Marcote, P. How Can Teachers Be Encouraged to Commit to Sustainability? Evaluation of a Teacher-Training Experience in Spain. *Sustainability* **2019**, *11*, 4309. [CrossRef]
9. Álvarez-García, O.; García-Escudero, L.Á.; Salvà-Mut, F.; Calvo-Sastre, A. Variables Influencing Pre-Service Teacher Training in Education for Sustainable Development: A Case Study of Two Spanish Universities. *Sustainability* **2019**, *11*, 4412. [CrossRef]
10. The Ministry of Education and Four Other Ministries. the Action Programme to Rejuvenate Education for the Teaching Staff 2018–2022. Available online: http://www.moe.gov.cn/srcsite/A10/s7034/201803/t20180323_331063.html (accessed on 15 September 2019).
11. China News. The Serious Shortage of High-Quality Educational Resources is the Major Contradiction of Chinese Education. Available online: http://edu.china.com.cn/2014-04/30/content_32246901.htm (accessed on 15 September 2019).
12. Song Chunpeng, Member of the CPPCC National Committee: The Situation of East is Strong, West is Weak in Universities Must Be Changed. Available online: http://www.moe.gov.cn/jyb_xwfb/xw_zt/moe_357/jyzt_2019n/2019_zt2/zt1902_dbwy/201903/t20190314_373410.html (accessed on 15 September 2019).
13. Ding, L.; Zeng, Y. Evaluation of Chinese higher education by TOPSIS and IEW — The case of 68 universities belonging to the Ministry of Education in China. *China Econ. Rev.* **2015**, *36*, 341–358. [CrossRef]

14. Johnes, J.; Yu, L. Measuring the research performance of Chinese higher education institutions using data envelopment analysis. *China Econ. Rev.* **2008**, *19*, 679–696. [CrossRef]

15. Hallinger, P.; Heck, R.H.; Murphy, J. Teacher Evaluation and School Improvement: An Analysis of the Evidence. *Educ. Assess. Eval. Account.* **2014**, *26*, 5–28. [CrossRef]

16. Herlihy, C.; Karger, E.; Pollard, C.; Hill, H.C.; Kraft, M.A.; Williams, M.; Howard, S. State and local efforts to investigate the validity and reliability of scores from teacher evaluation systems. *Teach. Coll. Rec.* **2014**, *116*, 1–28.

17. Papay, J. Refocusing the debate: Assessing the purposes and tools of teacher evaluation. *Harv. Educ. Rev.* **2012**, *82*, 123–141. [CrossRef]

18. Gagnon, D.J.; Hall, E.L.; Marion, S. Teacher evaluation and local control in the US: an investigation into the degree of local control afforded to districts in defining evaluation procedures for teachers in non-tested subjects and grades. *Assess. Educ. Princ. Policy Pract.* **2017**, *24*, 489–505. [CrossRef] [PubMed]

19. Ghosh, D.N. Analytic hierarchy process & TOPSIS method to evaluate faculty performance in engineering education. *Dipendra Nath Ghosh Et Al Uniascit* **2011**, *1*, 63–70.

20. Pavani, S.; Sharma, L.K.; Hota, H. A group expert evaluation for teachers by integrating fuzzy AHP and TOPSIS models. In Proceedings of the 2013 IEEE International Conference in MOOC, Innovation and Technology in Education (MITE), Jaipur, India, 20–22 December 2013; pp. 85–90.

21. Xu, X.; Wang, Y.; Yu, S. Teaching performance evaluation in smart campus. *IEEE Access.* **2018**, *6*, 77754–77766. [CrossRef]

22. Wang, W.; Liu, Z.; Yao, L.; Liu, G. Classroom teaching performance evaluation model of ideological and political education in colleges and universities based on TOPSIS and triangle fuzzy number. *Risti-Rev. Iber. Sist. E Tecnol. Inf.* **2016**, *2016*, 220–230.

23. Tseng, M.L. Using the extension of DEMATEL to integrate hotel service quality perceptions into a cause–effect model in uncertainty. *Expert Syst. Appl.* **2009**, *36*, 9015–9023. [CrossRef]

24. Kiani Mavi, R.; Standing, C. Cause and effect analysis of business intelligence (BI) benefits with fuzzy DEMATEL. *Knowledge Manag. Res. Pract.* **2018**, *16*, 245–257. [CrossRef]

25. Si, S.L.; You, X.Y.; Liu, H.C.; Zhang, P. DEMATEL technique: A systematic review of the state-of-the-art literature on methodologies and applications. *Math. Probl. Eng.* **2018**, *3696457*, 1–33. [CrossRef]

26. Liou, J.J.; Chuang, Y.C.; Zavadskas, E.K.; Tzeng, G.H. Data-driven hybrid multiple attribute decision-making model for green supplier evaluation and performance improvement. *J. Clean. Prod.* **2019**, *241*, 118321. [CrossRef]

27. Ding, J.F.; Kuo, J.F.; Shyu, W.H.; Chou, C.C. Evaluating determinants of attractiveness and their cause-effect relationships for container ports in Taiwan: users' perspectives. *Marit. Policy Manag.* **2019**, *46*, 466–490. [CrossRef]

28. Hu, H.Y.; Lee, Y.C.; Yen, T.M.; Tsai, C.H. Using BPNN and DEMATEL to modify importance–performance analysis model–A study of the computer industry. *Expert Syst. Appl.* **2009**, *36*, 9969–9979. [CrossRef]

29. Hu, H.Y.; Lee, Y.C.; Yen, T.M. Amend importance-performance analysis method with Kano's model and DEMATEL. *J. Appl. Sci.* **2009**, *9*, 1833–1846. [CrossRef]

30. Hu, H.Y.; Chiu, S.I.; Cheng, C.C.; Yen, T.M. Applying the IPA and DEMATEL models to improve the order-winner criteria: A case study of Taiwan's network communication equipment manufacturing industry. *Expert Syst. Appl.* **2011**, *38*, 9674–9683. [CrossRef]

31. Ho, L.H.; Feng, S.Y.; Lee, Y.C.; Yen, T.M. Using modified IPA to evaluate supplier's performance: Multiple regression analysis and DEMATEL approach. *Expert Syst. Appl.* **2012**, *39*, 7102–7109. [CrossRef]

32. Cheng, C.C.; Chen, C.T.; Hsu, F.S.; Hu, H.Y. Enhancing service quality improvement strategies of fine-dining restaurants: New insights from integrating a two-phase decision-making model of IPGA and DEMATEL analysis. *Int. J. Hosp. Manag.* **2012**, *31*, 1155–1166. [CrossRef]

33. Wilson, F.; Beaton, D. The theory and practice of appraisal: progress review in a Scottish university. *High. Educ. Q.* **1993**, *47*, 163–189. [CrossRef]

34. Tucker, A. *Chairing the Academic Department: Leadership Among Peers*; Oryx Press: Phoenix, AZ, USA, 1992.

35. Moses, I. Assessment and Appraisal of Academic Staff. *High. Educ. Manag.* **1996**, *8*, 79–86.

36. Michael, M.; Adrienne, E.H. Faculty Evaluation: A Prickly Pair. *High. Educ.* **1999**, *38*, 351.

37. Desselle, S.P.; Mattei, T.J.; Pete Vanderveen, R. Identifying and weighting teaching and scholarship activities among faculty members. *Am. J. Pharm. Educ.* **2004**, *68*, 1–11. [CrossRef]

38. Filipe, M.N.M.; Ferreira, F.A.F.; Santos, S.P. A multiple criteria information system for pedagogical evaluation and professional development of teachers. *J. Oper. Res. Soc.* **2015**, *66*, 1769–1782. [CrossRef]

39. Lane, J. Let's make science metrics more scientific. *Nature* **2010**, *464*, 488–489. [CrossRef]

40. Coccia, M. Measuring scientific performance of public research units for strategic change. *J. Informetr.* **2008**, *2*, 183–194. [CrossRef]

41. Bana e Costa, C.A.; Oliveira, M.D. A multicriteria decision analysis model for faculty evaluation. *Omega* **2012**, *40*, 424–436. [CrossRef]

42. Keeney, R.L.; See, K.E.; Von Winterfeldt, D. Evaluating academic programs: With applications to US graduate decision science programs. *Oper. Res.* **2006**, *54*, 813–828. [CrossRef]

43. Billaut, J.-C.; Bouyssou, D.; Vincke, P. Should you believe in the Shanghai ranking? An MCDM view. *Scientometrics* **2010**, 237. [CrossRef]

44. Oral, M.; Oukil, A.; Malouin, J.-L.; Kettani, O. The appreciative democratic voice of DEA: A case of faculty academic performance evaluation. *Soc.-Econ. Plan. Sci.* **2014**, *48*, 20–28. [CrossRef]

45. Hein, N.; Kroenke, A.; Júnior, M.M.R. Professor Assessment Using Multi-Criteria Decision Analysis. *Proc. Comput. Sci.* **2015**, *55*, 539–548. [CrossRef]

46. Chen, J.-F.; Hsieh, H.-N.; Do, Q.H. Evaluating teaching performance based on fuzzy AHP and comprehensive evaluation approach. *Appl. Soft Comput. J.* **2015**, *28*, 100–108. [CrossRef]

47. Chang, T.-C.; Wang, H. A Multi Criteria Group Decision-making Model for Teacher Evaluation in Higher Education Based on Cloud Model and Decision Tree. *Eurasia J. Math. Sci. Technol. Educ.* **2016**, *12*, 1243–1262. [CrossRef]

48. Dey Mondal, S.; Ghosh, D. An Integrated Approach of Multi-Criteria Group Decision Making Techniques to Evaluate the overall Performance of Teachers. *Int. J. Adv. Res. Comput. Sci.* **2016**, *7*, 38–45.

49. Nikolaidis, Y.; Dimitriadis, S.G. On the student evaluation of university courses and faculty members' teaching performance. *Eur. J. Oper. Res.* **2014**, *238*, 199–207. [CrossRef]

50. Lyde, A.R.; Grieshaber, D.C.; Byrns, G. Faculty Teaching Performance: Perceptions of a Multi-Source Method for Evaluation (MME). *J. Scholarsh. Teach. Learn.* **2016**, *16*, 82–94. [CrossRef]

51. Bi, H.H. A robust interpretation of teaching evaluation ratings. *Assess. Eval. High. Educ.* **2018**, *43*, 79–93. [CrossRef]

52. Lee, W.-S.; Tzeng, G.-H.; Cheng, C.-M. Using novel MCDM methods based on Fama-French three-factor model for probing the stock selection. In Proceedings of the 10th Asia Pacific Industrial Engineering and Management Systems Conference, APIEMS, Kitakyushu, Japan, 14–16 December 2012; pp. 14–16.

53. Gabus, A.; Fontela, E. World problems, an invitation to further thought within the framework of DEMATEL. *Battelle Geneva Res. Cent. Geneva Switz.* **1972**, 1–8.

54. Saaty, T.L. *Decision Making with Dependence and Feedback: The Analytic Network Process*; RWS Publ.: Casa Branca, Santo André-SP, Brazil, 1996; Volume 4922.

55. Chuang, Y.-C.; Hu, S.-K.; Liou, J.J.; Lo, H.-W. Building a Decision Dashboard for Improving Green Supply Chain Management. *Int. J. Inf. Technol. Decis. Mak.* **2018**, *17*, 1363–1398. [CrossRef]

56. Zhu, B.W.; Zhang, J.R.; Tzeng, G.H.; Huang, S.L.; Xiong, L. Public open space development for elderly people by using the DANP-V model to establish continuous improvement strategies towards a sustainable and healthy aging society. *Sustainability* **2017**, *9*, 420. [CrossRef]

57. Xiong, L.; Teng, C.L.; Zhu, B.W.; Tzeng, G.H.; Huang, S.L. Using the D-DANP-mV model to explore the continuous system improvement strategy for sustainable development of creative communities. *Int. J. Environ. Res. Public Health* **2017**, *14*, 1309. [CrossRef]

58. Feng, I.M.; Chen, J.H.; Xiong, L.; Zhu, B.W. Assessment of and improvement strategies for the housing of healthy elderly: Improving quality of life. *Sustainability* **2018**, *10*, 722. [CrossRef]

59. Liou, J.J.H.; Chuang, Y.C.; Tzeng, G.H. A fuzzy integral-based model for supplier evaluation and improvement. *Inf. Sci.* **2014**, *266*, 199–217. [CrossRef]

60. Gudiel Pineda, P.J.; Liou, J.J.H.; Hsu, C.-H.; Chuang, Y.-C. An integrated MCDM model for improving airline operational and financial performance. *J. Air Transp. Manag.* **2018**, *68*, 103–117. [CrossRef]

61. Shao, Q.G.; Liou, J.J.H.; Chuang, Y.C.; Weng, S.S. Improving the green building evaluation system in China based on the DANP method. *Sustainability* **2018**, *10*, 1173. [CrossRef]

62. Lu, M.T.; Liou, J.J.H.; Hsu, C.C.; Lo, H.W. A hybrid MCDM and sustainability-balanced scorecard model to establish sustainable performance evaluation for international airports. *J. Air Transp. Manag.* **2018**, *71*, 9–19. [CrossRef]

63. Sever, I. Importance-performance analysis: A valid management tool? *Tour. Manag.* **2015**, *48*, 43–53. [CrossRef]

64. Martilla, J.A.; James, J.C. Importance-performance analysis. *J. Mark.* **1977**, *41*, 77–79. [CrossRef]

65. Abalo, J.; Varela, J.; Manzano, V. Importance values for Importance–Performance Analysis: A formula for spreading out values derived from preference rankings. *J. Bus. Res.* **2007**, *60*, 115–121. [CrossRef]

66. Frauman, E.; Banks, S. Gateway community resident perceptions of tourism development: Incorporating Importance-Performance Analysis into a Limits of Acceptable Change framework. *Tour. Manag.* **2011**, *32*, 128–140. [CrossRef]

67. Cohen, J.F.; Coleman, E.; Kangethe, M.J. An importance-performance analysis of hospital information system attributes: A nurses' perspective. *Int. J. Med. Inf.* **2016**, *86*, 82–90. [CrossRef] [PubMed]

68. Boley, B.B.; McGehee, N.G.; Tom Hammett, A.L. Importance-performance analysis (IPA) of sustainable tourism initiatives: The resident perspective. *Tour. Manag.* **2017**, *58*, 66–77. [CrossRef]

69. Masoumik, S.M.; Abdul-Rashid, S.H.; Olugu, E.U. Importance-performance Analysis of Green Strategy Adoption within the Malaysian Manufacturing Industry. *Proc. CIRP* **2015**, *26*, 646–652. [CrossRef]

Article

Spanish Universities' Sustainability Performance and Sustainability-Related R&D+I

Daniela De Filippo [1,2,*], **Leyla Angélica Sandoval-Hamón** [1,3], **Fernando Casani** [1,3] and **Elías Sanz-Casado** [1,4]

1. Research Institute for Higher Education and Science (INAECU) (UAM-UC3M), 28903 Getafe, Spain; angelica.sandoval@uam.es (L.A.S.-H.); fernando.casani@uam.es (F.C.); elias@bib.uc3m.es (E.S.-C.)
2. Department of Library Science and Documentation, University Carlos III de Madrid, 28049 Madrid, Spain
3. Department of Business Administration, Autonoma University of Madrid, 28049 Madrid, Spain
4. Department of Library and Information Science, Carlos III University of Madrid, 28903 Getafe, Spain
* Correspondence: dfilippo@bib.uc3m.es

Received: 29 July 2019; Accepted: 8 October 2019; Published: 10 October 2019

Abstract: For its scope and the breadth of its available resources, the university system is one of the keys to implementing and propagating policies, with sustainability policies being among them. Building on sustainability performance in universities, this study aimed to: Identify the procedures deployed by universities to measure sustainability; detect the strengths and weaknesses of the Spanish university system (SUS) sustainability practice; analyse the SUS contributions to sustainability-related Research, Development and Innovation (R&D+I); and assess the efficacy of such practices and procedures as reported in the literature. The indicators of scientific activity were defined by applying scientometric techniques to analyse the journal (Web of Science) and European project (CORDIS) databases, along with reports issued by national institutions. The findings showed that measuring sustainability in the SUS is a very recent endeavour and that one of the strengths is the university community's engagement with the ideal. Nonetheless, high performance is still elusive in most of the items analysed. Whereas universities account for nearly 90 % of the Spanish papers published in the WoS subject category, Green and Sustainable Science and Technology, their contribution to research projects is meagre. A divide still exists in the SUS between policies and results, although the gap has been narrowing in recent years.

Keywords: university policies on sustainability; scientometric indicators; Web of Science; CORDIS; Spanish university system

---

1. Introduction

Universities as institutions make a significant social, economic, academic, scientific and technological contribution to their local and national environments. These contributions have been widely discussed in the literature [1–5] because universities are considered "the most prominent producers of fundamental knowledge, which has been argued to be one of the main drivers of economic growth" [6]. According to Eurostat [7], 19.6 million students and 1.5 million professors engaged in tertiary education in 2016, whilst public spending on higher education came to 1.2% of European GDP in 2015.

As those figures show, in light of its scope and the breadth of its available resources, the university system is a key agent in implementing and propagating all manner of policies, with sustainability being among them. Higher education institutions (HEIs) are consequently crucial to societal transformation [8].

One of the primary goals of the European Commission document, "Europe's 2020 Strategy for Smart, Sustainable and Inclusive Growth" is for 40% of the Union's population between the ages of 30 and 34 to have a post-secondary degree by the target year [9]. Higher education institutions, as knowledge and innovation centres, therefore play a significant role in furthering the migration to sustainable development models. At the same time, they are introducing change in their own processes to adapt to the new scenario, which impacts not only core education but research, institution management and community outreach [10].

As there are many higher education institutions, the world over has become increasingly aware of their impact on the environment. They have made substantial efforts to enhance their understanding of the environmental dimensions of their operations and the implications and impact of higher education activities [11]. HEIs therefore play a catalytic role in societies' engagement with sustainability. In light of that complex challenge, HEIs must pay greater heed to their internal and external modi operandi. This means that HEIs must realise that "the development must meet the needs of the present without compromising the ability of future generations to meet their own needs" [12]. In this line, the concept of a sustainable university is broad and includes consideration of "a sustainable university as a higher education institution, as a whole or as a part, that addresses, involves and promotes, on regional or global level, the minimization of environmental, economics, societal, and health negative effects in the use of their resources in order to fulfill its main functions of teaching, research, outreach and partnership, and stewardship among other as a way to helping society make the transition to sustainable life styles" [13]. Therefore, HEIs need to assume their responsibilities in education (ad hoc courses and curricula), research, on-campus operations and community outreach with greater integrity and transparency. This can be reflected, for instance, in resource allocation planning and commitment to sustainable development.

The push to make sustainability and sustainable development an overarching concern has become particularly prevalent since the last decades of the past century. Leal Filho, Brandli, Becke, Skanavis, Kounani, Sardi, and Raath have described the efforts of 35 universities (on five continents) to implement sustainability policies and procedures [14]. Other authors define a "sustainability or environmental policy" as "one element of HEIs' sustainability governance documents that also includes plans, strategies, and reports" [15].

Several national and international bodies have echoed these concerns and instituted sustainability-related statements, programmes, tools and systems. The Stockholm Declaration on the Human Environment (1972), the Tbilisi Declaration (1977) and the UNESCO's Magna Charta of European Universities (1988) constitute clear examples. Some of the most prominent programmes include the UN-supported Sustainable Development Solutions Network, the International Sustainable Campus Network (ISCN), the Association for the Advancement of Sustainability in Higher Education (AASHE) in the United States and the Environmental Association for Universities and Colleges (EAUC) in the United Kingdom. The rise in the number of scientific publications on the subject attests to growing research interest, prompting the Web of Science to establish a new subject category, Green and Sustainable Science and Technology (GSST), that groups journals dealing with ecology, energy, the environment, climate change, energy efficiency and related issues. The definition of the category set out in the WoS database reads: "This category covers resources that focus on basic and applied research on green and sustainable science and technology, including green chemistry; green nanotechnology; green building; renewable and green materials; sustainable processing and engineering; sustainable policy, management and development; environmental and agricultural sustainability; renewable and sustainable energy; and innovative technologies that reduce or eliminate damage to health and the environment" [16].

An analysis of universities' scientific activity in a given area may be broached from the perspective of their scientific and technological output. Scientometrics and bibliometrics are pivotal to the analysis, measurement and assessment of research activity [17]. Those disciplines frequently deal with the analysis of (researcher, group, institution, discipline or country) research and its impact on the scientific community. They nonetheless also embrace more innovative pursuits, such as the detection of new research fronts and emerging fields, network analysis and the identification of research niches.

In those endeavours, the information sourced from publications and patents may be supplemented by the analysis of research projects. When implemented under competitive research and development programmes, such projects furnish data apt for weighing basic against applied science and assessing the effort deployed in emerging fields or interdisciplinary or cross-border research, such as is often the case in environmental and socio-economic studies [18].

Further to this interest in sustainability, some authors have explored the field qualitatively [19,20]. Others have adopted a bibliometric approach to examine the development of sustainability science through the analysis of citations [21,22], journal interdisciplinarity [23] or social media repercussions [24,25]. Recent papers have analysed the dynamics and evolution of GSST (considering that it is an approach to sustainable research) [25,26] and others have focused on the specific study by research on higher education for sustainable development [27].

Earlier studies have also analysed scientific activity in related areas using European projects as a source of information to identify university activity and to analyse project content in depth [28]. The present authors developed a model based on project publications to relate research activity to the impact on the academic community as reflected in social networks [29].

As some authors mention, given the increasing interest on the sustainability in higher education, there is a clear need for a systematic review of the literature [30]. Then, HEIs around the world have become engaged in sustainable practices and methodologies. This is reflected in the "growing body of literature has investigated this trend" [31]. However, as is mentioned by several authors there are "a lack of studies analyzing impacts from a more holistic perspective" [4]. The measurement of these impacts also involves considering that "the tools generally lack indicators in research, education and community engagement areas" as was explained by authors as Yáñez, Uruburu, Moreno, and Lumbreras [10].

Against this backdrop, this study has focused on the Spanish case. This country was chosen because Spanish universities, through the Conference of Rectors of Spanish Universities (CRUE), show great interest in sustainability issues. In fact, CRUE had created a working group on environmental quality and sustainable development. Therefore, CRUE has had a very active role. As some authors mention "this working group compiles university experiences and the progress made in the area of environmental management and sustainability, promoting cooperation among universities in these areas" [32]. In addition, the Spanish university system is the third country in Europe in terms of the number of STEM graduates [33]. It is among the first in terms of the number of the adult population with a university education [34], and it is a clear reference also for the Ibero-American region. In this sense, according to data from the Ibero-American Network of Science and Technology Indicators, Spain is the number one country by number of doctors [35]. From the point of view of scientific production, its contribution is also very relevant since 75% of the country's publications come from the higher education sector [36]. Furthermore, it is the number one Ibero-American country in terms of both the total number of publications and publications per researcher [35].

Considering this scenario, the objectives pursued here are to:

- Identify and analyse sustainability measurement tools for universities;
- Reveal the strengths and weaknesses of the Spanish university system in terms of sustainability procedures;
- Detect the sustainability-related R&D activities in which universities participate (research projects, scientific papers);
- Analyse the relationship between tools and R&D+I activities conducted by universities.

2. Materials and Methods

Considering the importance of the analysis of scientific activity on sustainability in the higher education institutions, this work focuses on the study of reports, scientific papers and research projects.

In this line, the methodological framework used is the Information Metric Studies, which allow, through bibliometric techniques, the analysis of publications and projects on the proposed topic.

Information for this study was sourced from the following:

- Scientific publications and reports on sustainability measurement tools in universities;
- Sustainability reports authored by national bodies in Spain;

  - 'Evaluación de las políticas universitarias de sostenibilidad como facilitadoras para el desarrollo de los campus de excelencia internacional' [assessment of university sustainability policies in furthering the international campus of excellence programme] (2010) and;
  - 'Diagnóstico de la sostenibilidad ambiental en las universidades españolas' [diagnosis of environmental sustainability in Spanish universities] (2017);

- The Clarivate Analytics Web of Science is an international multidisciplinary database that has been indexing the most prominent scientific journals in science, technology, humanities and sociology since 1945, from which information was collected on Spanish university papers on sustainability.
- The CORDIS project database is the primary source of results from EU-funded projects since 1990 that carries information on the EU's framework programmes by call, country, subject and type of result: https://cordis.europa.eu/projects/es).

The study was broken down into the following stages:

1. A literature review of sustainability measurement in universities: Scientific publications and reports on sustainability measurement tools in universities were consulted and analysed considering the most frequent tools, dimensions and indicators included. A classification of the dimensions in categories was also carried out. The results are presented in Section 3.1.
2. An analysis of the Spanish data and identification of the strengths and weaknesses: Sustainability reports authored by national bodies were analysed. The results obtained in two surveys (2010 and 2017) were considered and the results are presented in Section 3.2.
3. The information retrieval in the subject category, Green and Sustainable Science and Technology, from Web of Science database. The following search strategy was used: "WC = Green and Sustainable Science and Technology AND CU = Spain". No cuts have been made by type of document or date when collecting all the publications in journals included in this subject.
4. The obtention of the main bibliometric indicators: Output by country, institution, discipline and year. By identifying production by institution, it has been possible to detect documents from Spanish universities and calculate the contribution of the higher education system. The activity index has been calculated to measure the intensity of the production related to sustainability, both in Spain and in the higher education system and in each of the Spanish universities. The results are presented in Section 3.3.
5. The identification of European projects on related subjects in CORDIS database. The selection of the Seventh Framework Programme projects because they are the most numerous and the call has already been completed. The obtention of information on the participation by Spanish institutions. The results are presented in Section 3.4.

3. Results

The findings are set out below.

3.1. Sustainability Measurement in Universities

Integrating sustainability in universities entails creating tools that enable institutions to assess their engagement with the economic, social and environmental dimensions of sustainability and to continually improve their performance in those realms. Measuring sustainability remains a complex and challenging process for higher education institutions, however, especially institutions in the early stages of their sustainable development programmes [37].

The three resources used to analyse and measure sustainability in universities were accounts; narratives such as reports and similar; and indicators [38]. The proliferation in recent decades of papers of the many tools in place attests to the interest in such measurements. Some of the most popular publications analyzing and comparing sustainability assessment tools were published by authors such as Shriberg [39]; Alshuwaikhat and Abubakar [40]; Yarime & Tanaka [41]; Sayed, Kamal and Asmuss [42]; Gómez, Sáez-Navarrete, Lioi, and Marzuca [37]; Salvioni, Franzoni, and Cassano [43]; Berzosa, Bernaldo and Fernández-Sanchez [44]; Findler, Schönherr, Lozano and Stacherl [45] or Parvez and Agrawal [46]. Then, the tools have been described in the international literature to quantify sustainability-related activities in higher education institutions. According to Alghamdi [47], the ones most commonly used include the following:

- Sustainability Assessment Questionnaire (SAQ) (2001)
- Graphical Assessment of Sustainability in University (GASU) (2006)
- Sustainable University Model (SUM) (2006)
- University Environmental Management System (UEMS) (2008)
- Assessment Instrument for Sustainability in Higher Education (AISHE) (2009)
- Benchmarking Indicator Questions – Alternative University Appraisal (BIQ-AUA) (2009)
- Unit-based Sustainability Assessment Tool (USAT) (2009)
- The Green Plan (2012)
- Sustainable Campus Assessment System (SCAS) (2014)
- Adaptable Model for Assessing Sustainability in Higher Education (AMAS) (2014)
- Sustainability Tracking, Assessment and Rating System (STARS) (2014)
- Green Metric – UI's GreenMetric World University Ranking (GM) (2014)

Some of the most prominent features (dimensions and indicators) of the afore mentioned 12 tools are set out in Table 1. The links and references consulted to extract the information are shown in the table.

These tools differ in typology, the number of indicators and methodology for determining the integration of sustainability in university activities. By way of example of the details for some of these tools, the parameters addressed (and their relative weights) in sustainability tracking, assessment and rating systems (STARS) include academic courses (AC) (28%), engagement (EN) (20%), operations (OP) (35%), planning and administration (PA) (15%), innovation and leadership (IN) (2%). The academic dimension covers indicators associated with research and projects, with three sub-sections: support for research, research and scholarship and open access to research, further broken down into items.

Sustainability assessment questionnaires (SAQs), in turn, comprise eight dimensions: curriculum, research and scholarship, operations, faculty and staff, development and reviews, outreach and services, student opportunities, and administration, mission and planning They are designed to stimulate discussion and further assessment by campus representatives knowledgeable of and responsible for the activities specified and include a specific dimension for assessing research on sustainability.

Green Metrics, another tool, encompasses dimensions such as education (ED) (18%), (setting and infrastructure (SI) (15%), energy and climate change (EC) (21%), waste (WS) (18%), water (WR) (10%) and transportation (TR) (18%). The education dimension covers research and project indicators and more specifically, total research funds (in USD) dedicated to environmental and sustainability research and the number of scholarly papers published on the environment and sustainability.

Table 1. Tools for measuring sustainability in universities: Key features.

Tools	Dimensions	Indicators	References/Link
SAQ	Curriculum. Research and scholarship. Operations. Faculty and Staff. Development and Reviews. Outreach and services. Student Opportunities. Administration, mission and planning.	35	(SAQ 2001) http://ulsf.org/sustainability-assessment-questionnaire/.
GASU	Economic. Environmental. Social. Education (curriculum and research).	59	Lozano, R. (2006)
SUM	Education. Research. Outreach and partnership. Sustainability on campus.	23	Velazquez, L., Munguia, N., Platt, A., & Taddei, J. (2006).
UEMS	University EMS (environmental management and improvement. Green campus). Public participation and social responsibility. Sustainability teaching and research.	27	Alshuwaikhat, H.M., & Abubakar, I. (2008)
AISHE	Operation. Education. Research. Society. Identity.	30	Roorda, N. (2004).
BIQ-AUA	Governance. Education. Research. Outreach.	30	Gómez *et al.* (2015)
USAT	Teaching and research community service. Operations and management. Student involvement. Policy and written statements.	75	PSPE (2012)
THE GREEN PLAN	Strategy and governance. Teaching and training. Research. Environmental management. Social policy and regional presence.	44	Green Plan (2010),
SCAS	Management. Education and research. Environment. Local community. Special reporting.	48	Alghamdi, N., den Heijer, A., & de Jonge, H. (2017)
AMAS	Institutional commitment. Example setting. Advancing sustainability (education, research and public engagement).	25	Gómez *et al.* (2015)
STARS 2.0	Academic courses. Engagement. Operations. Planning and administration. Innovation.	74	(STARS 2014) https://reports.aashe.org/accounts/login/?next=/tool/
GM	Setting and infrastructure. Energy and climate change. Waste. Water. Transportation. Education	33	(GM 2019) http://greenmetric.ui.ac.id

In short, sustainability measurement in universities embraces a variety of realms, some internal and others external to the university community, as reflected in the diversity of parameters defined in these tools and the number of indicators applied to each.

These 12 tools consider indicators that can be grouped within each of the categories (curricula and competences, research, campus operations, community outreach, university and governance, sustainable assessment and reporting) that have been collected in the model for sustainable development in universities by authors as Alonso-Almeida et al [48] and Lozano et al [49,50]. However, as mentioned above, each tool uses different indicators and weights for each dimension (Table 2).

Table 2. Categories and dimensions of tools.

Categories	Dimensions
Curricula and competences	Curriculum. Education. Sustainability teaching. Teaching and training. Academic courses
Research	Research and scholarship
Campus operations	Operations. Sustainability on campus. Green campus. Setting and infrastructure. Energy and climate change. Waste. Water. Transportation.
Community outreach	Development and Reviews. Outreach and services. Outreach and partnership. Public participation and social responsibility. Local community. Social policy and regional presence. Public engagement Society.
University and governance	Administration, mission and planning. Student Opportunities. Economic. Identity. Faculty and Staff. Governance. Strategy and governance. Management. Policy and written statements. Institutional commitment. Economic. Environmental. Social. Innovation.
Sustainable assessment and reporting	Special reporting

The research dimension is explicitly included in all tools, except STARTs and GM which include it in academics and education, respectively. However, the sustainability research dimension uses a small number of indicators compared to the total of the indicators of each tool, mainly related to the number of publications, number of projects, funding support, institutional support, staff (teachers and/or students) involved in this type of research. Some examples are: In the GM tool of 33 indicators, there are only two focusing on sustainability research (number of academic publications and funds); and in the SAQ tool that has 35 indicators and only three are related to research (projects and professors focused on this subject, as well as the presence of multidisciplinary or interdisciplinary structures for research on sustainability issues).

3.2. Sustainability in Spanish Universities

The Spanish university system presently comprises 83 institutions (50 public and 33 private). Together they account for 27% of total national R&D spending and employing 36.8% of the personnel engaging in research [33]. The intensity of their activity is attested to by the findings, for Spanish universities author 75% of the country's scientific publications listed in international databases, such as the Web of Science.

Most Spanish universities conduct some manner of sustainability-related activity (administration, education, research), although they only began to address this issue in the nineteen nineties, somewhat later than countries, such as America and other parts of Europe. Those initial endeavours around university sustainability in Spain were induced by inter-university projects and attendance at meetings or seminars on the subject. Against that backdrop, in 2002 the Conference of Spanish University Vice-Chancellors (Spanish initials, CRUE) created a working group on environmental quality and sustainable development to further pro-sustainability action in Spanish universities. In July 2008, the group was restructured into the Sectoral Commission on Environmental Quality, Sustainable Development and Risk Prevention in Universities (Spanish initials, CADEP) [51]. In 2009, the commission was renamed CRUE-Sustainability. At least 68 universities have participated in some of its events. In June 2007, the University of Santiago de Compostela hosted a standing seminar for university environmental action entitled "Indicators and Sustainability in Universities". The conclusions included a proposal to create a technical working group to establish a system for assessing sustainability performance in Spanish universities. That project envisaged the definition of system of indicators to assess the universities' sustainability policies, compile models for implementing those policies and identify good practice in connection with international campus of excellence projects. The result was a system of indicators for assessing university sustainability agreed to and tested by most of Spain's university system. On an initiative authored by several universities, that led 2 years later to the creation of the CRUE Sectoral Commission-Sustainability to compile the experience of higher education institutions on environmental management, progress in heightening environmental awareness in the university community and risk prevention. It was also intended to further cooperation and exchange in these areas and establish good practice.

Those commissions and working groups organised a series of sustainability-related activities in Spanish universities and authored reports on the subject. In 2010, a questionnaire was circulated among all higher education institutions, to which 30 public (62% of the Spanish University System) and one private university responded. The findings were written up in a report [52] diagnosing university sustainability policy in Spain and defining the realms and indicators to be used to measure Spanish universities' contribution to sustainability. Those indicators afford an assessment framework for the progress made in sustainability policies, rendering progress more visible for the university community and society at large. The 31 universities responding to the questionnaire were consulted about the structure of the report. All were visited to discuss their inquiries and suggestions and obtain a first-hand view of sustainability programmes in the Spanish university system. The report revealed that the universities analysed engaged most intensely in areas such as environmental awareness, waste management and sustainability courses and least in green procurement, water management, social responsibility and environmental impact assessment. The follow-up on that first report (2010–2011) included the review of the tool used by the CADEP-CRUE working groups and the one employed by the group on university sustainability assessment (GESU) from November 2014 to April 2015. The outcome was a new document entitled "Sistema de evaluación ambiental de la Universidad Española- GESU-CRUE v3", Spanish university environmental assessment system-GESU-CRUE v3.

A second version of the questionnaire formulated in 2017 and responded to by 33 universities spawned the report "Diagnóstico de la sostenibilidad ambiental en las universidades españolas' (diagnosis of environmental sustainability in Spanish universities). One of its major conclusions was that the universities studied had made environmental improvements in terms of organisation, with the highest mean scores reported for environmental policy, awareness and engagement. Nearly all the respondent institutions had environmental policy officers and had attained a degree of engagement deemed as acceptable by the university community. Education and research, areas where implementation was essentially nil, scored the lowest. The findings on green campus management showed that universities had made a substantial effort in controlling environmental parameters such as water, energy, waste and biodiversity, although the implementation of improvements in those areas was scantly systematised. The difficulties were also observed in the adoption of measures to enhance green procurement. Many universities had developed plans for improving sustainable transportation, although most were still in the initial phases [51].

According to the results of the two surveys (Figure 1), the universities listed in the 2017 report scored higher in sustainability policy, water management and environmental impact assessment than the institutions included in the 2010 report. In En relación a los demás aspectos analizados, cabe destacar que una de las fortalezas de las universidades que se recoge en los dos informes es la implicación de la comunidad universitaria, en este ámbito se incluyen: another vein, one of the strengths recorded in both reports was university community engagement, including cursos, jornadas, noticias, congresos, difusión de actividades de la sociedad en general, etc. courses, seminars, press releases, congresses and dissemination of activities to society at large. The 2010 and 2017 reports also revealed considerable room for improvement in connection with environmental impact assessment, social responsibility, green procurement, education and research.

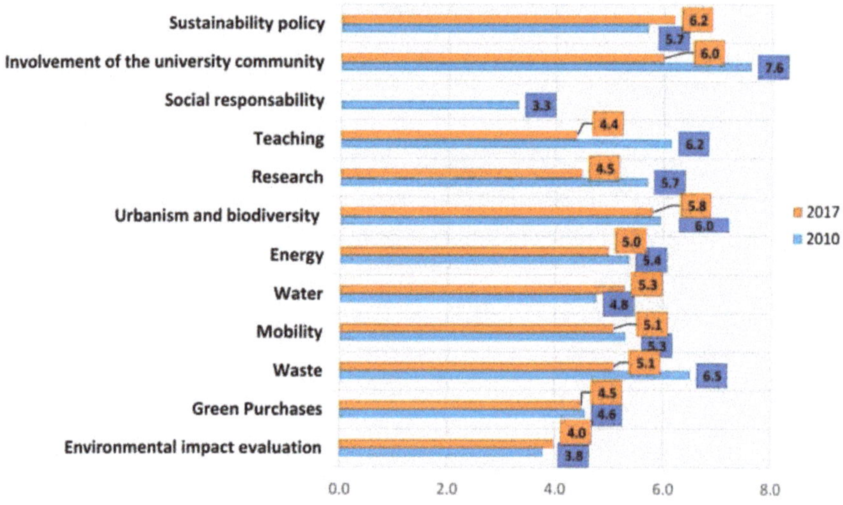

Figure 1. Conference of Rectors of Spanish Universities (CRUE) report findings, 2010 and 2017.

The differences between the two reports were not significant due to most of the areas considered maintain very similar results. With the exception of the involvement of the university community, teaching and research which, although they have reduced their scores (even some by 1.8 points) continue to be above the average of the scores. These differences are related to the improvement adjustment of the report, which led to modify some aspects of the questionnaire and simplify the total of questions (from 176 to 140) but keeping most of the questions the same as in the 2010 version so that the information can be comparable. In the case of the social responsibility aspect in the 2017 report, no data is available because it was not considered.

In short, in the results of the 2010 and 2017 reports, Spanish universities stand out in the score of sustainable policies and the involvement of the university community, but they must improve in the other areas that correspond to the effective implementation of sustainability, as well as in the monitoring and evaluation of the environmental impact generated by HEIs.

3.3. Publications on Sustainability

The Web of Science database lists 79,014 papers in Green and Sustainable Science and Technology, GSST (through 2018). Considering the whole period analysed, the average annual increase was 17.3%.

Although documents dated as early as 1994 were listed, the number of publications was higher in the last 5 years. As can be seen in Figure 2, a 66% of production is concentrated between 2014 and 2018. It is important to note that the decrease in the number of documents in 2018 is due to the updating of the database.

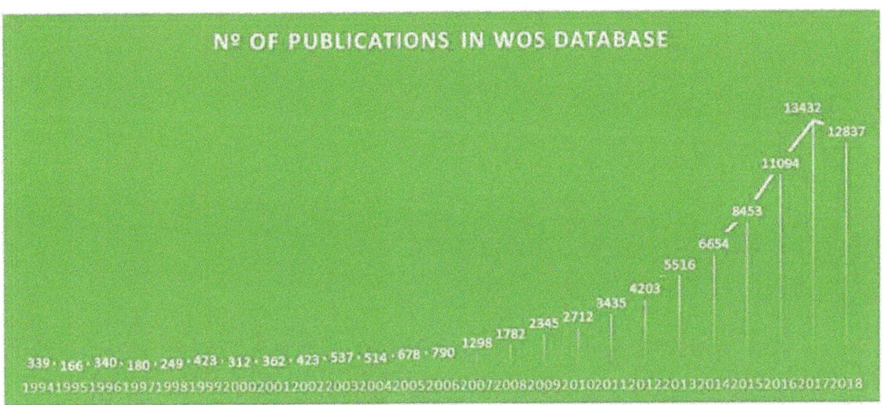

Figure 2. Annual evolution of Green and Sustainable Science and Technology (GSST) publications in WoS database. Source: Compiled by authors based on Web of Science information.

With 20% of the total, China had the highest output, followed by United States, India and England. Whilst the major countries obviously had the highest output, those values were normalised to accommodate size with the activity index (AI), which compares a country's output in a given area (GSST in this case) to its total output in the WoS database. An AI of > 1 is indicative of greater than expected output in a given field. As Table 3 shows, Netherlands, Belgium, Austria, Malaysia and Australia had particularly high AI values in GSST. In contrast, United States, England, Germany, France and Japan published fewer papers on sustainability than expected.

As Table 3 shows, Spain (with 4179 papers) ranked fifth by output, accounting for 5.3% of all world papers and exhibiting the mean yearly growth of 21.8% (3.3 points more than world growth). As shown in Figure 3, Spanish production has grown steadily and 70% is concentrated in the last 5 years. The importance of this production is also evident in the increase in the world's contribution (2% in 1994 and 5.8% in 2018).

Table 3. GSST output by countries (> 1% of total) and activity index.

Country	No. Papers	%	Activity Index (AI)
PEOPLE'S R CHINA	15,911	20.12	2.39
USA	11,226	14.20	0.47
INDIA	5411	6.84	2.72
ENGLAND	5093	6.44	0.91
SPAIN	4179	5.28	1.80
GERMANY	3985	5.04	0.77
ITALY	3635	4.60	1.23
AUSTRALIA	3606	4.56	5.35
CANADA	3010	3.81	0.92
SOUTH KOREA	2885	3.65	1.59
FRANCE	2656	3.36	0.74
NETHERLANDS	2525	3.19	> 10
JAPAN	2487	3.15	0.56
MALAYSIA	2302	2.91	9.12
IRAN	2190	2.77	3.12
SWEDEN	2041	2.58	1.73
BRAZIL	1995	2.52	1.38
TURKEY	1811	2.29	1.86
TAIWAN	1432	1.81	1.42
PORTUGAL	1213	1.53	2.59
DENMARK	1156	1.46	1.62
SWITZERLAND	1036	1.31	0.81
GREECE	1016	1.28	1.99
BELGIUM	986	1.25	> 10
FINLAND	976	1.23	1.76
SAUDI ARABIA	967	1.22	3.54
NORWAY	917	1.16	1.84
SCOTLAND	864	1.09	1.07
AUSTRIA	807	1.02	> 10

Source: Compiled by authors based on Web of Science information.

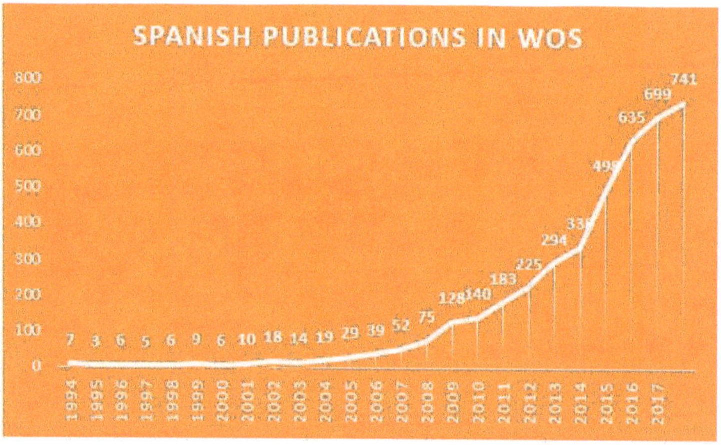

Figure 3. Number of Spanish GSST papers listed in WoS, 1994–2017. Source: Compiled by authors based on Web of Science information.

Whilst all the papers listed dealt with GSST, the publishing journals were also indexed under other subject categories. According to Figure 4, which depicts Spain's and the world's output by subject area, Spain followed the same pattern as the world mean, with a slightly higher proportion in environment-related disciplines only.

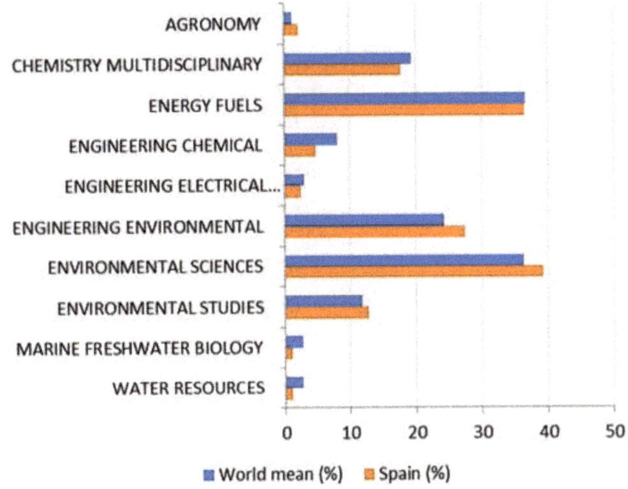

Figure 4. Number of papers by topic (Spain and world mean). Source: Compiled by authors based on Web of Science information.

Spanish output was concentrated in major cities such as Barcelona, Madrid and Valencia, although some activity was recorded in all 17 of the country's regions. This is logical, given that these cities concentrate a significant proportion of research centers and universities that are the main producers of the papers analyzed. In the distribution shown on the map in Figure 5, the size of the nodes denotes the output volume.

Figure 5. Distribution by city of GSST papers authored in Spain. Source: Compiled by authors based on Web of Science information.

Interestingly, Spanish output was highly concentrated in its universities, with 91% of the papers authored by at least one higher education institution.

The National Research Council (CSIC: no breakdown by institution is provided) produced more papers than any other organisation, whilst the most active universities were the Autonomous University of Barcelona and the Technical Universities of Valencia, Madrid and Catalonia (Table 4). As shown in Table 3, polytechnic universities have intensive production in the GSST field (AI > 2), as do other small or medium-sized universities such as La Rioja (AI = 4.1), Lleida (AI = 3.7), Almería (AI = 3.4) and Jaen (AI = 3.1). On the other hand, the large universities, being generalists, present a lower intensity of production than expected.

Spanish universities partnered most intensely with institutions in UK, Italy, USA, Portugal, France and Netherlands. Figure 6 shows the number of documents in co-authorship between Spanish universities and foreign partners (the size of the node is proportional to the number of documents). The main partners are the European and North American leaders in the topic studied (green circles in Figure 6). However, there is not such an intense relationship with large producers from other regions such as China, India, Australia, and South Korea.

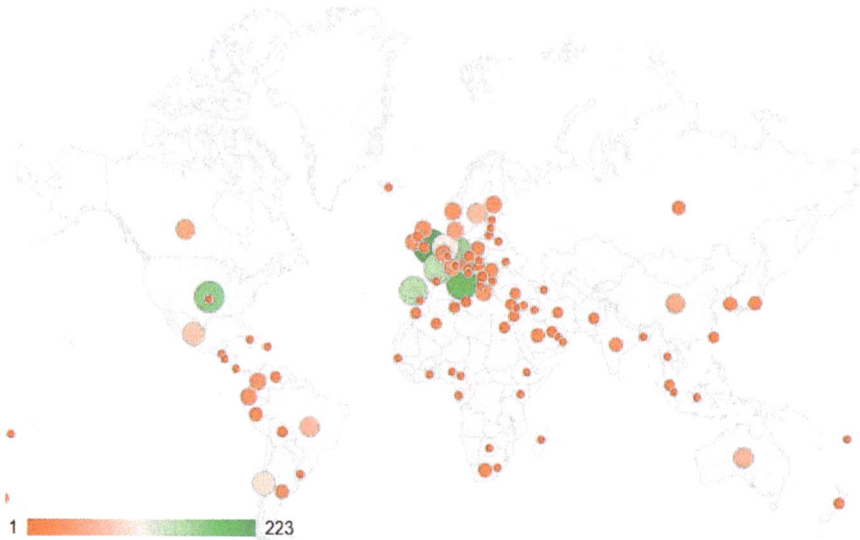

Figure 6. Countries of origin of Spanish universities' major partners. Source: Compiled by authors based on Web of Science information.

Following an analysis of the content of GSST papers, 12 of them mention tools for measuring sustainability in universities. All their authors were affiliated with universities: five in Canada, two in Saudi Arabia, two in USA and one each in Chile, UK, India and Mexico. The journals publishing the largest number of such papers were: *Journal of Cleaner Production; Sustainability; Environment Development and Sustainability; International Journal of Sustainability in Higher Education.*

Table 4. Output by institution (> 2% of documents).

Institution	No. Papers	%	Activity Index (AI)
NATIONAL RESEARCH COUNCIL (CSIC)	475	11.36	0.78
AUTONOMOUS UNIVERSITY OF BARCELONA	235	5.62	0.92
UNIVERSITAT POLITECNICA DE VALENCIA	217	5.19	2.09
POLYTECHNIC UNIVERSITY OF MADRID	213	5.09	2.22
POLYTECHNIC UNIVERSITY OF CATALONIA	203	4.85	1.85
UNIVERSITY OF BASQUE COUNTRY	189	4.52	1.32
UNIVERSIDAD DE CORDOBA	168	4.02	2.68
UNIVERSITY OF ZARAGOZA	166	3.97	1.36
UNIVERSITY OF SEVILLA	158	3.78	1.12
UNIVERSITY OF SANTIAGO DE COMPOSTELA	136	3.25	1.16
UNIVERSITY OF GRANADA	131	3.13	0.77
UNIVERSIDAD DE CASTILLA LA MANCHA	126	3.01	2.02
UNIVERSITAT ROVIRA I VIRGILI	125	2.99	1.91
UNIVERSITY OF OVIEDO	118	2.82	1.21
UNIVERSITAT DE LLEIDA	108	2.58	3.74
UNIVERSIDAD DE JAEN	106	2.53	3.11
UNIVERSITY OF VIGO	103	2.46	1.57
UNIVERSITY OF BARCELONA	101	2.41	0.24
UNIVERSIDAD DE ALMERIA	99	2.36	3.40
COMPLUTENSE UNIVERSITY OF MADRID	95	2.27	0.39
UNIVERSIDAD DE EXTREMADURA	84	2.01	1.69
UNIVERSITAT JAUME I	80	1.91	2.05
IRTA	73	1.74	4.60
UNIVERSITAT D ALACANT	73	1.74	1.19
UNIVERSITY OF VALENCIA	68	1.62	0.33
CSIC UPV INSTITUTO DE TECNOLOGIA QUIMICA ITQ	65	1.55	6.74
UNIVERSIDAD DE CANTABRIA	64	1.53	1.18
UNIVERSIDAD DE VALLADOLID	64	1.53	1.02
AUTONOMOUS UNIVERSITY OF MADRID	60	1.43	0.25
CIEMAT	60	1.43	4.77
UNIVERSIDAD DE LA RIOJA	59	1.41	4.15
UNIVERSIDADE DA CORUNA	57	1.36	1.23
BARCELONA INSTITUTE OF SCIENCE TECHNOLOGY	56	1.34	0.99
UNIVERSIDAD REY JUAN CARLOS	56	1.34	1.70
UNIVERSITAT DE GIRONA	50	1.19	1.02

Source: Compiled by authors based on Web of Science information.

3.4. Sustainability Projects

European calls for projects have become one of the main avenues for scientific and technological activity and a major source of funding for Spanish institutions. For the analysis of projects related to sustainability, the call of the 7th Framework Programme has been chosen and information retrieved from the CORDIS database confirmed Spain's significant participation in this call (Spanish institutions took part in a quarter of the total projects) (Table 5).

Table 5. Distribution of projects under Framework Programmes, by call (all topics).

Call	No. of Projects	Spanish Participation	% Spanish Participation/Total Projects
H2020 (2014–2020)	5348	1530	28.61
FP7 (2007–2013)	25,630	6334	24.71
FP6 (2002–2006)	10,102	2822	27.94
FP5 (1999–2002)	17,202	3710	21.57
Total	58,282	14,396	24.70

Source: Compiled by authors based on CORDIS information

According to a report by the Centre for Technological Development in Industry, in the Seventh Framework Programme (FP7, 2007–2013), Spain ranked in the same sixth place in funding received (after Germany (17.8%), United Kingdom (17.2%), France (12.5%), Italy (9.3%) and Netherlands (8.4%)) as it had under FP6. Qualitatively speaking, the results attained in FP7 were substantially better than in the preceding edition, where Spain headed 10.7% of the projects, compared to 6.0% in FP6.

The Seventh Framework Programme comprised 23 calls in different areas. Spain participated in over 50% of the projects awarded under eight of those calls. It was particularly active in FP7-SME (Specific Programme-Capacities: Research for the benefit of SMEs), designed to strengthen small and medium-sized enterprises; FP7-KBBE (Specific Programme-Cooperation: Food, Agriculture and Biotechnology); FP7-INFRASTRUCTURES (Specific Programme-Capacities: Research Infrastructures) and PF7-NMP (Specific Programme-Cooperation: Nanosciences, Nanotechnologies, Materials and New Production Technologies).

Six of the subjects addressed by those projects were associated with sustainability. Environmental protection attracted the largest number of projects (823), whilst 264 addressed energy savings. Spain had a significant presence in all six, with participation ranging from 34% in biofuels to 73% in waste management (Figure 7).

Universities were scantly involved in these projects, contributing to 16% at most. The highest rates of participation were observed in projects on sustainable development and environmental protection. The latter was the area where the largest number of higher education institutions participated (Table 6).

Table 6. Spanish university participation in FP7 sustainability-related projects.

Area	Spanish University Participation		
	No. of Projects	% Projects	No. of Universities
Energy Savings	25	9.47	15
Biofuels	2	5.26	2
Sustainable Development	13	16.25	9
Renewable Energies	9	13.43	8
Waste Management	1	6.67	1
Environmental Protection	127	15.43	37

Source: Compiled by authors based on CORDIS information.

A total of 40 Spanish universities, 39 public and one (Pontifical University of Comillas) private, participated in European sustainability-related projects. Technical universities participated most intensely, with the Technical University of Madrid heading the list. Taken together, however, the Catalonian institutions (Autonomous University of Barcelona, the Technical University of Catalonia and the University of Barcelona) had a significant presence in these calls (Table 7).

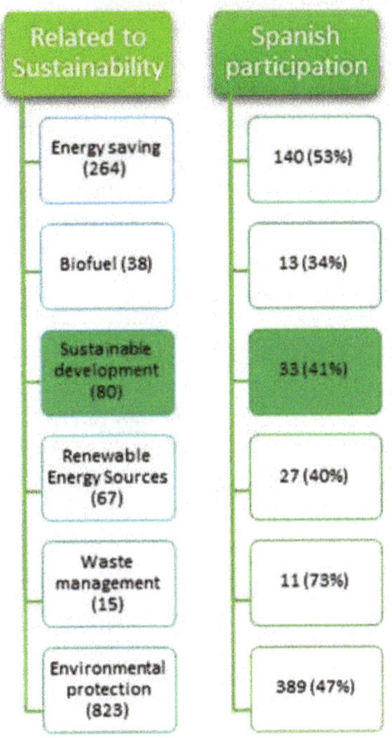

Figure 7. Total sustainability-related projects and Spanish participation. Source: Compiled by authors based on CORDIS information.

Table 7. Universities participating in FP7 university-related projects (>2 projects).

University	No. of Projects
Univ. Politécnica de Madrid	19
Univ. Autónoma de Barcelona	16
Univ. Politécnica de Cataluña	12
Univ. Politécnica de Valencia	11
Univ. de Barcelona	9
Univ. Complutense de Madrid	8
Univ. de Santiago de Compostela	7
Univ. de Córdoba	6
Univ. de Granada	6
Univ. de Cantabria	5
Univ. del País Vasco	5
Univ. Pontificia de Comillas	5
Univ. de Almería	4
Univ. de Oviedo	4
Univ. de Sevilla	4
Univ. de Valencia	4
Univ. de Valladolid	4
Univ. de Zaragoza	4
Univ. Carlos III de Madrid	3
Univ. de Alcalá de Henares	3
Univ. de Alicante	3
Univ. de Castilla la Mancha	3
Univ. Pablo de Olavide	3
Univ. Politécnica de Cartagena	3

Source: Compiled by authors based on CORDIS information.

4. Discussion

The assessment mechanisms are useful tools for diagnosing universities' sustainability performance. The most prominent international studies conducted on the subject have as a rule used information delivered by green metrics, a tool that processes data very generically and yields no details on research particulars such as project participation or paper write-up. Using this tool to analyse nine Indian HIEs, Parvez and Agrawal [46] found that only two had made formal sustainability progress in terms of education. The specific indicators studied were academic courses on sustainable development and the existence of a website on sustainability. They did not, however, analyse research findings or papers on sustainability. Similarly, Marrone, Orsini, Asdrubali and Guattari [53] analysed the green metrics scores for universities in several countries and regions (India, Indonesia, Japan, USA, Canada, Africa, Middle East, South America and European Union). The findings were also given generally, by category only and focusing on inter-country comparisons, with no information on indicators. In education, Canada and the European Union scored highest. Suwartha and Sari [54], in turn, studied 25 US universities that used the tool in 2011, analysing the scores for each by indicator, although in this case the authors grouped education under the category, setting and infrastructure.

Beringer, Wright and Malone [55] studied the state of sustainability in higher education (SHE) in Atlantic Canada, in which the tool of choice was SAQ. They found that the majority of higher education institutions in Atlantic Canada were engaged in sustainable development work, most notably in the area of curriculum. Sustainability research and scholarship is spread amongst faculty and students. Many institutions have inter- or multi-disciplinary research structures to address sustainability questions across campus and in collaboration with community partners. They nonetheless acknowledged that the dimension to be still only moderately developed and identified student commitment to research projects as an avenue for speedier progress.

Consernign SAQ, a paper by Lidstone, Wright and Sherren [15] on sustainability research reviewed 21 Canadian HEIs that used this measuring tool. Their findings showed that 50% of these universities' research plans included sustainability goals. The particulars most intensely studied by these universities included funding, interdisciplinarity and structures (organisation and other university resources). This tool was also analysed by Parvez and Agrawal [46] in their study of nine HIEs in India. The findings on research delivered by STARS are somewhat more detailed, for they include three groups of research-related indicators: research and scholarship, support for research and open access to research, with at least 10 items under each. The latter authors concluded that two of the nine Indian universities, in addition to offering courses on sustainability, undertook initiatives to establish new courses and mechanisms to support and fund research.

Likewise using STARS data, Salvioni, Franzoni and Cassano [43] analysed three groups of universities in the 2015 international Academic Ranking of World Universities (ARWU) Top 500: The first twenty best positions in the Top 500, an intermediate group formed by the last twenty universities classified among the Top 100 and the last twenty positions in the Top 500. The authors acknowledged that universities in the first group assumed sustainable culture more effectively in areas such as research. They stressed that "there is higher integration and inclusion of the sustainability theme in the institutional, managerial, research and teaching activities of universities placed in the Top 5 rankings compared to universities in subsequent position".

In another study, Alshuwaikhat, Adenle and Saghir [56] using SAQ, found public universities in Saudi Arabia to be in the initial stage of integrating sustainability in both the curriculum and research, stating "sustainability-related projects are not prioritised within universities and sustainable financial management practices are not significant".

It is important to note that all these sustainability assessment tools are used by universities on a voluntary basis. However, public authorities in most countries are promoting the transparency and accountability of universities through more information about their activities to society, which may include sustainability reports. In the area of public funding of universities, the use of allocation mechanisms at least partially performance-based for teaching and research is spreading [57], but they still do not include sustainability indicators. In many countries, the most important outputs taken into consideration for performance-based funding are those related to research, which also do not introduce criteria that promote sustainability research.

The case of Spain, as mentioned above, is interesting because of the relevant role played by the higher education system. It also has its own initiatives for the analysis and measurement of sustainability in universities. When analyzing the results obtained in this study, it can be observed that in terms of measures for the development of sustainability, universities scored highest in items relating to university community engagement (7.6), waste management (6.5) and campus grounds and biodiversity (6.0) and lowest in environmental impact assessment and green procurement (<5.0). The 2010 and 2017 reports also revealed considerable room for improvement in all areas, especially environmental impact assessment, social responsibility, green procurement, education and research. Although environment-related degrees have been in place in the last two areas, sustainability has yet to be integrated holistically across all university activities, rather than as a separate chapter or stand-alone curricular content.

The Spanish experience has also served as a reference for other regions. This is why a similar study was carried out in Latin America under the title "Definición de indicadores para la evaluación de las políticas de sustentabilidad en universidades latinoamericanas" (definition of indicators for assessing sustainability policies in Latin American universities). According to the findings, the region's countries scored lower than Spain. With a score of 6.1 (higher than Spain's 3.3), social responsibility was the sole exception and the only item attesting to significant commitment to sustainability [58]. In all the other items analysed (sustainability policy, university community engagement, education, research, urban planning, energy, water, transportation, waste management, green procurement, environmental impact assessment), Latin American universities scored lower than their Spanish counterparts, in particular in connection with green procurement and transportation. The report consequently identified a pressing need to develop and integrate sustainability in all aspects of university life.

Spanish governing bodies introduced a new way to allocate public universities funding based on performance criteria. By funding universities according to their outputs, rather than inputs, state policy makers in Spain believe they are providing an incentive for universities to improve their quality management and accountability [59]. In other words, it is about universities optimizing their management and activities considering ethical values and transparency to achieve a greater positive impact on all aspects of society. One of the main outcomes of this performance funding system has been the greater number of strategic plans or sustainability reports articulated and published online by Spanish universities in recent years [60].

Reviewing R&D+I from a bibliometric perspective, the number of scientific papers constitutes a good measure of the scientific community's interest in a given subject. The creation in Journal Citation Reports of the subject category Green and Sustainable Science and Technology (GSST) laid the ground for measuring the impact of sustainability-related research.

Although the earliest papers on the subject date back to the nineteen nineties, the number of articles has risen substantially in the last 5 years, especially at a yearly rate of over 17%.

Whilst research majors (China, USA, India, etc.) account for the highest output in GSST, smaller countries (Netherlands, Belgium, Austria and Malaysia) devote more effort to the area relative to the total research effort deployed.

Spain ranks fifth worldwide by total number of articles on the subject and its output is growing faster than the world average (21.8% versus 17.0%). The intensity of the country's activity in the area is also attested to by its activity index which, at 1.8, denotes greater devotion to the area than expected on the grounds of its overall WoS-listed output. Further WoS evidence for that assertion lies in Spain's ninth place in overall output compared to its fifth place in papers relating to sustainability.

Another feature of Spain's activity in the field is the concentration of sustainability research in three of its major cities (Barcelona, Madrid and Valencia), cities that concentrate the great universities. The presence of powerful university systems in all three is consistent with the fact that universities account for 91% of Spain's output in this subject category and with the community's interest in sustainability and related issues.

Although the country's most productive institution is the National Research Council, four universities also lay claim to significant output figures: The Autonomous University of Barcelona and the Technical Universities of Valencia, Madrid and Catalonia. Together, those five institutions account for over 20% of Spain's entire production in this area of research. Likewise, it has been shown how some small and medium sized universities present an intense productive activity in subjects related to sustainability.

Spain also plays a significant role in European projects on the subject, with a success rate of over 30% in the most prominent sustainability-related calls organised under the Seventh Framework Programme. One less favourable aspect of that success is that universities participate only marginally in such calls, where the private sector prevails. Earlier studies have shown that this is not uncommon in several other areas [21]. In any event, the same four universities found to reign in scientific paper output also rank highest by the number of Seventh Framework Programme projects, accounting for over 30% of the projects awarded in the area to the SUS as a whole.

Spanish universities' scant presence in sustainability projects should prompt academic management to change their research strategies for the present findings confirm the HEI community's interest in engaging in science on the subject. One possible way to raise participation would be through partnering with the private sector, for companies have acquired an ever more significant role, comparable to their predominance in patent applications in areas such as renewable energies.

It is evident that the use of scientometric tools, such as the analysis of publications or projects, has certain limitations. Among them is the difficulty of accurately defining the specific area of sustainability, so all quantitative studies are an approximation. However, the use of tools external to the information provided by the universities themselves can contribute to reducing the biases of interpretation (or manipulation) of the data. Likewise, the use of absolute indicators, such as the number of projects or publications, combined with relational indicators, such as the activity index, can offer a good measure of the effort of each of the institutions. In this sense, the case of Spain shows that the limitation raised in the literature on the scarcity of indicators on research in universities (compared to the total of the indicators of each tool) can be approached in a complementary way from a scientometric perspective.

5. Conclusions

After the development of this study, some relevant aspects should be mentioned as conclusions.

It has been detected that the interest in sustainability has been growing and this is evident in publications and projects on the subject as well as in the initiatives developed in different environments to promote it. The higher education sector is presented as one of the most fertile spaces for the development of measures in this field.

The methodology used in this study, which combines bibliometric techniques with the analysis of institutional documents, may be useful to identify a country's commitment to the development of initiatives, such as those related to sustainability.

The Spanish case can be taken as a reference to analyze the situation of other European and Latin American countries and detect points and strengths with respect to the implementation of measures on sustainability in universities.

The most common tools for measuring sustainability in universities use a small number of generic indicators (compared to the total of the indicators of each tool) of scientific output on sustainability. The presented study of the Spanish case shows the possibility of including several more specific indicators to analyse the topic of university research in sustainability (thematic specialization and AI). It would be very interesting for the different tools to include some of these indicators collected externally with an objective criterion, which would allow studies to be carried out with more reliable data in order to make international comparisons and facilitate the university's accountability in this area.

Regarding public policies that aim to promote the sustainability of universities, it should be noted that although the use of the tools studied is voluntary for the institutions, actions in terms of transparency and accountability can help to promote measures that encourage information on the impact of their actions on society. In the field of research, the incorporation of indicators on sustainability research, analysed for the Spanish case, in the performance-based funding models would provide an important incentive to promote research in this field.

Author Contributions: Conceptualisation, all authors; methodology, D.D.F.; data cleansing and formal analysis, D.D.F. and L.A.S.-H.; writing—original draft preparation, D.D.F. and L.A.S.-H.; writing—review and editing, F.C. and E.S.-C.; project administration and fund raising, D.D.F. and E.S.-C.

Funding: This project received funding from the European Union's Horizon 2020 Research and Innovation Programme under grant 741657, SciShops.eu. The content of this article does not reflect the official opinion of the European Union. Responsibility for the information and views expressed in the article lies entirely with the authors.

Acknowledgments: The methods used were developed on the framework of the project entitled 'Detection of new research and innovation fronts. Analysis of knowledge flows in the scientific domain, industry and society in the field of energy efficiency' (ref.: CSO2014-51916-C2-1-R), funded by the Spanish Ministry of the Economy and Competitiveness.

Conflicts of Interest: The authors declare no conflicts of interest.

References

1. Agasisti, T.; Barra, C.; Zotti, R. Research, knowledge transfer and innovation: The effect of Italian universities' efficiency on local economic development 2006–2012. *J. Reg. Sci.* **2019**, 1–31. [CrossRef]
2. Dagiliūtė, R.; Liobikienė, G. University contributions to environmental sustainability: Challenges and opportunities from the Lithuanian case. *J. Clean. Prod.* **2015**, *108*, 891–899. [CrossRef]
3. Barra, C.; Zotti, R. Investigating the human capital development-growth nexus does the efficiency of universities matter? *Int. Reg. Sci. Rev.* **2017**, *40*, 638–678. [CrossRef]
4. Findler, F.; Schönherr, N.; Lozano, R.; Reider, D.; Martinuzzi, A. The impacts of higher education institutions on sustainable development: A review and conceptualization. *Int. J. Sustain. High. Educ.* **2019**, *20*, 23–38. [CrossRef]
5. Fitzgerald, H.E.; Bruns, K.; Sonka, S.T.; Furco, A.; Swanson, L. The centrality of engagement in higher education. *J. High. Educ. Outreach Engagem.* **2016**, *20*, 223–244.
6. Carree, M.; Malva, A.D.; Santarelli, E. The contribution of universities to growth: Empirical evidence for Italy. *J. Technol. Transf.* **2014**, *39*, 393–414. [CrossRef]
7. EUROSTAT. Tertiary Education Statistics. Online Publications. 2019. Available online: https://ec.europa.eu/eurostat/statistics-explained/index.php?title=Tertiary_education_statistics (accessed on 20 September 2019).
8. Lozano, R.; Lozano, F.J.; Mulder, K.; Huisingh, D.; Waas, T. Advancing higher education for sustainable development: International insights and critical reflections. *J. Clean. Prod.* **2013**, *48*, 3–9. [CrossRef]
9. European Commission. Europe 2020: A Strategy for Smart, Sustainable and Inclusive Growth: Communication from the Commission. Publications Office of the European Union. 2010. Available online: https://ec.europa.eu/eu2020/pdf/COMPLET%20EN%20BARROSO%20%20%20007%20-%20Europe%202020%20-%20EN%20version.pdf (accessed on 14 May 2019).
10. Yáñez, S.; Uruburu, Á.; Moreno, A.; Lumbreras, J. The sustainability report as an essential tool for the holistic and strategic vision of higher education institutions. *J. Clean. Prod.* **2019**, *207*, 57–66. [CrossRef]
11. Carpenter, D.; Meehan, B. Mainstreaming environmental management: Case studies from Australasian universities. *Int. J. Sustain. High. Educ.* **2002**, *3*, 19–37. [CrossRef]
12. WCED. *Our Common Future*; Oxford University Press: Oxford, UK, 1987.
13. Velazquez, L.; Munguia, N.; Platt, A.; Taddei, J. Sustainable University: What can be the Matter? *J. Clean. Prod.* **2006**, *14*, 810–814. [CrossRef]
14. Leal Filho, W.; Brandli, L.L.; Becker, D.; Skanavis, C.; Kounani, A.; Sardi, C.; Raath, S. Sustainable development policies as indicators and pre-conditions for sustainability efforts at universities: Fact or fiction? *Int. J. Sustain. High. Educ.* **2018**, *19*, 85–113. [CrossRef]

15. Lidstone, L.; Wright, T.; Sherren, K. Canadian STARS-rated campus sustainability plans: Priorities, plan creation and design. *Sustainability* **2015**, *7*, 725–746. [CrossRef]

16. Clarivate Analytics. *Web of Science*; Clarivate Analytics: Philadelphia, PA, USA, 2019.

17. Callon, M.; Courtial, J.P.; Penan, H. *Cienciometría: La Medición de la Actividad Científica: De la Bibliometríaa la Vigilancia Tecnológica*; Trea: Gijón, Spain, 1995.

18. Plaza, L. Obtención de indicadores de actividad científica mediante el análisis de proyectos de investigación. In *Indicadores Bibliométricos en Iberoamérica*; En Albornoz, M., Ed.; RICYT: Buenos Aires, Argentina, 2001; pp. 63–70.

19. Jerneck, A.; Olsson, L.; Ness, B.; Anderberg, S.; Baier, M.; Clark, E.; Hickler, T.; Hornborg, A.; Kronsell, A.; Lovbrand, E.; et al. Structuring sustainability science. *Sustain. Sci.* **2011**, *6*, 69–82. [CrossRef]

20. Kajikawa, Y.; Ohno, J.; Takeda, Y.; Matsushima, K.; Komiyama, H. Creating an academic landscape of sustainability science: An analysis of the citation network. *Sustain. Sci.* **2007**, *2*, 221–231. [CrossRef]

21. Kajikawa, Y.; Tacoa, F.; Yamaguchi, K. Sustainability Science: The changing landscape of sustainability research. *Sustain. Sci.* **2014**, *9*, 431–438. [CrossRef]

22. Buter, R.K.; Van Raan, A.F.J. Identification and analysis of the highly cited knowledge base of sustainability science. *Sustain. Sci.* **2014**, *8*, 253–267. [CrossRef]

23. Bettencourt, L.M.A.; Kaur, J. Evolution and structure of sustainability science. *Proc. Natl. Acad. Sci. USA* **2011**, *108*, 19540–19545. [CrossRef]

24. Serrano-López, A.; De Filippo, D. Do scientifically relevant subjects arouse general interest? Bibliometric and altmetric analysis of green and sustainable science and technology output. In Proceedings of the 22 International Conference on Science and Technology Indicators, Paris, France, 6–8 September 2017.

25. De Filippo, D.; Serrano-Lopez, A. Academic impact and social media presence in Green and Sustainable Science and Technology publications. In Proceedings of the 23th International Conference on Science and Technology Indicators, Leiden, The Netherlands, 12–14 September 2018.

26. Pandiella-Dominique, A.; Bautista-Puig, N.; De Filippo, D. Trends in sustainability research output: Analysis of "Green and sustainable science and technology" category. *J. Innov. Sustain. Dev.* **2019**, in press.

27. Hallinger, P.; Chatpinyakoop, C. A Bibliometric Review of Research on Higher Education for Sustainable Development, 1998–2018. *Sustainability* **2019**, *11*, 2401. [CrossRef]

28. García-Zorita, C.; Marugán-Lázaro, S.; De Filippo, D. Aplicación de minería de texto para el análisis métrico sobre ahorro energético en proyectos del Séptimo Programa Marco. *Bibliotecas Anales de Investigación* **2018**, *14*, 149–163.

29. De Filippo, D.; Serrano-López, A.E. From Academia to Citizenry. Study of the Flow of Scientific Information from Projects to Scientific Journals and Social Media in the Field of "Energy Saving". *J. Clean. Prod.* **2018**, *199*, 248–256. [CrossRef]

30. Figueiró, P.S.; Raufflet, E. Sustainability in higher education: A systematic review with focus on management education. *J. Clean. Prod.* **2015**, *106*, 22–33. [CrossRef]

31. Lo, K. Campus sustainability in Chinese higher education institutions: Focuses, motivations and challenges. *Int. J. Sustain. High. Educ.* **2015**, *16*, 34–43. [CrossRef]

32. Larrán, M.; Andrades, F.J. Implementing Sustainability and Social Responsibility Initiatives in the Higher Education System: Evidence from Spain. In *Integrative Approaches to Sustainable Development at University Level*; Springer: Cham, Switzerland, 2015; pp. 113–128.

33. Instituto Nacional de Estadística (INE). España En Cifras. 2018. ISSN 2255-0410. Available online: http://www.ine.es/prodyser/espa_cifras (accessed on 03 March 2019).

34. CRUE. La Universidad Española en Cifras 2016–2017. 2018. Available online: http://www.crue.org/SitePages/La-Universidad-Espa%C3%B1ola-en-Cifras.aspx (accessed on 08 January 2019).

35. RICYT. Indicadores. 2019. Available online: http://www.ricyt.org/indicadores (accessed on 23 March 2019).

36. Sanz-Casado, E.; Bautista-Puig, N.; De Filippo, D.; Mauleón, E.; de Souza, C.; Lascurain, M.L.; García-Zorita, C.; Serrano López, A.; Marugán, S. Informe IUNE 2019. Actividad Investigadora de las Universidades Españolas (VI Edición). 2019. Available online: www.iune.es (accessed on 12 August 2019). [CrossRef]

37. Gómez, F.U.; Sáez-Navarrete, C.; Lioi, S.R.; Marzuca, V.I. Adaptable model for assessing sustainability in higher education. *J. Clean. Prod.* **2015**, *107*, 475–485. [CrossRef]

38. Lozano, R. A tool for a Graphical Assessment of Sustainability in Universities (GASU). *J. Clean. Prod.* **2006**, *14*, 963–972. [CrossRef]

39. Shriberg, M. Institutional assessment tools for sustainability in higher education: Strengths, weaknesses, and implications for practice and theory. *High. Educ. Policy* **2002**, *15*, 153–167. [CrossRef]
40. Alshuwaikhat, H.M.; Abubakar, I. An integrated approach to achieving campus sustainability: Assessment of the current campus environmental management practices. *J. Clean. Prod.* **2008**, *16*, 1777–1785. [CrossRef]
41. Yarime, M.; Tanaka, Y. The issues and methodologies in sustainability assessment tools for higher education institutions: A review of recent trends and future challenges. *J. Educ. Sustain. Dev.* **2012**, *6*, 63–77. [CrossRef]
42. Sayed, A.; Kamal, M.; Asmuss, M. Benchmarking tools for assessing and tracking sustainability in higher educational institutions: Identifying an effective tool for the University of Saskatchewan. *Int. J. Sustain. High. Educ.* **2013**, *14*, 449–465. [CrossRef]
43. Salvioni, D.M.; Franzoni, S.; Cassano, R. Sustainability in the higher education system: An opportunity to improve quality and image. *Sustainability* **2017**, *9*, 914. [CrossRef]
44. Berzosa, A.; Bernaldo, M.O.; Fernández-Sanchez, G. Sustainability assessment tools for higher education: An empirical comparative analysis. *J. Clean. Prod.* **2017**, *161*, 812–820. [CrossRef]
45. Findler, F.; Schönherr, N.; Lozano, R.; Stacherl, B. Assessing the Impacts of Higher Education Institutions on Sustainable Development—An Analysis of Tools and Indicators. *Sustainability* **2019**, *11*, 59. [CrossRef]
46. Parvez, N.; Agrawal, A. Assessment of sustainable development in technical higher education institutes of India. *J. Clean. Prod.* **2019**, *214*, 975–994. [CrossRef]
47. Alghamdi, N.; den Heijer, A.; de Jonge, H. Assessment tools' indicators for sustainability in universities: An analytical overview. *Int. J. Sustain. High. Educ.* **2017**, *18*, 84–115. [CrossRef]
48. Alonso-Almeida, M.M.; Marimon, F.; Casani, F.; Rodriguez-Pomeda, J. Diffusion of sustainability reporting in universities: Current situation and future perspectives. *J. Clean. Prod.* **2015**, *106*, 144–154. [CrossRef]
49. Lozano, R. Incorporation and institutionalization of SD into universities: Breaking through barriers to change. *J. Clean. Prod.* **2006**, *14*, 787–796. [CrossRef]
50. Lozano, R.; Young, W. Assessing sustainability in university curricula: Exploring the influence of student numbers course credits. *J. Clean. Prod.* **2013**, *49*, 134–141. [CrossRef]
51. CRUE. *Informe 2017. Diagnóstico de la Sostenibilidad Ambiental en las Universidades Españolas*; CRUE: Madrid, Spain, 2018.
52. Benayas, J. *Evaluación de las Políticas Universitarias de Sostenibilidad como Facilitadoras para el Desarrollo de los Campus de Excelencia Internacional*; CRUECADEP: Madrid, Spain, 2011.
53. Marrone, P.; Orsini, F.; Asdrubali, F.; Guattari, C. Environmental performance of universities: Proposal for implementing campus urban morphology as an evaluation parameter in Green Metric. *Sustain. Cities Soc.* **2018**, *42*, 226–239. [CrossRef]
54. Suwartha, N.; Sari, R.F. Evaluating UI GreenMetric as a tool to support green universities development: Assessment of the year 2011 ranking. *J. Clean. Prod.* **2013**, *61*, 46–53. [CrossRef]
55. Beringer, A.; Wright, T.; Malone, L. Sustainability in higher education in Atlantic Canada. *Int. J. Sustain. High. Educ.* **2008**, *9*, 48–67. [CrossRef]
56. Alshuwaikhat, H.; Adenle, Y.; Saghir, B. Sustainability assessment of higher education institutions in Saudi Arabia. *Sustainability* **2016**, *8*, 750. [CrossRef]
57. Estermann, T.; Pruvot, E.B.; Claeys-Kulik, A.L. *Designing Strategies for Efficient Funding of Higher Education in Europe*; DEFINE Interim Report; European University Association (EUA): Brussels, Belgium, 2013.
58. RISU (Red de Indicadores de Sostenibilidad Universitaria). *Definición de Indicadores para la Evaluación de las Políticas de Sustentabilidad en Universidades Latinoamericanas*; Publicación Resumen del Proyecto; Universidad Autónoma de Madrid: Madrid, Spain, 2014.
59. Vilardel, I.; Álvarez, M. Algunas reflexiones sobre la planificación estratégica de la universidad: El caso de la Universitat Autònoma de Barcelona (1998–2010). *Investigaciones de Economía de la Educación* **2014**, *5*, 350–368.
60. Larran, M.; Madueño, J.H.; Cejas, M.Y.C.; Peña, F.J.A. An approach to the implementation of sustainability practices in Spanish universities. *J. Clean. Prod.* **2015**, *106*, 34–44. [CrossRef]

Article

The Motivation of Students at Universities as a Prerequisite of the Education's Sustainability within the Business Value Generation Context

Alzbeta Kucharcikova *, Martin Miciak, Eva Malichova, Maria Durisova and Emese Tokarcikova

Department of Macro and Microeconomics, Faculty of Management Science and Informatics, University of Zilina, Univerzitna 8215/1, 01026 Zilina, Slovakia; Martin.Miciak@fri.uniza.sk (M.M.); Eva.Malichova@fri.uniza.sk (E.M.); Maria.Durisova@fri.uniza.sk (M.D.); Emese.Tokarcikova@fri.uniza.sk (E.T.)
* Correspondence: Alzbeta.Kucharcikova@fri.uniza.sk; Tel.: +421-41-513-4426

Received: 16 September 2019; Accepted: 8 October 2019; Published: 10 October 2019

Abstract: The aim of this article is to identify substantial factors affecting the motivation of universities' students to be actively engaged in the education process and define recommendations for the increase of this motivation. As a result, the sustainability of education at universities will be supported, contributing to the increase of the value of human capital of students and, subsequently, to the generation of value for the stakeholder groups in those enterprises where the graduates will be employed. The research hypothesis is focused on the presence of differences in students' motivation in relation to their gender, study program, and the year of study. To effectively achieve this aim, the analysis, comparison, and the synthesis of the theoretical background was performed, using available sources of secondary data found in the pieces of domestic and foreign professional literature. The pieces of knowledge obtained were supplemented and combined with pieces of information acquired from the questionnaire survey conducted, focusing on the motivation of students of informatics and management at a university in the Slovak Republic. As tools of statistical analysis, tests of independence suitable for nominal categorical data were applied. It was revealed that young people are motivated to study at a university, specifically at the Faculty of Management Science and Informatics, mainly by the prospect of better chances in the labor market, the possibility of getting a higher salary, and higher qualification. The motivation to study at a university in order to improve the opportunity of getting employed in the labor market was more frequently perceived by women. Despite the fact that the level of teaching is considered to be high by almost 50% of the students regardless of their gender, study program, or the year of study, their motivation also stems from their expectations related to their future jobs. The students of informatics expect to have a team of friendly colleagues, delightful and stimulating working conditions, and the opportunity to do meaningful work. Among the students of management, meaningful work was replaced by the opportunity for self-fulfillment. When focusing on other factors, the differences based on the gender, study program, or the year of study were not statistically significant. Based on these findings, specific measures for the faculty's management were proposed.

Keywords: motivation; human capital investments; generation of business value; sustainability; universities

---

1. Introduction

The current time is characterized by open markets and rapid changes in production and sales conditions. At the same time, the risk of new economic crises emerging is still rising. To strengthen the competitiveness of whole economies but also of the individual enterprises, it is necessary to increase the performance and efficiency, which requires the implementation of new technologies and changes in the business processes, the orientation on value management, education, and the emphasis put on the increase of the value of human capital because people represent the most important driver in enterprises and in society.

Human capital includes all the innate and during-life-acquired knowledge, skills, experience, and talent of a person [1].

The scope, structure, and focus of the investment in human capital are affected by macroeconomic as well as microeconomic factors. These include, for example, the society-wide or an enterprise's strategic goals [2], current state within the industry and the enterprise's position in the given industry [3–5], corporate culture [6], the enterprise's orientation on social responsibility [7,8], or the support of sustainability in the long term [9].

Based on the research conducted earlier [10], it was revealed that the enterprises as well as society consider the investment in education to be the most familiar form of investment in human capital. This investment serves for the increase of the level of knowledge and skills of individuals, and for the desirable change in their attitudes.

Education is provided by various institutions. These also include universities [11]. For universities to be able to provide high-quality education for the needs of the practice, they need to identify and affect the factors that motivate young people to enroll for a specific university program and to study. To secure the sustainability of education at universities [12] as well as within the implementation of the value management in enterprises [13], it is also necessary to know the wishes and expectations of students [14] related to their employment in the labor market [15], or more specifically, within the practical operation of enterprises. Accordingly, universities can plan and implement adequate measures afterwards [16]. While increasing the value of human capital via university education [17], the success achieved is influenced by multiple factors. These encompass the phase of the economic cycle, historical development, and the engagement of the country in international structures, measures of economic policy [18,19], attitude of society towards education [20], demand for the highly-qualified workforce on the side of enterprises and the overall situation in the labor market, the level of science and technology in the country [21], quality and reputation of a particular institution providing the education, quality and attractiveness of the study programs and their alignment with the current needs in the labor market, quality and attitude of the teachers to the education process, teaching methods being applied, motivation of students, etc.

A pivotal element in the process of education is represented by the students themselves. There are numerous factors affecting students during their studies, but the most important one is their motivation for studying and for becoming proficient experts within specific needs of the practice. The motivation for studying is a prerequisite for successful achievement of expected results within the education process as well as for the sustainability of this process.

This will also contribute to the increase in value generation in enterprises that employ successful graduates with high level of their human capital.

This was the reason why this particular research was focused on the identification of substantial factors that affect the motivation of students to start studying at a university and actively participate in the education process as well as their expectations related to their future jobs.

Via the identification of these factors of motivation together with the designed recommendations, this research will contribute to the solving of the issue of securing the education's sustainability (which contributes to the increase in the value of people's human capital), while respecting the requirement of enterprises for the increase in the generation of value for the stakeholder groups within the implementation of the value management concept.

The logical connection of the above-mentioned professional and scientific areas is depicted in Figure 1, and it is more thoroughly elaborated in the following sections. However, this is a typical situation happening in research when the real system is quite complex and consists of an abundance of elements. One possible solution to coping with this issue is the application of abstraction.

Following this method, the research presented in this article works only with selected aspects within the motivational readiness of university students. Subsequently, the research tries to describe the link between motivational readiness (motivation for studying) of students and the sustainability of the education system. Finally, the sustainability of the education system is connected to the increase in the human capital value of university graduates. After being employed in various enterprises, these graduates, with enhanced levels of their human capital, become valuable contributors to the value creation in those enterprises.

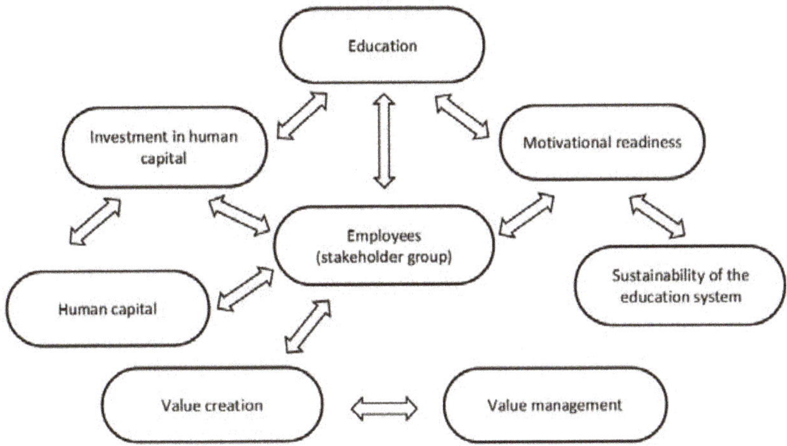

Figure 1. The logical structure of the studied topic, consisting of individual elements and their relationships.

2. Literature Review

2.1. Education and the Generation of Value

In the time of globalization and dynamic changes, education is becoming the decisive force of successful enterprises. More than ever before, it is necessary to maintain the knowledge and skills of employees at a desirable level.

The stakeholder groups of an enterprise are represented by persons, groups, or organizations that are able to affect the enterprise's existence, or that can be affected by the enterprise's activities. Such groups include the owners, creditors, employees, suppliers, customers, competitors, the government, local authorities, non-governmental organizations and pressure groups, communities, and the media. The stakeholder groups defined this way are related to the value orientation of the business. A common goal of all stakeholder groups is the long-term operation and prosperity of the enterprise. Therefore, the aspect of sustainability is important here as well. Employees, as one of the stakeholder groups, embody the carriers of human capital. Via the generation of value for the employees, the enterprise's human capital is growing. This can be determined as the intangible assets [22]. Additionally, the intangible assets represent a factor contributing to the increase of quality of internal processes, which subsequently generate the value. Other assets (machines, equipment, hardware, software, patents, copyrights, trademarks) are the sources of value generation via human capital that is in possession of the employees. Without employees and their knowledge, skills, experience, abilities, creativity, and personal traits, other assets would only be static, idle things or rights.

The generation of value for the employees, which results in the increase of human capital, is closely related to the concept of value management. This is a style of management focused especially on people, acquisition of skills, and support of synergy and innovation with the aim to maximize the total performance of an organization [23]. Value management is a process consisting of managerial activities performed via approaches and methods used at various managerial levels of enterprises, focusing on the value.

When the principles, methods, and activities of the value management are applied, the value is being generated for the stakeholder groups, which results in their strengthened loyalty to the enterprise—an owner is willing to invest additional capital, a customer becomes a long-term customer, an employee directs his/her activities towards the fulfillment of the goals set, other stakeholder groups spread the enterprise's good name, etc. [24]. According to Obeng, it is necessary to manage relationships with the stakeholder groups, while the strategy can be focused on all or only on a few selected groups. The author uses the concept of the stakeholder concentration index [25].

The procedure of generation of value for employees starts with the expression of this value, continues with its measurement and its transformation into financial and non-financial indicators, and it ends with the determination of activities for the generation of this value. The value for employees can be expressed as an adequate basic salary, a complex system of additional benefits, job security, respecting of payday deadlines, career development and satisfaction of the needs for self-fulfillment, and so on. The measurement of this value can be expressed via the salary, bonuses, an average length of employment according to categories of employees, the number of official praises, the number of opinions expressed, and the number of proposals. The managerial activities generating the value for employees can include the offer of educational and training activities (ranging from practical training to advanced managerial courses) [26], communication of the employees' representatives with the management, increasing of the quality of working environment and positive influence on motivation, performance, and the general interest in employees within the enterprise [27]. The increase in the value is a result of an optimal combination of numerous activities and factors that need to be monitored, analyzed and incorporated in the designing of variants for the solution with the subsequent selection of the optimal one [28].

Another specific feature of human capital in enterprises is the fact that on the one side it represents a part of the enterprise's market value [29], and on the other side, it is closely connected to one of the stakeholder groups within the concept of value management, and thus the employees. The employed people expect that a rich portfolio of motivational tools will be provided for them [30–33], including, for example, meaningful work assignments, an opportunity for self-fulfillment, professional development, fair remuneration, delightful surroundings, job security, friendly atmosphere at the workplace, career advances, fringe benefits, professional management, and others. Via the fulfillment of these expectations, the value is generated for the employees, which develops the human capital available.

Among the internal stakeholders, the employees have considerable importance. Employees primarily acquire knowledge and skills from the education system. The education shall be focused on the creativity of students, independent problem solving, integration of pieces of knowledge obtained from multiple subjects studied and from multiple scientific fields [34]. Within the education process, the cooperation of three elements is necessary: The teacher, the student, and the content of the syllabus.

In the process of education, a university teacher [35,36] is a mediator of the content, who helps a student embrace the knowledge, using the selected, adequate teaching method to familiarize the student with new pieces of knowledge via the application of them in examples. The teacher also points out the common features and relationships within the content and explains the logical structure and connections of individual concepts being taught to the student. The student embraces the pieces of knowledge via his/her active approach to studying. The degree of his/her active approach is directly proportional to the teaching method the teacher selected, which considerably affects the success of the goals' fulfillment. The way of the teacher's activity influences the way of the student's activity in the process of embracing the knowledge.

Subsequently, in enterprises, the value is generated only by those employees (once being the students themselves) who

- participate in continuous education,

- have knowledge gained within multiple scientific fields,
- are able to address the managers with the products that will solve their problems and contribute to the growth of revenues, reduction of costs, increase of the productivity of labor, and to the generation of the enterprise's profit,
- are able to look at an issue from the technological as well as from the business perspective.

The education process shall react to the continuous connection of education and the employment of the graduates in the practice. The graduates will add value to the pieces of information via their interpretation and identification of connections. A precondition for potentially smart employees, managers, or owners of enterprises is represented by well-educated students. These are able to perform a survey or research, seek and select the electronically processed pieces of information needed, and synthesize them into coherent pieces of knowledge. The emphasis is put on the building of the ability of students to get oriented in the plethora of new pieces of information and the ability to utilize them.

An effective way of teaching uses progressive didactic methods and cooperative forms of work. The primary goal is to actively engage the student in the cognitive process. This is based on the cognition via activity, an active relationship of the student with the natural, economic, or social environment in which the issue is being solved [37]. This approach is oriented on the experience gained by the student, and it develops the mental structures in connection with the corresponding processes. In the case of conventional teaching methods, the student is being presented with particular, mutually isolated pieces of knowledge. The most emphasis shall be put on the creativity of students and their own solutions of the assignments, on the integration of the knowledge, and experience from multiple subjects and from real life during the solving of a particular issue. The students need to learn the abilities of self-presentation, effective communication, and self-sufficiency. Subsequently, this will be reflected in their willingness and flexibility for the solving of various problematic situations [38].

In relation to the generation of value for employees as one of the stakeholder groups, it is meaningful for the enterprise to establish cooperation with universities. The alignment and connection of the needs in the practice with the education at universities lowers the need of an employer to organize other forms of education within the enterprise, both with the general or specialized focus. The enterprise's internal education needs to be evaluated together with the strategy of employees' engagement and with the Critical Success Factors [39].

Within the concept of value management, in concordance with the theory of human capital, it is important to focus the generation of value for employees on the elimination of unwillingness of people to share the knowledge and undergo changes in their routine procedures and methods. Education significantly contributes to this. This is the case of education in enterprises but mainly of education at universities, before an individual becomes an employee.

2.2. Education and Sustainability

Numerous research studies were performed focusing on the sustainability in education and on the education for the continuous sustainable development at universities [40,41]. This research is focused on the identification of expectations and motivational factors of students at a university, specifically on the factors that can influence their decision-making on enrolling, successful studying, and finishing the studies at a university. This will support the sustainability of education at universities and contribute to the increase in the quality and value of human capital that is in possession of the graduates. When the graduates become employed, the research's results can also contribute to the increase of value of enterprises within the concept of value management.

The basic and conventional mission of the university education process is to perform professional preparation and form human capital via the connection between the education process, research, and the requirements of the labor market. Education plays a key role in the development of a personality and engagement of individuals (as a stakeholder group), and it shall not only support the increase in general and specific theoretical knowledge and the knowledge of the world but also the effective transition of the knowledge to the surroundings and the practical application in personal and professional life. This is an important part of the university's mission since the knowledge and skills of the graduates, based on which they perform their decisions, will be able to considerably influence the quality of life of future generations from the ecological, environmental, social, cultural, economic, and personal perspectives.

The level and quality of education contributes to the sustainable development of the economy at the international as well as at the national, regional, and local level. The results of research studies point out the positive economic effect and the ability to secure sustainable development in those cities and regions in which there are universities and their graduates [42]. In addition to that, the mutual cooperation of universities with enterprises results in various innovation activities in the form of applied new technologies and patents [43]. In a synergic way, the education emphasizing the practical side of the application of the students' knowledge in the business practice increases their professional and technical competencies, cognitive and non-cognitive abilities, which enables them to get decent and adequately remunerated jobs.

Sustainable development is a broad concept that has many forms of application and that is related to the sustainability of the university education itself. For the education to be the driver of development and prosperity of economies, it is necessary to meet these conditions:

- Universities themselves need to apply the principles and goals of sustainable development and set a good example because they play an important role in society and they represent very influential parties [44]. It is also important for the management of a university to publicly declare and integrate sustainability into the strategic plans. This will lead and motivate them to actually implement different dimensions of sustainable development into practice [45,46]. Students and other university stakeholder groups often evaluate its quality based on the position in international rankings. The systems of evaluation, which focus on the current state and also monitor and quantify the effort towards sustainability of universities, include the system of evaluation according to the UI Green Metric World University Ranking (UIGM). This system evaluates the performance in the categories including the infrastructure, energy and climatic change, waste management, transport management, education, and research [47,48]. Besides that, other authors also created and described frameworks and tools for universities focused on the measurement and strengthening of the resilience of their infrastructure, culture, and systems, and on the ways of contributing to the resilience of communities [49].

- Universities need to educate their students in the field of sustainable development. This commitment explicitly means the securing of innovative educational approaches and education of a high quality with the emphasis being put on international cooperation within the field of education (study programs, mobility programs), creation of excellence in the field of skills development, support of life-long learning, and the access to education for everyone. Within this field, literature research on the key competencies and solutions for sustainable development can be useful, especially including the specific case studies focused on the form of education within the field of sustainable development [50–52] and the measurement of success of the education process [53,54].

- Universities need to operate in the conditions where the graduates can find employment in enterprises in which the management is oriented on the value and the managers want to achieve sustainable development. With this approach, the enterprise is able to solve the needs and expectations of its stakeholder groups in the long run, and it is able to create social values and support the best possible utilization of the limited resources available [55]. The sustainable management reaches its goals more effectively and more efficiently, while it is able to direct the performance with regard to the generation of value [56]. The increasing values of economic indicators of sustainability have a multiplicative effect on the creation of new opportunities for the development of enterprises, which are affected by the regional specialization [57].

Within the economy's cycle, education of high quality, based on the principles of sustainable development, has a considerable impact on the increase of the quality of human capital, which enables the strengthening of sustainable development of enterprises and increases their value and economic prosperity of the whole country. On the other hand, this enables sustainable investment with a substantial impact in the field of university education, contributing to the sustainable development of the universities themselves.

For enterprises and universities/faculties to be able to meet the requirement of the education's sustainability and of the increase in the generation of value for the stakeholder groups, they need to identify and subsequently influence the factors that motivate students to study. This motivation represented the topic on which the questionnaire survey for the students of the selected faculty was oriented.

2.3. Motivation for Studying

A motive represents an internal reason, which causes a change in the person's behavior and leads to the fulfillment of his/her needs. The motivation for studying or for learning can be perceived in two perspectives. One is created by the motivation of the participants of particular educational activities being performed, which follow the content and the structure of the study program (internal motivation). The second perspective is represented by the motivation for the learning itself, which is affected by the expected benefits from the education (external motivation).

The key element within the realization of the education process is represented by the students themselves. Their motivational readiness for studying depends, among other things, on their emotional state, cultural and educational background, or the physical conditions in the classrooms.

There is an abundance of motivational factors. Within this research, they were divided into three groups:

- Expectations before starting the studies at a university,
- quality of realization of educational activities during the studies at a university,
- expectations and vision related to a future career.

These factors are closely related to each other and they are in a relationship of mutual influence. Therefore, the strength of motivation is the intersection of all these factors (Figure 2).

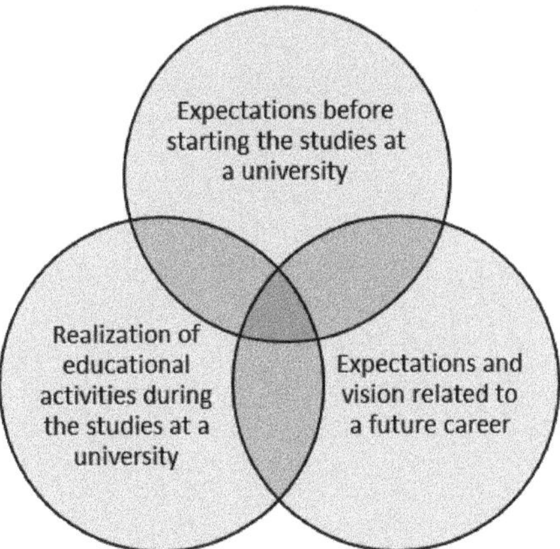

Figure 2. The factors of students' motivation at universities.

Revealing which specific elements create these three groups of motivational factors will help solve the issue of securing the education's sustainability, respecting the requirement of enterprises for the increase in the generation of value for the stakeholder groups. This can be considered to be an important contribution of this article to application of the research results into practice.

Motivation is the most important factor of university education since the studying at universities is beyond the compulsory school attendance, and thus the motivation of a university student is primarily dependent on the student's personality. In Slovakia, there is also a still present pressure on the successful completion of this type of education from the family members of a student. However, there is a difference between the motivation for studying (learning) and the motivation to only obtain the diploma or the university degree itself. In the first case, the student realizes the desirability and usefulness of the knowledge for his/her future employment and for the personal life, he/she has a strong self-motivation for the advances and cognition. In the second case, the motivation is based on the misconception that a "piece of paper" without the knowledge obtained at a university will automatically help him/her find a job in the labor market. Alternatively, it is based on the assumption that the status of a university student helps postpone the start of the working life. (In Slovakia, even though many students work part-time or have temporary jobs during the studies, the parents still tend to support them financially until they finish their studies.) This perspective can be considered mistaken, leading to incompetence in future jobs. The universities try to correct it via asking for feedback from enterprises and they try to flexibly adapt the content and scope of accredited mandatory and optional subjects.

In general, it can be assumed that the internal motivation of a university student includes the usefulness of the knowledge gained for life, obtaining the qualification for future jobs, curiosity, desire to learn something new, getting the social status of an educated person, and the value system of the student. When talking about the internal motivation, the most frequently used terms shall be the need of cognition, self-fulfillment, self-transcendence, but, at present, also the terms such as the pro-sociality and the job/mission, and the need for life-long learning.

The external motivation can encompass the expectations placed on the student, i.e., the stimuli from the labor market or the demands of employers [58], expected future salary in the given field, professional status and working conditions, social status in an explicit way, i.e., the society's perception of an educated person, the influence of the family and surroundings, or the influence of teachers at lower levels of the education system, prestige of a university, the overall environment at the university (including supportive teachers), subjects with appropriate content and purpose, motivational scholarships or job offers for the best students, lectures done by experts from the practice, internships in enterprises, job fairs, a possibility to participate in international student mobility programs, the need to be in a young and inspiring environment, and the interpersonal relationships [59,60].

The development of information-communication technology (ICT) enabled the increase of the students' motivation as well. It offers various new tools supplementing and supporting the educational process that facilitates studying and increases student engagement [61,62]. The experts then point out the fact that many students, regardless of the chosen direction of their studies, seek studying subjects that can increase their business skills and education [63,64], while these pieces of knowledge help them find better employment in the labor market or start their own businesses. The expectations of a future career are one of the motivating factors for young people to study at a university. By employing highly motivated students, enterprises will attract employees with a high level of human capital. This will also contribute to increasing business value generation for the stakeholders.

The students' motivation for studying at a particular university and their expectations about future careers was addressed in the questions asked within the questionnaire survey focused on the students.

3. Materials and Methods

The aim of the article is to identify substantial factors influencing the motivation of students at universities to actively engage in the education process and define the recommendations for the increase of this motivation so that the sustainability of this form of education is supported in the long term. This will contribute to the increase of the value of human capital of students, and, subsequently, also to the generation of value for the stakeholder groups in the enterprises in which the graduates of universities will be employed.

The effective attainment of this goal starts with the analysis, comparison, and synthesis of the theoretical background of the studied issues, based on the available pieces of secondary data found in the domestic and foreign professional literature. The studied topic overlaps several theoretical concepts and approaches. The central element is represented by the process of education (specifically university education) as a specific form of investment in human capital. Such investment increases the value of the human capital available and prolongs the period during which it is usable, as it is described in the concept of human capital management. Another perspective is added by the value management concept where education directly affects the value that the enterprise can create for the stakeholder groups. Within this concept, the employees themselves are one of the stakeholder groups. Since students will become employees in the future, it is efficient to deal with the sustainability of their motivation already at the phase of their university studies. The logic behind the theoretical background of this topic together with the interrelationships of individual elements are depicted in Figure 1.

Another step leading to the attainment of the aim is the collection of the primary data. The data points were collected via the method of sociological inquiry with the application of the questionnaire technique. The questionnaire was anonymous, focused on the students of informatics and management of the University of Zilina, Faculty of Management Science and Informatics in the Slovak Republic.

The survey was conducted in 2018, including students from the first and the second grade at the Faculty of Management Science and Informatics. The population consisted of 577 students. A total of 306 filled questionnaires were collected. This sample size put the margin of error to 3.84% at the confidence level of 99%. This survey represents a starting/preparatory exploration within the conditions of the faculty. In the future, it is planned to broaden the scope of the research, including more students at the faculty and at the partnered faculties in Slovakia and abroad.

The purpose of the inquiry was to reveal the current state of the motivation of students for studying at university and of their expectations about their future careers. The identification of students' expectations in relation to their employment in the labor market contributes to the creation of the situation where education at a university helps students obtain the qualification needed for getting jobs that will meet their wishes. The identification features of respondents included the gender, study program, and the year of study. Based on these features, the basic description of the research sample was elaborated. In the questionnaire, three questions were used in which multiple choices could be selected by a respondent. In nine questions, only one choice could be selected. Within the total number of 12 questions included in the questionnaire, in four of them the respondents could also freely express/add their own opinions. Particular questions in the questionnaire were connected to three basic groups of factors affecting the motivation of students (Figure 2). Specifically, the questions were focused on the reasons leading young people to start studying at a university (expectations before starting the studies), the perceived level of teaching (realization of teaching activities), and the expectations related to their future careers. These elements represent basic variables to be used to confirm or reject the main hypothesis underlying the whole research. The content of questions in the questionnaire follows the findings from the literature review and it is also inspired by a previous study conducted in the Czech Republic [65]. Therefore, the validity of the specific questionnaire applied in this research is supported by the application of a similarly constructed tool used in the above-mentioned study. Overall validity was enhanced by the fact that the questionnaire and its final form was checked by several experts within the field of human capital and higher education. To make sure that the tool applied in the research was inherently consistent, Cronbach's alpha was calculated. The result value of 0.73 shows that the tool's consistency is acceptable. However, for future research projects, the tool can be altered to achieve an even higher value.

The research hypothesis was defined as follows: The motivation of students for studying differs in relation to their gender, study program, or the year of study. Here, the motivation consists of three elements included in the aforementioned variables. This way the research hypothesis indicates its decomposition into particular statistical hypotheses (e.g., the reasons leading young people to start studying at a university differ in relation to their gender). In the article, the individual sections of the results part are structured accordingly, with the questions in the questionnaire encompassing the variables applied in the hypotheses' testing. The assessment of the hypotheses is not explicitly listed, only the test results and their interpretation are included in the article.

Within the processing and interpretation of the results obtained from the primary data, specific forms and techniques of exploratory analysis were applied. The processed data outputs were appropriately listed in tables and depicted via histograms for the support of interpretation of the results achieved. The nature of the data entries themselves (categorical data) determined the application of relevant methods of statistical analysis. Depending on the number of categories, the statistically significant differences were detected via suitable methods for the testing of independence of two variables, including the Pearson's chi-squared test and the z-score related to the group of chi-squared methods, at the significance level of $\alpha = 0.05$. The Pearson's chi-squared test was applied in cases when one of the variables had more than two categories. This test identifies the presence of statistically significant differences based on the comparison of the actually observed and expected frequencies [66,67]. This method of statistical analysis is often used by researchers in various fields, for example, in the field of management, marketing, and business [68–73]. The method is also used in research projects focused on the field of education [74–77]. In cases of the questions in which the respondents could choose more than one answer, the responses were evaluated separately. This way, multiple association tables were created, encompassing the frequencies for dichotomic nominal variables. From these, z-scores were calculated to be applied in the statistical testing of hypotheses. Other research works, including those focused on the field of education, often utilize other statistical methods as well [78–82], but their application is based on working with the numeric data type.

Based on the results of the statistical analysis in combination with the findings extracted from the review of the professional literature, finally, the recommendations leading to the increase of the motivation of students for an active participation in the process of education were defined and described in the article, within the context of sustainability of the whole system.

4. Results

The results obtained from the processing of the primary data points contribute to the identification and description of the current state of the motivational readiness of students of universities for studying. High motivational readiness is a fundamental precondition for the achievement of high-quality education because it is interconnected with the effort of students to reach great studying results during their studies. Students who actively participate in the education process become qualified job seekers in the labor market after finishing their studies, they have high values of their human capital, and they subsequently become valuable employees of enterprises, contributing to the value generation in those enterprises. Only if these three phases are connected and sufficiently aligned, the sustainability of the education system in the long run can be achieved.

A basic outlook on the processed data can be gained from the description of the research sample captured in Table 1. In terms of gender, men had the majority position within the sample. In terms of the study program, the majority position was held by the students of informatics. In terms of the last identification feature represented by the year of study, the students from the first year of university studies prevailed in the research sample. Even though the proportions of individual groups in the sample are not equal, the numbers of respondents within these groups enable the application of methods of statistical analysis for the detection of differences between these groups ($n > 20$).

Table 1. Characteristics of the research sample.

Gender	Number	%
Male	206	67.32
Female	100	32.68
Study Program	**Number**	**%**
Informatics	193	63.07
Management	113	36.93
Year of Study	**Number**	**%**
First year	270	88.24
Second year	36	11.76

The results obtained from the survey are structured in accordance with three basic groups of factors affecting motivational readiness of students for studying. The first group of factors includes the reasons of students for starting to study at a university. The second group is connected to the realization of educational activities during the studies at a university. The last group of factors gives an account of the students' expectations in relation to their future work positions and assignments.

4.1. Reasons Leading to Studying at a University

The reasons causing students to start studying at a university represent the elemental source of their motivation during the whole studies. The aggregated results of the corresponding question from the questionnaire are captured in Figure 3 in a graphical way to enhance their interpretation.

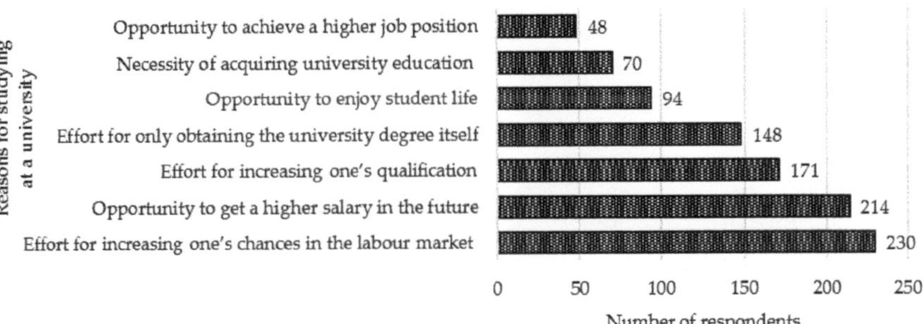

Figure 3. Reasons leading to studying at a university.

The results show that the two most frequent sources of motivation for starting to study at a university are the effort for increasing one's chances in the labor market (75.16% of respondents) and the opportunity to get a higher salary in the future (69.93% of respondents). An interesting fact is that almost half the students chose the reason of only trying to get the university degree itself for the sake of having it. Contrary to the effort for increasing one's qualification, the desire to only acquire the university degree represents only a superficial interest in the studies themselves, which implies less attention paid to the content of the subjects being taught and lower interest in the active participation in the education process. The opportunity of getting a higher position was chosen only by 15.69% of respondents. This result implies that students do not fully realize that due to the university education they will be able to achieve higher job positions later in their careers. This finding thus creates space for future improvement. Implementation of appropriately directed recommendations for the process of university education can increase the perception of this advantage by the students, which will contribute to higher motivation for achieving superb study results.

Then, within the reasons leading to studying at a university, the presence of statistically significant differences based on the student's gender was analyzed. The purpose was to reveal whether there occurs significant differentness of motives for studying at a university between men and women. Table 2 shows the results of a statistical analysis based on the testing of independence using the z-score together with the p-value (at the significance level of $\alpha = 0.05$) enabling the interpretation of these results.

Table 2. Reasons leading to studying at a university in relation to gender.

Reason	Men %	Women %	z-Score	*p*-Value
Effort for increasing one's chances in the labor market	69.90	86.00	3.057	0.002
Opportunity to get a higher salary in the future	68.45	73.00	0.815	0.415
Effort for increasing one's qualification	54.85	58.00	0.520	0.603
Effort for only obtaining the university degree itself	46.12	53.00	1.130	0.258
Opportunity to enjoy student life	30.58	31.00	0.074	0.941
Necessity of acquiring university education	19.90	29.00	1.777	0.076
Opportunity to achieve a higher job position	15.53	16.00	0.105	0.916

A statistically significant difference was corroborated by the test only for the motive of the effort for increasing one's chances in the labor market. Women chose this motive more often (86% of women) in comparison with men (69.90% of men). Overall, the motives of women and men were not considerably different. However, the significant difference was identified for the most frequent motive. Since we consider this motive to be a strong factor affecting the effort for achieving great study results, there is space for directing the recommendations toward the increase of perception of this form of motivation among the male students. One of the possible reasons for the difference identified is the socially-conditioned state when women can perceive their chances in the labor market as lower than those of men, and they try to get an advantage via higher education.

Then the attention was paid to the identification of significant differences in the perception of the motives based on the study program. The results of this analysis were reached via the calculation of the z-score again, and they are listed in Table 3 in a structured way, together with the corresponding p-values (at the significance level of $\alpha = 0.05$).

Table 3. Reasons leading to studying at a university in relation to the study program.

Reason	Informatics	Management	z-Score	*p*-Value
Effort for increasing one's chances in the labor market	74.09	76.99	0.566	0.571
Opportunity to get a higher salary in the future	69.43	70.80	0.252	0.801
Effort for increasing one's qualification	54.92	57.52	0.442	0.658
Effort for only obtaining the university degree itself	47.67	49.56	0.319	0.750
Opportunity to enjoy student life	29.53	32.74	0.587	0.557
Necessity of acquiring university education	21.24	25.66	0.888	0.374
Opportunity to achieve a higher job position	11.40	23.01	2.695	0.007

Within the research sample, there were 79.13% of male respondents studying the study program informatics, and there were 70% of women studying the study program management. Therefore, it is an interesting finding that the results of identification of differences based on the study program as the distinguishing feature did not copy the previous results obtained from the identification of differences based on the student's gender. This is supported by the fact that, in this case, the difference was not detected in the perception of the motive represented by the effort for the increase in one's chances in the labor market. The only statistically significant difference identified from the perspective of the study program was the opportunity for getting a higher job position later in the career. This motive is more often perceived by the students of management (23.01% of students of management in comparison with 11.40% of students of informatics). Such situation can actually be expected since the graduates in informatics aspire mainly to get the positions of programmers and they want to become experts in their professional field. On the other hand, the graduates in management as a study program are expected to get higher up the career ladder to the managerial positions in enterprises after some time.

The last identification feature, whose effect on the results obtained was tested via the techniques of statistical analysis, was the year of study. The purpose of this focus was the evaluation of changes in the perceived motives over time, as a result of maturing, or as a consequence of realized educational activities. In this case, the z-score was calculated once again, following the dichotomic nominal data type. The test's results with the corresponding p-values (at the significance level of $\alpha = 0.05$) and with the relative frequencies of the responses for the individual years of studies are listed in Table 4.

Table 4. Reasons leading to studying at a university in relation to the year of study.

Reason	First Year (%)	Second Year (%)	z-Score	p-Value
Effort for increasing one's chances in the labor market	74.44	80.56	0.797	0.425
Opportunity to get a higher salary in the future	70.37	66.67	0.455	0.649
Effort for increasing one's qualification	57.41	44.44	1.471	0.141
Effort for only obtaining the university degree itself	47.78	52.78	0.564	0.573
Opportunity to enjoy student life	31.11	27.78	0.407	0.684
Necessity of acquiring university education	23.33	19.44	0.522	0.602
Opportunity to achieve a higher job position	13.70	30.56	2.612	0.009

In the case of differences based on the year of study, studying the research sample, similar results were obtained to those reached when focusing on the differences in relation to the study program. This can be connected to a more detailed structure of the sample itself. The respondents from the first year of study were mainly from the study program informatics (69.63%), and, on the contrary, the students from the second year of study were mainly the students of the study program management (86.11%). Such structure of the research sample skewed the explanatory power of the differences related to the year of study.

When summarizing the partial findings within the reasons for starting to study at a university, the following conclusions were drawn that will help direct the recommendations focused on the support of the students' motivation with the aim to strengthen the sustainability of the system in the long run. The most frequent motive for starting to study at a university is the effort for the increase in one's chances in the labor market, which can be considered a very positive motive. However, within this motive, a significant difference was detected between men and women. Overall, it was revealed that the study program has a negligible effect on the reasons for starting to study at a university.

4.2. The Level of Teaching at a University

In the questionnaire survey conducted, the group of factors focused on the quality of the realization of educational activities during study at a university was represented by the question about perceived level of teaching. The aggregate results of this question are listed in Table 5.

Table 5. The perceived level of teaching at a university.

Level	Number of Respondents	%
High	167	54.75
Medium	136	44.59
Low	2	0.66

Within the research sample, 54.75% of respondents considered the level of teaching to be high and 44.59% of respondents considered it to be medium. This result can be perceived as generally positive, but there is still considerable space for future improvement. Since the realized educational activities during the studies can change and influence the direction as well as the strength of the students' motivation, this section was thoroughly examined in the research.

Again, the examination started with the identification of statistically significant differences in the perception of the level of teaching in relation to the students' gender. The number of categories of the relevant variables implied the application of the Pearson's chi-squared test and its interpretation based on the corresponding *p*-value (at the significance level of $\alpha = 0.05$). The results of this test are listed in Table 6.

Table 6. The perceived level of teaching at a university in relation to gender.

Level of Teaching	Men %	Women %	Pearson's Chi-Squared Test
High	54.15	56.00	1.518
Medium	44.88	44.00	*p*-value
Low	0.98	0.00	0.678

The test's result was calculated for three degrees of freedom, with the critical value being $C = 7.815$ (at the significance level of $\alpha = 0.05$). In accordance with the corresponding *p*-value, the result is negative. This means that the dependence between the perceived level of teaching and the respondent's gender was not identified. Men and women perceived the level equally, which implies that during the designing and implementation of the measures for the increase of the perceived level of teaching, it is not necessary to take the students' gender into account.

The impact of the study program on the perceived level of teaching was analyzed as well. The results of the statistical testing, together with the relative frequencies, are listed in Table 7.

Table 7. The perceived level of teaching at a university in relation to the study program.

Level of Teaching	Informatics %	Management %	Pearson's Chi-Squared Test
High	50.00	62.83	5.6554
Medium	49.48	36.28	*p*-value
Low	0.52	0.88	0.130

With three degrees of freedom and the critical value of $C = 7.815$, at the significance level of $\alpha = 0.05$, the test did not confirm the presence of a statistically significant difference in the perceived level of teaching in relation to the study program. However, when looking at the relative frequencies, it can be seen that a slightly higher level was perceived by the students of management. To corroborate this tendency, it would be possible to broaden the research in the future, including a larger sample of students in it. An additional piece of information, shedding more light on the findings, is the fact that the students of management and informatics have several common subjects during the first year of their studies. Therefore, in future research, it would also be possible to filter out the impact of the common subjects on the perceived level of teaching. Subsequently, the finding obtained this way would be compared with the results of this research.

Finally, the influence of the year of study on the studied variable was analyzed. If the results from the previous section are followed, there is an assumption that while focusing on the impact of the study program and the year of study, the results obtained will be similar. The actually achieved results are listed in Table 8. The types of data entries and the number of categories led to the application of the Pearson's chi-squared test.

Table 8. The perceived level of teaching at a university in relation to the year of study.

Level of Teaching	First Year (%)	Second Year (%)	Pearson's Chi-Squared Test
High	55.93	45.71	9.171
Medium	43.33	54.29	*p*-value
Low	0.74	0.00	0.027

Due to the same number of categories in the input data, the number of degrees of freedom as well as the critical value for the test did not change (three degrees of freedom, $C = 7.815$, significance level of $\alpha = 0.05$). Based on the *p*-value reached, the test's result shows the presence of the statistically significant dependence between the perceived level of teaching and the year of study. This means that, in this case, the results differ when compared with those reached for the previous impact studied. The students of the first year perceive the level of teaching to be higher than the students of the second year. Within the overall concept of this research, these results represent a negative direction of the impact of the realized educational activities on the students' motivational readiness. From the perspective of the sustainability of the whole system, it would be desirable to focus on the causes of this effect more closely.

The aggregate results within the group of factors focused on the quality of the educational activities and their effect on the students' motivational readiness include these points:

- Only slightly above 50% of students consider the level of the educational activities to be high, which represents potential for further improvement,
- the dependence between the perceived level of teaching and the respondents' gender or their study program was not confirmed,
- the testing for the dependence between the perceived level of teaching and the year of study implies a negative impact of time and experience on the given variable, which means that it is desirable to direct the recommendations for future improvement toward this area.

4.3. The Students' Expectations Related to Their Future Jobs

The last group of factors affecting the students' motivational readiness is represented by their expectations and wishes related to their future careers. This group of factors is considered to be especially important since it is placed at the boundary between the education system and the labor market. In a similar way to the previous sections, the results obtained are at first presented aggregately, then they are structured into sub-sections created by focusing on the dependence between the given variable and the identification features of respondents. The aggregate results are captured in Figure 4.

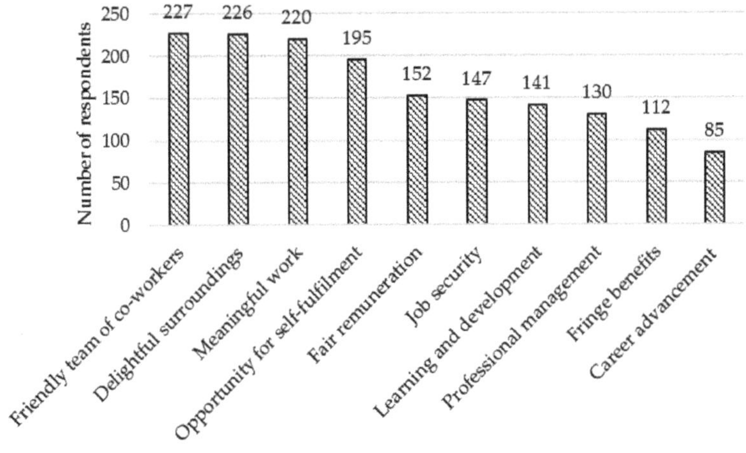

Figure 4. The students' expectations related to their future jobs.

Based on the absolute frequency of the responses, the students most frequently expect their future employment to provide them with a friendly team of co-workers, delightful surroundings at the workplace, meaningful work and work tasks, and sufficient opportunities for their self-fulfillment.

On the other hand, the least frequent expectations in relation to the future jobs were represented by the career advancement and by fringe benefits. The aggregate results reflect the opinions of the current students who, after finishing their studies, will enter the labor market. Therefore, the employers can use them as inspiration, for example, while creating and promoting new job offers.

Other findings are created by the analysis of the results from the perspective of detection of the dependence between the students' expectations and the identification features. The first identification feature was gender of respondents, the same as in the previous sections. Based on the type of the data and the number of individual categories, the z-score was calculated here. Its results, together with the relative frequencies, are listed in Table 9.

Table 9. The students' expectations related to their future jobs, with the dependence on gender being studied.

Expectations	Men %	Women %	z-Score	p-Value
Friendly team of co-workers	72.33	78.00	1.063	0.288
Delightful surroundings	69.90	82.00	2.259	0.024
Meaningful work	70.87	74.00	0.571	0.568
Opportunity for self-fulfillment	60.19	71.00	1.844	0.065
Fair remuneration	45.63	58.00	2.030	0.042
Job security	44.66	55.00	1.698	0.089
Learning and development	46.12	46.00	0.019	0.985
Professional management	40.78	46.00	0.867	0.386
Fringe benefits	32.52	45.00	2.125	0.034
Career advancement	24.76	34.00	1.693	0.090

A statistically significant dependence was detected for the following expectations: Delightful surroundings, fair remuneration, and fringe benefits. In all three cases, the expectations are more frequent among women. In a similar way to the differences identified within the reasons for starting to study at a university, the causes of the situation here can also include the socially-conditioned different behavior of women and men. The possible implications of these findings are applicable on the side of employers. In the case when the employers want to attract more women to secure the diversity of working teams, they can focus on the identified factors in their job offers.

Another part is focused on the impact of the study program on the expectations related to future jobs. The nature of the study programs themselves implies certain estimated differences. Confirmation of these differences based on the z-score and the corresponding p-values (at the significance level of $\alpha = 0.05$), together with the relative frequencies, are captured in Table 10.

Table 10. The students' expectations related to their future jobs, with the dependence on the study program being studied.

Expectations	Informatics %	Management %	z-Score	p-Value
Friendly team of co-workers	72.02	77.88	1.130	0.259
Delightful surroundings	72.02	76.99	0.955	0.340
Meaningful work	73.06	69.91	0.591	0.555
Opportunity for self-fulfillment	58.03	73.45	2.708	0.007
Fair remuneration	49.22	50.44	0.206	0.837
Job security	50.78	43.36	1.253	0.210
Learning and development	45.60	46.90	0.221	0.825
Professional management	42.49	42.48	0.002	0.999
Fringe benefits	32.64	43.36	1.879	0.060
Career advancement	23.32	35.40	2.277	0.023

The test results confirm the estimated different expectations, specifically within the opportunity for self-fulfillment and career advancement, which are more often expected among the students of management. These two factors are more important for future managers regarding the character of this work. An important finding is that independently from the study program, the opportunity for further learning and development is not often expected by the students. In relation to the current global environment, where the emphasis is being put on the life-long learning, this result again opens space for strengthening the motivational readiness of students as a consequence of suitably designed recommendations for the changes in the education process.

Finally, the dependence between the expectations related to future jobs and the year of study was analyzed. Since the data had the same type as the data entries in the previous case, the same procedure was applied again. Its results, accompanied by the *p*-values (at the significance level of $\alpha = 0.05$) and the relative frequencies, are listed in Table 11.

Table 11. The students' expectations related to their future jobs, with the dependence on the current year of study.

Expectations	First Year (%)	Second Year (%)	z-Score	*p*-Value
Friendly team of co-workers	73.33	80.56	0.930	0.352
Delightful surroundings	71.85	88.89	2.185	0.029
Meaningful work	71.48	75.00	0.441	0.659
Opportunity for self-fulfillment	63.33	66.67	0.391	0.696
Fair remuneration	48.52	58.33	1.106	0.269
Job security	48.15	47.22	0.104	0.917
Learning and development	46.67	41.67	0.565	0.572
Professional management	43.70	33.33	1.182	0.237
Fringe benefits	36.30	38.89	0.303	0.762
Career advancement	28.15	25.00	0.396	0.692

The results show that, once again, the situation is not exactly the same as it was described for the impact of the factor represented by the study program. This supports the importance of studying the dependence of the results on the year of study separately. Based on the z-score calculated, the dependence between the expectation in the form of the delightful surroundings and the year of study was identified. Specifically, this expectation was more often among the students of the second year (88.89% of them) in comparison with the students of the first year (71.85% of them). This makes the particular expectation at the top of the list among students of the second year. This expectation can be considered to be one of the additional ones that the job seeker shall focus on only after satisfying his/her fundamental needs related to the work itself and the career path. This unexpected result, when the delightful surroundings are expected more often than e.g., the fringe benefits or the opportunities for the career advancement, elicits the need for deeper examination of the causes of this state in future research.

Overall, the results within the last group of factors affecting the motivational readiness of students revealed the following facts:

- In relation to their future jobs, students emphasize a good atmosphere at the workplace and the working conditions and interpersonal relationships, only then they focus on the work itself and on the opportunities for self-fulfillment,
- there were socially conditioned differences between genders identified in relation to the job expectations,
- the occurrence of the estimated differences based on the study program was corroborated in relation to the nature of the programs themselves,
- the results showing the differences based on the year of study do not exactly copy the situation within the study program, which confirms that this group of factors affecting the motivational readiness can change over time as well.

Based on these results, the research hypothesis was confirmed, following the specific differences described in the corresponding sections. These findings were taken into consideration while designing the recommendations for future improvement.

5. Discussion

The students' motivation to study at a university is affected by various expectations, motives and factors, including, for example, the opinions about the quality of education at a particular university, place of permanent residence [83], and others. This topic was also studied in the work done by Weberova et al. [84]. Based on the research presented in this article it was revealed that the most frequent motive for starting to study at the particular university is the effort for the increase in one's chances in the labor market. This is also corroborated by the statement of Chodasova et al. [85], saying that in the current time of globalization, education as a form of investment in human capital is an ideal platform for the improvement of the position of an unemployed person in the labor market, including its effect in the long run. Women in the survey chose this motive more often than men, which can be caused by the fact that, in the Slovak Republic, women are considered to be a disadvantaged group within the labor market, and they are trying to enhance their position via education.

On the one hand, the level of teaching is being affected by the students' approach to studying, but on the other hand, the key role is played by the teacher, including his/her competencies, willingness to continue learning new things [35,86], his/her approach to students, willingness to implement the latest knowledge into the subjects' content, and to use modern teaching methods. Teaching methods represent an important tool for the realization of the education process. The selection and suitable application of a method shall reflect the students' needs and it shall also react to the current society-wide trends in technical [87] as well as economic development. The selection of appropriate methods is also determined by various factors, such as the number of students in the study group, spatial and technical conditions, motivation of students for studying, professional level and experience of the teachers, and last but not least, by the quality and accessibility of the didactic tools and the supporting studying materials. Within this research, it was revealed that only slightly above 50% of students consider the level of education activities to be high (regardless of their gender, study program or the year of study). Therefore, there is still a huge potential for further improvement of the teacher's work and for the improvement of the content as well as the form of education and its particular activities.

It was revealed that among the students of informatics, regarding their expectations in relation to their future career, the most motivating factor is represented by a friendly team of co-workers, followed by delightful surroundings and meaningful work and work tasks. Among the students of management, meaningful work was substituted by the opportunities for self-fulfillment, which is probably a consequence of their ambition of getting managerial positions after finishing the studies. It seems that the issue of remuneration and fringe benefits is not that important for current students. This can be caused by the fact that the students of the first two years of the studies participated in the research, for whom the question of independent funding of their own needs is still a bit distant from the time perspective. Another reason can be the fact that despite the low average salary level in Slovakia, the graduates from universities, who find employment as IT professionals and managers, can achieve remuneration that is above average.

Within a wider context, the macroeconomic effects from the investment in individual components of human capital, for example, in the form of university education, lie in the increase and the sustainability of multi-factor productivity and macroeconomic performance, increase in incomes for the public budgets, growth of the life standard of citizens, and finally, in the growth of the knowledge level of people in the whole country [65,88]. In the case when the country does not pay sufficient attention to the education of citizens, economic inefficiency occurs [89], accompanied by the dissatisfaction of citizens and their possible emigration abroad.

Based on the research presented in this article, performed using the technique of a questionnaire which was created utilizing the inspiration from a particular study [65], it was revealed that young people are motivated to study at the particular faculty due to a better chance of finding employment and getting a higher salary and qualification. The motivation to study at the faculty due to better employment in the labor market was more often perceived among women. Despite the fact that the level of teaching is considered to be high by almost 50% of the students, regardless of their gender, study program or the year of study, the motivation to study is also supported by the students' expectations related to the jobs they will do in the future. The students of informatics expect mainly a friendly team of colleagues, delightful and stimulating working conditions, and the possibility of doing meaningful work. For the students of management, the third most often expectation is the opportunity for self-fulfillment. In relation to other factors, the differences based on the gender, study program, or the year of study were not statistically significant.

The greatest motivation for young people to study at the particular faculty is a better chance of getting employment in the labor market. For better employment of graduates in the practice, it is necessary for the students to have the opportunities to connect knowledge with practice during their studies. This is an interesting source of motivation for them to learn because it contributes to a more accurate idea about their future. In teaching, the teachers need to connect the latest results of science and research from the world and from the university with examples and forms of their application in the practice even more. The students are also more inclined to accept and absorb the information from a teacher who has experience also from the outside of the school environment, or who is working on the projects assigned from the business environment.

Based on the results obtained, several measures for the Faculty of Management Science and Informatics are designed within this research, which can serve as an inspiration for other faculties and universities, even beyond the borders of the Slovak Republic. The implementation of the measures recommended can increase the motivation of students to study at the university, and specifically at the particular faculty, which will secure the sustainability of education and strengthen the reputation of the faculty among the public audience. By employing highly motivated graduates, enterprises will acquire employees with a high level of human capital. This will also contribute to increasing business value generation for the stakeholders.

If students perceive that they can achieve an advantaged position in the labor market and that their possibility of getting favorable employment is strengthened, their motivation to start studying and to study successfully, even at the particular faculty, will be increased. This will create synergy among all three areas studied: Value generation in an enterprise (as an employer), the sustainability of education, and the students' motivation.

6. Conclusions

Education, as one of the forms of investment in human capital, improves the starting position and the negotiation power of individuals in the labor market, enables them to get a higher salary, or achieve higher performance at work. The effects from the employment of highly qualified people or from the investment in human capital realized at the level of enterprises lie, for example, in the increase in the production's quality, or in the increase of the productivity and performance. The investment in human capital and its value, and the ability to efficiently utilize the human capital available, also represents an important aspect of value creation and an enterprise's competitiveness in the current dynamic, open markets. The increased care of the employees via the investment leads to higher satisfaction with the work an employee does. For the employee, this represents higher motivation for performing his/her work tasks with higher quality, and it increases his/her loyalty to the enterprise. The enterprise gets qualified, healthy, and educated employees, which increases the value of human capital of the employees and the value for the stakeholder groups of the enterprise. This way the enterprise acquires an advantage against its competitors in the market space. All of this again contributes to the sustainability of education in the country.

When approaching this from a more specific perspective, attention can be paid to the motivation of students at universities. This is being influenced by various factors. The satisfaction of students and the sustainability of education depends on the fulfillment of these factors. Student's motivation can be affected by experience from the past, opinions of friends and acquaintances, reputation of the university, situation in the labor market, etc. The article was focused on the reasons leading the students to study at a university, the level of the teaching, and the students' expectations related to their future careers.

The recommendations for the faculty include paying heed to continuous updating of the content of the study programs being provided and to their consistent connection with the actual requirements of the labor market. These are the requirements of the enterprises as future employers of the graduates. These enterprises have a chance to get employees with high level of human capital, which will increase their value in the market. Since this factor was statistically higher among women, there is a chance to increase the motivation of girls for studying informatics at the faculty. The increase of the number of girls studying informatics currently belongs to the aims of other faculties too, and it is also a national as well as an international effort.

The level of teaching affects the resulting quality of education at universities. During the education process, the students directly interact with the teachers. The teacher is supposed to be competent and able to appropriately (e.g., via experiences and examples) explain a specific topic to a group of students. Teachers can apply various teaching styles in the process of education. Another factor of the education's quality is the portfolio of the methods utilized. When utilizing a suitable combination of different modern teaching methods accompanied by the right alignment of the learning styles of the students with the teaching styles of the teachers, the interest of the students in studying is encouraged, together with their creativity, and their critical thinking and expert argumentation are improved. This way, it is possible to enhance the quality and attractiveness of the university education as well as the readiness of the students not only for the successful passing of the exam within the given subject and the defense of the thesis, but also for the solving of common-life and work situations. This will create a precondition for the successful employment of students outside the school, thus in the practice. On the one hand, this will contribute to the sustainability of education at the faculty. On the other hand, successful graduates with high level of human capital will be a valuable asset for the enterprises in which they will be employed, and they will be able to considerably contribute to the generation of value for the enterprises' stakeholder groups. It is necessary to implement all of this within the conditions of the faculty and the university as well. It is helpful if the teachers themselves work on increasing their qualification, professional level and competencies, and they learn new and modern methods of teaching. It is important that the teacher is able to identify his/her preferred style of teaching, and after considering its advantages and disadvantages, he/she is able to use other styles as well, reacting to the learning styles of the students. It is also necessary to thoroughly and adequately often update the content of the subjects in the study program, in harmony with the world-wide trends in the specific professional fields.

Another recommendation focused on the increase in the quality of education at the faculty is to establish a program of regular training courses, focusing on new, progressive, and participative education methods, which shall be mandatory for all teachers. All of this will contribute to the sustainability of education at the faculty.

The motivational factors of the students also include the expectations related to their future jobs, being connected to a friendly working team, working conditions, meaningful work, and the opportunity for self-fulfillment. Therefore, the recommendation for the faculty's management and its employees is to regularly present these expectations at common work meetings, during specialized activities as well as at scientific conferences. These meetings include the gatherings of the IT cluster, in which the faculty represents a respected member, together with important employers and institutions of the labor market within the field of information technology. If personal managers can attract clever university graduates and create suitable working conditions aligned with their expectations, it will result in the increase in value generation in enterprises and in the increase in their competitiveness in the market.

Future research of the students' motivation can be focused on other factors and the research sample can be broadened by including students from higher grades. Since the situation in the labor market is constantly developing over time and the conditions in schools as well as in enterprises are changing, it is planned to repeat the survey regularly, after a certain period of time. Subsequently, in the future, it will be possible to perform the survey and comparison including students from partnered faculties or other universities in different countries as well.

Author Contributions: A.K. conceptualization, validation, project administration, writing the paper, writing—review and editing, visualization of the paper. M.M. data curation, formal analysis, visualization of the paper, writing the paper, writing—review and editing. E.M. methodology, data curation, formal analysis, visualization of the paper. M.D. investigation, resources, writing the paper. E.T. investigation, resources, writing the paper.

Acknowledgments: This work was supported by project APVV-16-0297 Updating of anthropometric database of Slovak population, Grant system of the University of Zilina, project VEGA 1/0382/19 Building a sustainable relationship with stakeholders of enterprise through value creation using ICT.

Conflicts of Interest: The authors declare no conflict of interest.

References

1. Armstrong, M. *Human Resource Management. Modern Concepts and Procedures*, 13th ed.; Grada Publishing: Praha, Czech Republic, 2015; p. 928.
2. Cahyaningsih, E.; Sensuse, D.I.; Arymurthi, A.M.; Wibowo, W.C.; Sari, W.P. The cycle of knowledge in government human capital management. In Proceedings of the 3rd International Conference on Information Technology Systems and Innovation, ICITSI 2016, Bandung-Bali, Indonesia, 24–27 October 2016. [CrossRef]
3. Gejdoš, M.; Danihelová, Z. Valuation and timber market in the Slovak Republic. *Procedia Econ. Financ.* **2015**, *34*, 697–703. [CrossRef]
4. Hitka, M.; Lorincová, S.; Ližbetinová, L.; Bartáková Pajtinková, G.; Merková, M. Cluster Analysis Used as the Strategic Advantage of Human Resource Management in Small and Medium-sized Enterprises in the Wood-Processing Industry. *BioResources* **2017**, *12*, 7884–7897. [CrossRef]
5. Ližbetinová, L.; Štarchoň, P.; Lorincová, S.; Weberová, D.; Pruša, P. Application of cluster analysis in marketing communications in small and medium-sized enterprises: An empirical study in the Slovak Republic. *Sustainability* **2019**, *11*, 2302. [CrossRef]
6. Stacho, Z.; Potkány, M.; Stachová, K.; Marcineková, K. The Organizational Culture as a Support of Innovation processes' Management. *Int. J. Qual. Res.* **2017**, *10*, 769–784. [CrossRef]
7. Socoliuc, M.; Grosu, V.; Hlaciuc, E.; Stanciu, S. Analysis of social responsibility and reporting methods of Romanian companies in the countries of the European Union. *Sustainability* **2018**, *10*, 4662. [CrossRef]
8. Vetráková, M.; Hitka, M.; Potkány, M.; Lorincová, S.; Smerek, L. Corporate sustainability in the process of employee recruitment through social networks in conditions of Slovak small and medium enterprises. *Sustainability* **2018**, *10*, 1670. [CrossRef]
9. Lorincová, S.; Hitka, M.; Štarchoň, P.; Stachová, K. Strategic Instrument for Sustainability of Human Resource Management in Small and Medium-Sized Enterprises Using Management Data. *Sustainability* **2018**, *10*, 3687. [CrossRef]
10. Kucharčíková, A.; Mičiak, M. Human capital management in transport enterprises with the acceptance of sustainable development in the Slovak Republic. *Sustainability* **2018**, *10*, 2530. [CrossRef]
11. Hiadlovsky, V.; Dankova, A.; Gundova, P.; Vinczeova, M. University as Innovative Organization in the Era of Globalization. In Proceedings of the 16th International Scientific Conference on Globalization and its Socio-Economic Consequences, Rajecke Teplice, Slovakia, 5–6 October 2016; pp. 646–653.
12. Ramísio, P.J.; Costa Pinto, L.M.; Gouveia, N.; Costa, H.; Arezes, D. Sustainability Strategy in Higher Education Institutions: Lessons learned from a nine-year case study. *J. Clean. Prod.* **2019**, *222*, 300–309. [CrossRef]
13. Pereira, C.; Ferreira, C.; Amaral, L. An IT Value Management Capability Model for Portuguese Universities: A Delphi Study. *Procedia Comput. Sci.* **2018**, *138*, 612–620. [CrossRef]
14. Srivastava, A.P.; Venkatesh, M.; Yadav, M. Evaluating the implications of stakeholder's role towards sustainability of higher education. *J. Clean. Prod.* **2019**, *240*, 118270. [CrossRef]

15. Stuss, M.M.; Szczepańska-Woszczyna, K.; Makieła, Z.J. Competences of Graduates of Higher Education Business Studies in Labor Market I (Results of Pilot Cross-Border Research Project in Poland and Slovakia). *Sustanability* **2019**, *11*, 4988. [CrossRef]

16. Filho, W.L.; Skanavis, C.; Kounani, A.; Brandli, L.L.; Shiel, C.; do Paco, A.; Pace, P.; Mifsud, M.; Beynaghi, A.; Price, E.; et al. The role of planning in implementing sustainable development in a higher education context. *J. Clean. Prod.* **2019**, *235*, 678–687. [CrossRef]

17. Shabrova, N.; Kuzminchuk, A. Students' Educational Activity Stimulation as the Development Factor of their Human Capital. In Proceedings of the 11th International Days of Statistics and Economics, Prague, Czech Republic, 14–16 September 2017; pp. 1391–1399.

18. Vîrlănuţă, F.; Schin, G.; Stanciu, S. Study on the evolution of financial insurance in Romania. In Proceedings of the 29th International Business Information Management Association Conference-Education Excellence and Innovation Management through Vision 2020, Vienna, Austria, 3–4 May 2017. Code 129797.

19. Klučka, J.; Hajek, P.; Vit, O.; Basova, P.; Krijt, M.; Paszekova, H.; Souckova, O.; Mudrik, R. Risks of Regulation in Network Industry-Case of the Slovak Republic. In Proceedings of the International Conference of Central-Bohemia-University (CBUIC)-Innovations in Science and Education, Prague, Czech Republic, 22–24 March 2017. [CrossRef]

20. Ďuračík, M.; Kršák, E.; Hrkút, P. Current Trends in Source Code Analysis, Plagiarism Detection and Issues of Analysis Big Datasets. *Procedia Eng.* **2017**, *192*, 136–141. [CrossRef]

21. Cahyaningsih, E.; Sensuse, D.I.; Arymurthi, A.M.; Wibowo, W.C. Knowledge Management Strategy of Government Human Capital Management. In Proceedings of the 14th IEEE Student Conference on Research and Development (SCOReD), Kuala Lumpur, Malaysia, 13–14 December 2016.

22. Svejby, K.E. Methods for Measuring Intangible Assets. 2010. Available online: http://www.sveiby.com/articles/EmergingStandard.html (accessed on 4 May 2019).

23. EN 1325–2014 Value Management. Available online: https://infostore.saiglobal.com/en-gb/Standards/EN-1325-2014-339930_SAIG_CEN_CEN_779173/ (accessed on 14 May 2019).

24. Ďurišová, M. *Value and Its Expression in the Company*; EDIS: University of Žilina, Žilina, Slovak, 2017; p. 124.

25. Obeng, E. Bullseye: An argument for effectively managing retail stakeholder relationships. *J. Retail. Consum. Serv.* **2019**, *49*, 327–335. [CrossRef]

26. Sekerin, V.D.; Gaisina, L.M.; Shutov, N.V.; Abdrakhmanov, N.K.; Valitova, N.E. Improving the Quality of Competence-Oriented Training of Personnel at Industrial Enterprises. *Qual.-Access Success* **2018**, *19*, 68–72.

27. Ližbetinová, L.; Hitka, M.; Li, C.; Caha, Z. Motivation of employees of transport and logistics companies in the Czech Republic and in a Selected Region of the PRC. *MATEC Web Conf.* **2017**, *134*, 00032. [CrossRef]

28. Davies, R.H. *Value Management: Translating Aspirations into Performance*; Routledge: New York, NY, USA, 2016; p. 7.

29. Edvinsson, L.; Malone, M.S. *Intellectual Capital: Realizing Your Company's True Value by Finding its Hidden Roots*; Harper Collins Publishers: New York, NY, USA, 1997.

30. Hitka, M.; Lorincová, S.; Gejdoš, M.; Klariá, K.; Weberová, D. Management approach to motivation of white-collar employees in forest enterprises. *BioResources* **2019**, *14*, 5488–5505. [CrossRef]

31. Ližbetinová, L. Satisfaction with the motivational level of Czech employees. In Proceedings of the IBIMA 2018: Innovation Management and Education Excellence through Vision 2020, Milan, Italy, 25–26 April 2018; pp. 3859–3865.

32. Stachová, K.; Stacho, Z.; Blšťáková, J.; Hlatká, M.; Kapustina, L.M. Motivation of employees for creativity as a form of support to manage innovation processes in transportation-logistics companies. *Nase More* **2018**, *65*, 180–186. [CrossRef]

33. Chlpeková, A.; Večera, P.; Šurinová, Y. Enhancing the effectiveness of problem-solving processes through employee motivation and involvement. *Int. J. Eng. Buss. Manag.* **2014**, *6*, 31. [CrossRef]

34. Săvoiu, G.; Vasile, D.; Tăchiciu, L. An Inter-, Trans-, Cross- and Multidisciplinary Approach to High Education in the Field of Business Studies. *Amfiteatru Econ.* **2014**, *16*, 707–725.

35. Sharma, P.; Pandher, J.S. Quality of teachers in technical higher education institutions in India. *High. Educ. Ski. Work-Based Learn.* **2018**, *8*, 511–526. [CrossRef]

36. Yinon, H.; Orland-Barak, L. Career stories of Israeli teachers who left teaching: A salutogenic view of teacher attrition. *Teach. Teach. Theory Pract.* **2017**, *23*, 914–927. [CrossRef]

37. Mushynska, N.; Kniazian, M. Social innovations in the professional training of managers under the conditions of knowledge economy development. *Balt. J. Econ. Stud.* **2019**, *5*, 137–143. [CrossRef]

38. Xie, L.; Guan, X.; Huan, T.C. A case study of hotel frontline employees' customer need knowledge relating to value co-creation. *J. Hosp. Tour. Manag.* **2019**, *39*, 76–86. [CrossRef]

39. Aquilani, B.; Silvestri, C.; Ruggieri, A. Sustainability, TQM and Value Co-Creation Processes: The Role of Critical Success Factors. *Sustainability* **2016**, *8*, 995. [CrossRef]

40. Farinha, C.S.; Azeiteiro, U.; Caeiro, S.S. Education for sustainable development in Portuguese universities: The key actors' opinions. *Int. J. Sustain. High. Educ.* **2018**, *19*, 912–941. [CrossRef]

41. Leal Filho, W.; Wu, Y.C.J.; Brandli, L.L.; Avila, L.V.; Azeiteiro, U.M.; Caeiro, S.; Madruga, L.R.D.G. Identifying and overcoming obstacles to the implementation of sustainable development at universities. *J. Integr. Env. Sci.* **2017**, *14*, 93–108. [CrossRef]

42. Corejova, T.; Rostasova, M. University–Industry partnership in the context of regional and local development. In Proceedings of the 15th International Conference on Information Technology Based Higher Education and Training-ITHET 2016, Istanbul, Turkey, 8–10 September 2016; p. 7760698.

43. Madudova, E.; Majercakova, M. The influence of university-firm cooperation on firm value chain. In Proceedings of the 16th Internat. Confer. on Information Technology Based Higher Education and Training-ITHET 2016, Ohrid, Macedonia, 10–16 June 2017; p. 8067819.

44. Torabian, J. Revisiting Global University Rankings and Their Indicators in the Age of Sustainable Developement. *Sustain. J. Rec.* **2019**, *12*, 167–172. [CrossRef]

45. Farinha, C.; Caeiro, S.; Azeiteiro, U. Sustainability strategies in Portuguese higher education institutions: Commitments and practices from internal insights. *Sustainability* **2019**, *11*, 3227. [CrossRef]

46. Kadoic, N.; Redep, N.B.; Divjak, B. A new method for strategic decision-making in higher education. *Central Europ. J. Oper. Res.* **2018**, *26*, 611–628. [CrossRef]

47. Suwartha, N.; Sari, R.F. Evaluating UI Green Metric as a tool to support green universities development assessment of the year 2001 ranking. *J. Clean. Prod.* **2013**, *61*, 46–53. [CrossRef]

48. Lauder, A.; Sari, R.F.; Suwartha, N.; Tjahono, G. Critical review of a global sustainability ranking: GreenMetric. *J. Clean. Prod.* **2015**, *108*, 852–8633. [CrossRef]

49. Storms, K.; Simundza, D.; Morgan, E.; Miller, S. Developing a Resilience Tool for Higher Education Institutions: A Must-Have in Campus Master Planning. *J. Green Build.* **2019**, *14*, 187–197. [CrossRef]

50. Duffin, M.; Perry, E.E. Regional Collaboration for Sustainability via Place-Based Ecology Education: A Mixed–Methods Case Study of the Upper Vallez Teaching Place Collaborative. *Educ. Sci.* **2018**, *9*, 6. [CrossRef]

51. Jeong, J.S.; González-Gómez, D.; Cañada-Cañada, F. Prioritizing elements of science education for sustainable development with MCDA-FDEMATEL method using flipped e-learning scheme. *Sustainability* **2019**, *11*, 3079. [CrossRef]

52. Silvius, G.; Schipper, R. Exploring Responsible Project Management Education. *Educ. Sci.* **2018**, *9*, 2. [CrossRef]

53. Durek, V.; Kadoic, N.; Begicevic, N. Assessing the digital maturity level of higher education institutions. In Proceedings of the 41st Inter. Convention on Information and Communication Technology, Electronics and Microelectronics-MIPRO 2018, Opatija, Croatia, 21–25 May 2018; pp. 671–676.

54. Balaban, I.; Redjep, N.B.; Calopa, M.K. The Analysis of Digital Maturity of Schools in Croatia. *Int. J. Emerg. Technol. Learn.* **2018**, *13*, 4–15. [CrossRef]

55. Barabanova, M.; Lebedeva, L.; Rastova, Y.; Uvarov, S. Use of system tools in value-oriented approach in management. *Econ. Ann.-Xxi* **2018**, *173*, 32–37. [CrossRef]

56. Medvecká, I.; Biňasová, V.; Kubinec, L. Planning and Performance Evaluation of the Manufacturing Organizations. *Procedia Eng.* **2017**, *192*, 46–51. [CrossRef]

57. Madudova, E.; Corejova, T.; Valica, M. Economic Sustainability in a Wider Context: Case Study of Considerable ICT Sector Sub-Divisions. *Sustainability* **2018**, *10*, 2511. [CrossRef]

58. Calopa, M.K.; Horvat, J.; Kuzminski, L. The Role of Human Resource Management Practice Mediated by Knowledge Management (Study on companies from ICT sector, Croatia). *Tem J.* **2015**, *4*, 178–186.

59. Choi, J. Sustainable Behaviour: Study Engagement and Happiness among University Students in South Korea. *Sustainability* **2016**, *8*, 599. [CrossRef]

60. Sharok, V.V. Emotional and Motivational Factors of Satisfaction with University Education. *Sib. Psikhol. Zhur.-Sib. J. Psychol.* **2018**, *69*, 33–45. [CrossRef]
61. Lee, J.; Song, H.D.; Hong, A.J. Exploring Factors, and Indicators for Measuring Students' Sustainable Engagement in e-Learning. *Sustainability* **2019**, *11*, 985. [CrossRef]
62. Kim, H.J.; Hong, A.J.; Song, H.D. The Relationships of Family, Perceived Digital Competence and Attitude, and Learning Agilitz in Sustainable Student Engagement in High Education. *Sustainability* **2018**, *10*, 4635. [CrossRef]
63. Vukovic, K.; Kedmenec, I.; Korent, D. The Impact of Exposure to Entrepreneurship Education on Student Entrepreneurial Intentions. *Croatian J. Educ.* **2015**, *17*, 1009–1036. [CrossRef]
64. Fang, C.; Chen, L.W. Exploring the Entrepreneurial Intentions of Science and Engineering Students in China: A Q Methodology Study. *Sustainability* **2019**, *11*, 2751. [CrossRef]
65. Šafránková, J.M.; Šikýř, M. Sustainable development of the professional competencies of university students: Comparison of two selected cases from the Czech Republic. *J. Secur. Sustain. Issues* **2017**, *7*, 321–333. [CrossRef]
66. Loisel, S.; Takane, Y. Partitions of Pearson's Chi-square statistic for frequency tables: A comprehensive account. *Comput. Stat.* **2016**, *31*, 1429–1452. [CrossRef]
67. Rahardia, D.; Yang, Y.; Zhang, Z.W. A Comprehensive Review of the Two-Sample Independent or Paired Binary Data, with or without Stratum Effects. *J. Mod. Appl. Stat. Methods* **2016**, *15*, 215–223. [CrossRef]
68. Kljucnikov, A.; Popesko, B. Export and its Financing in The SME Segment. Case Study from Slovakia. *J. Compet.* **2017**, *9*, 20–35. [CrossRef]
69. Mura, L.; Haviernikova, K.; Machova, R. Empirical Results of Entrepreneurs' Network: Case Study of Slovakia. *Serb. J. Manag.* **2017**, *12*, 121–131. [CrossRef]
70. Oboh, C.S.; Ajibolade, S.O. Strategic management accounting and decision making: A survey of the Nigerian Banks. *Future Bus. J.* **2017**, *3*, 119–137. [CrossRef]
71. Velcovska, S.; Krbova, P.K. Consumer Attitudes towards Food Quality Labels in Selected European Union Countries. In Proceedings of the 3rd International Conference on European Integration, Ostrava, Czech Republic, 19–20 May 2016; pp. 1068–1077.
72. Okreglicka, M. Internal Innovativeness and Management of Current Finances of Enterprises in Poland. In Proceedings of the Business Challenges in the Changing Economic Landscape-14th Conference of the Eurasia-Business-and-Economics-Society, Barcelona, Spain, 23–25 October 2014; pp. 225–237. [CrossRef]
73. Radosavljevic, Z.; Cilerdzic, V.; Dragic, M. Employee Organizational Commitment. *Int. Rev.* **2017**, *12*, 18–26. [CrossRef]
74. Mesicek, L.; Petrus, P.; Kovarova, K. What Can We Learn from Students' Tests Results in Subject Management? In Proceedings of the 14th International Conference Efficiency and Responsibility in Education, Prague, Czech Republic, 8–9 June 2017; pp. 247–254.
75. Almalki, S.A.; Almojali, A.I.; Alothman, A.S. Burnout and its association with extracurricular activities among medical students in Saudi Arabia. *Int. J. Med. Educ.* **2017**, *8*, 144–150. [CrossRef] [PubMed]
76. Pooler, J.A.; Morgan, R.E.; Wong, K.R.; Wilkin, M.K.; Blitstein, J.L. Cooking Matters for Adults Improves Food Resource Management Skills and Self-confidence among Low-Income Participants. *J. Nutr. Educ. Behav.* **2017**, *49*, 545. [CrossRef]
77. Sneck, S.; Saarnio, R.; Isola, A.; Boigu, R. Medication competency of nurses according to theoretical and drug calculation online exams: A descriptive correlational study. *Nurse Educ. Today* **2016**, *36*, 195–201. [CrossRef] [PubMed]
78. Tanaka, Y. *Economics of Cooperative Education: A Practitioners' Guide to the Theoretical Framework and Empirical Assessment of Cooperative Education*; Routledge: London, UK, 2014; pp. 61–78. [CrossRef]
79. Thakur, A.; Kaur, R. Relationship between Self-concept and Attitudinal Brand Loyalty in Luxury Fashion Purchase: A Study of Selected Global Brands on the Indian Market. *Manag. J. Contemp. Manag. Issues* **2015**, *20*, 163–180.
80. Ankad, R.B.; Shashikala, G.V.; Herur, A.; Manjula, R.; Chinagudi, S.; Patil, S. PowerPoint presentation in learning physiology by undergraduates with different learning styles. *Adv. Phys. Educ.* **2015**, *39*, 367–371. [CrossRef]
81. Park, H.L.; Shrewsbury, R.P. Student Evaluation of Online Pharmaceutical Compounding Videos. *Am. J. Pharm. Educ.* **2016**, *80*, 30. [CrossRef] [PubMed]

82. Zucolot, M.L.; de Oliveira, V.; Maroco, J.; Campos, J.A.D.B. School engagement and burnout in a sample of Brazilian students. *Curr. Pharm. Teach. Learn.* **2016**, *8*, 659–666. [CrossRef]

83. Reznik, S.D.; Vdovina, O.A. Regional University Teacher: Evolution of Teaching Staff and Priority Activities. *Eur. J. Contemp. Educ.* **2018**, *7*, 790–803. [CrossRef]

84. Weberova, D.; Starchon, P.; Lizbetinova, L. Comparison of Motivational Preferences of University Students and Employees. In Proceedings of the 30th International Business-Information-Management-Association Conference, Madrid, Spain, 8–9 November 2017; pp. 4184–4193.

85. Chodasová, Z.; Tekulová, Z.; Hľušková, L.; Jamrichová, S. Education of students and graduates of technical schools for contemporary requirements of practice. *Procedia-Soc. Behav. Sci.* **2015**, *174*, 3170–3177. [CrossRef]

86. Adnan Suwandi, S.; Nurkamto, J.; Setiawan, B. Teacher Competence in Authentic and Integrative Assessment in Indonesian Language Learning. *Int. J. Instr.* **2019**, *12*, 701–716. [CrossRef]

87. Azeiteiro, U.M.; Filho, W.L.; Caeiro, S. *E-Learning and Education for Sustainability*; Elsevier: Amsterdam, The Netherlands, 2015; p. 290. [CrossRef]

88. Starecek, A.; Vranakova, N.; Koltnerova, K.; Chlpekova, A.; Caganova, D. Factors affecting the motivation of students and their impact on academic performance. In Proceedings of the 14th International Conference on Efficiency and Responsibility in Education (ERIE), Prague, Czech Republic, 8–9 June 2017; pp. 396–407.

89. Martin-Sardesai, A.; Guthrie, J. Human capital loss in an academic performance measurement system. *J. Intellect. Cap.* **2018**, *19*, 53–70. [CrossRef]

 *sustainability*

Article

A Cooperative Interdisciplinary Task Intervention with Undergraduate Nursing and Computer Engineering Students

Pilar Marqués-Sánchez [1], Isaías García-Rodríguez [2,*], José Alberto Benítez-Andrades [3], Mari Carmen Portillo [4], Javier Pérez-Paniagua [1] and María Mercedes Reguera-García [1]

[1]   SALBIS Research Group, Department of Nursing and Physiotherapy Health Science School, University of León, Campus de Ponferrada s/n, C.P. 24401 León, Spain; pilar.marques@unileon.es (P.M.-S.); jperep06@estudiantes.unileon.es (J.P.-P.); mercedes.reguera@unileon.es (M.M.R.-G.)
[2]   SECOMUCI Research Group, Escuela de Ingenierías Industrial e Informática, Universidad de León, Campus de Vegazana s/n, C.P. 24071 León, Spain
[3]   SALBIS Research Group, Department of Electric, Systems and Automatics Engineering, University of León, Campus of Vegazana s/n, C.P. 24071 León, Spain; jbena@unileon.es
[4]   Wessex ARC NIHR. School of Health Sciences, University of Southampton, University Road, Southampton SO17 1BJ, UK; M.C.Portillo-Vega@soton.ac.uk
*   Correspondence: Isaias.garcia@unileon.es

Received: 16 September 2019; Accepted: 9 November 2019; Published: 11 November 2019

Abstract: This study proposed a collaborative methodology among university students in different grades in order to find sustainable strategies that are an added value for students, teachers, and society. In daily professional practice, different professionals must develop skills to collaborate and understand each other. For that reality to be sustainable, we believe that experiences must begin in the context of higher education. Social network analysis offers a new perspective on optimizing relationships between university students. The main goal of this study was to analyze students' behavior in their networks following an educational intervention and the association with academic performance, resilience and engagement. This was a descriptive quasi-experimental study with pre–post measures of a cooperative interdisciplinary intervention. Participants comprised 50 nursing and computer engineering students. We measured help, friendship, and negative network centrality, engagement, resilience, and academic performance. No significant differences were observed between pre–post-intervention centrality measures in the negative network. However, the help and friendship networks presented statistically significant differences between inDegreeN, OutDegreeN and EigenvectorN on the one hand, and resilience and engagement—but not academic performance—on the other. Academic performance was solely associated with the team to which participants belonged. Cooperative interdisciplinary learning increased the number of ties and levels of prestige and influence among classmates. Further research is required in order to determine the influence of engagement and resilience on academic performance and the role of negative networks in network formation in education. This study provides important information for proposals on sustainable assessments in the field of higher education.

Keywords: academic performance; cooperative learning; engagement; engineering; interdisciplinary learning; nursing; resilience; social network analysis; students

---

1. Introduction

The university context is essential for university students to be sensitive to behaviors aligned with sustainability [1]. Universities have considered joining the the Sustainable Development Goals (SDGs) as a strategic factor. They have carried this out through multiple organizations, such as the

Sustainable Development Solutions Network (SDSN), the Environmental Association for Universities and Colleges (EAUC), the Association for the Advancement of Sustainability in Higher Education (AASHE), and the Australasia Campus towards Sustainability (ACTS). The SDGs are a set of priorities and aspirations to guide all countries to address the most pressing challenges in the world, including health and social welfare issues [2]. In this context and aligned with the definition of sustainable development [3], the concept of "sustainable assessment" emerges in the early years of the 21st century. Sustainable assessment is defined as an assessment "that meets the needs of the present and [also] prepares students to meet their own future learning needs" [4].

The main idea behind sustainable assessment is to prepare students to undertake assessment tasks that they will have to face during their lives [4]. Learning cannot be sustainable if it requires continuing information from teachers on student's work [5]. University students must build the capacity to become judges of their own learning; this includes self-assessment but also peer and collaborative assessment. In this sense, assessment must go beyond the idea of getting a mark for a given course: It must be seen as an educational tool and not as a simple measure or learning outcomes. In relation to sustainable assessment, the concept of "evaluative judgement" has gained attention today in higher education contexts, being defined as "the capability to make decisions about the quality of work of self and others" [6].

Learning in a collaborative environment is one of the scenarios where the ideas from sustainable assessment can be more beneficial, beyond the traditional ideas of summative and formative assessment [7]. Self-, peer-, and lecturer-based assessment can be used in conjunction in order to obtain a sustainable assessment system, as in the case of the authentic assessment for sustainable learning model [8].

Sustainable assessment is a field of current active research that needs to explore a plethora of possibilities; however, there are few applied studies on interventions that develop such sustainable practices [5]. We consider that studying networking in a collaborative and interdisciplinary learning experience is a good proposal within the sustainable assessment approach, as interdisciplinary collaboration during the future working lives of the students will be an everyday issue. To perform this kind of studies, teachers must design teaching–learning strategies that assess these collaborations. This research explores learning and assessment approaches based on interdisciplinary and collaborative work among students from the degrees of nursing and computer engineering, with an emphasis on studying the social relationships that are established during this collaboration and how these social interactions affect engagement and academic outcomes.

The period of adolescence is accompanied by changes in the socioaffective process that affect perception, ties between peers, and inclusion in social groups [9]. During this process, social influence between peers reaches its zenith, while the same influence exerted by parents begins to decline [10]. Peer relations in the classroom can promote the knowledge, skills, and social capital necessary to successfully transition from adolescence to adulthood [11]. In addition, the development of classroom relationships is intimately related to students' academic performance [12].

Understanding engagement and resilience in the classroom helps to elucidate the formation of ties between students. When forming working groups, the selection of team members by the students themselves improves engagement and motivation [13]. Furthermore, promoting engagement has positive effects on academic performance and reduces dropout [14]. Moreover, resilience helps students to solve problems [15] and enhances subjective wellbeing [16]. It is highly important to instil resilience in future health professionals [17]. Interpersonal behaviors can affect engagement and resilience as a result of the status generated and reputation processes within groups.

Classroom relationships generate a rich ecosystem of social ties that requires a theoretical research framework to gain understanding. Social network analysis (SNA) comprises a method for analyzing the structure of ties within a network, which is its main difference from other methods of analysis [18]. SNA is based on the idea that the ties between network participants are meaningful. Hence, these ties are analyzed to elucidate their significance [19]. A social network is a set of nodes, some of which are linked by lines. The nodes represent individuals or groups, and the lines indicate that the nodes are connected among them, generating a social structure [20]. This network of relationships or networks transfers resources inherent to the structure generated among individuals [21]. SNA studies the contact that exists not only between the actors but also between their goals and objectives, since their objectives are achieved through connections and relational behaviors [18,22,23].

Friendship and help networks have been among the most frequently analyzed networks in SNA, while more recently, the negative network has also emerged as a subject of study. Analyses of the friendship network examine friendship ties between nodes and the degree centrality of their intensity [24]. Centrality is defined as the position of the actor in the network [25]. Findings are useful to determine the influence of friendships and ties between peers on the acquisition of new values and behavior modification [14] or the importance of the most popular students and their impact on the dynamics of relationships [26]. Analyses of the help network assess the connection and intensity between nodes when problem-solving or seeking advice [27]. Previous studies have confirmed the utility of the help network when seeking prenatal information in the absence of formal resources [28] or receiving support and acceptance from students in the case of speech difficulties [29]. In relation to negative networks, it has been demonstrated that nodes avoid interacting with other nodes [30], and researchers have underlined the influence of negative ties in the workplace [31,32]. Negative relationships refer to the intensity of disgust established between two nodes and to whether the person knows that the other person dislikes him or her [31].

The structural analysis of networks can be applied to various fields of study, such as business relationships [33], tourist travel intentions [34], resilience in disasters [35], and mental health [36]. In the field of education, the context of the present study, SNA has been applied to explore (i) the social influence of ties on adolescents' mental health [37], (ii) the social dynamics of groups in educational camps [38], and (iii) the formation of friendships between students from different ethnic groups [39].

Among engineering students, SNA has been used to determine the influence of networks on performance, demonstrating that a higher number of ties is associated with better academic outcomes [40]. In nursing students, SNA has been used to explore the influence of networks on engagement and resilience [41] and determine their role in the development of technological competence [42]. It has been found that the network perpective is suitable for the analysis of resilience, since it is developed in sociological systems [43]. The network perspective is focused on the study of the structure of these sociological systems, and the ability of that structure to be resilient. A command of information and communication technologies is not among the competences instilled in nursing. Gamification is considered a novel and interesting approach to the development of competencies in new technologies [44] and its use to instil computing skills in the nursing profession has proven effective as a means to achieve a better command of these technologies [45].

These issues motivated the present study of how the dynamics of social interaction are associated with classroom behavior when sharing ideas in order to achieve a good academic outcome. The interdisciplinary intervention aimed at determining nursing and computer engineering students' behavior that has not previously been analyzed and remains unexplored in the literature. Notably, our SNA included an analysis of negative networks between students, which has not previously been examined in this population. In particular, the recent literature contains few studies that have focused on students and the impact of engagement and resilience [41,46], and none of them included academic performance among their variables. The present study is the first to include SNA in an interdisciplinary intervention with university students.

The study objectives were:

- To quantify pre–post changes in behavior following a cooperative task intervention, analyzing centrality variables in the help, friendship, and negative networks of computer engineering and nursing students engaged in the fields of information, communications, and health science.

- To graphically represent pre–post changes following a cooperative task intervention in the help, friendship, and negative networks of computer engineering and nursing students engaged in the fields of information, communications, and health science.
- To determine pre–post changes in engagement and resilience following a cooperative task intervention.
- To determine the relationship between academic performance and centrality, engagement, resilience, and sociodemographic variables.

The objectives allowed us to propose a cooperative and multidisciplinary framework, which will be sustainable for Nursing and Computer Engineering degrees.

2. Materials and Methods

This was a quasi-experimental descriptive study with pre–post-intervention measures.

2.1. Sample Description

Participants were recruited using convenience sampling [47]. Students from two undergraduate courses were approached and presented the project. One of the courses belongs to the fourth year of the Computer Engineering degree and the other one to the third year of the Nursing degree. These degree courses were taught on different campuses located 113 kilometers apart. As shown in Table 1, 26 students from the Nursing degree course and 24 students from the Computer Engineering degree agreed to participate in the project after being informed. Interdisciplinary work groups were formed with randomly selected students, so that there would be a similar number of nursing and engineering students in each work team.

Table 1. Descriptive data and comparison of pre–post-intervention centrality variables (N = 50).

Degree	Sex								
	Men	N	(%)	Women	N	(%)	Total	(%)	
Nursing	4	15.4	(%)	22	84.6	(%)	26	100	(%)
Computer engineering	19	79.2	(%)	5	22.08	(%)	24	100	(%)
Total		23			27		50	100	(%)

The sample consisted of 50 students taking two different degree courses at a public university in Spain. All the individuals participated voluntarily in the study after being informed.

2.2. Variables

The variables analyzed were as follows (Table 2):

Table 2. Variables and conceptual variables.

Metrics	Conceptual Definition
Degree studied, sex, team	Descriptive variables of the sample.
Engagement	Work-related, positive or satisfactory, persistent cognitive affective state. It is composed of three basic dimensions: absorption (concentration), vigor (tenacity, effort), and dedication (enthusiasm, inspiration, pride, defiance) [41,48].
Resilience	Individual's capacity to respond to stress in a healthy manner, such that they can achieve goals at the lowest physical and psychological cost [41,46,49,50].
Academic performance	Students 'knowledge of a cooperative interdisciplinary task [12].
Centrality structural variables	IndegreeN (degree of received ties surrounding the individual), OutDegreeN (degree of emitted ties), EigenvectorN (degree of prestige or influence), BetweennessN (degree of intermediation [19,41].

2.3. Instruments Used to Collect Data

Data on variables were collected by means of an online questionnaire viewable on any device (desktop, laptop, mobile device or tablet). The questionnaire was accessed via a URL, entering a username and password, and incorporated an automatic anonymization system. The server ensured secure data transfer via SSL encryption and HTTPS (Hypertext Transfer Protocol Secure).

The questionnaire included the following:

- Each student's degree, sex, and team;
- The UWES-S scale (Utretch Work Engagement Scale-Students) adapted to measure the level of engagement in university students [48,51,52] and validated for the Spanish population [41,45]. This scale consists of 17 items scored on a scale from 0 (never) to 6 (always or every day);
- The Connor–Davidson scale, version CD-RISC, validated in Spanish in 2011, used to measure resilience. This consists of 10 items scored using a Likert scale from 0 (never true) to 4 (almost always true) [49,50];
- Academic performance was measured based on the mark obtained in the subject, scored from 0 to 10. This mark included individual marks for team work and a written test;
- Centrality structural variables were measured using a 5-point Likert scale to assess the sociocentric networks of all study participants. The networks assessed were (a) friendship network: Which of the following classmates do you consider a friend? [24]; (b) help network: Which of the following classmates do you ask for help when you have a problem/doubt/difficulty regarding course work? [27] and (c) which of the following classmates do you avoid interacting with? [30]

2.4. Procedure

Data were collected on two occasions: first in the initial face-to-face session and again on the day when the completed task was presented.

Descriptive variables, engagement, resilience, and marks were processed using Microsoft Excel.

Structural variables of sociocentric network centrality were analyzed using square matrices for each network. It was necessary to dichotomize the data using intermediate encoding of the friendship, help, and negative networks. We used normalized data for centrality values in accordance with UCINET v. 6.666 [53] (Table 3).

Table 3. Dichotomization of network interactions.

Network	Centrality Variable	Values
Help	Without support	0, 1
	With support	2, 3, 4
Friendship	Without friendship	0, 1
	With friendship	2, 3, 4
Negative	Without avoidance	0, 1
	With avoidance	2, 3, 4

2.5. Intervention

The interdisciplinary intervention consisted of dividing the class into nine teams of five or six students each from both degree courses; these teams were required to carry out a cooperative task on applications in the field of health. Participants were taught at different campuses located 113 kilometers apart, so they only met face-to-face at the initial session when the teachers presented the cooperative task. Their subsequent contacts took place via online networks.

The cooperative task involved three stages: (a) an initial face-to-face session to explain the task objectives and method and establish personal contact during a 5-hour session in a non-academic environment (a cafeteria with brunch included); (b) implementation of the task over the course of 40 days subsequent to the initial session, communicating by means of mobile phones, emails, and instant messaging; (c) presentation of the completed work by all members of the team via videoconference between the two campuses.

To complete the task, the nursing students had to explain a healthcare need, and the engineering students had to formulate a technological solution for the identified healthcare need. In their oral presentations, all team members were required to present part of the completed work but with the caveat that the nursing students had to present the technological solution and the engineering students had to talk about the healthcare need.

Assessment was achieved by both teacher and peer evaluation of the collaborative work. The nursing degree students evaluated the computing engineering students' presentations (except those from their own working group) and vice-versa: The computer engineering students evaluated the nursing students' talks. Academic performance was measured as the individual mark awarded for the results of the cooperative task (evaluated by the teacher), a multiple-choice test (for the peer evaluation previously described), and classroom participation.

2.6. Ethical Considerations

Participants were informed of the study objectives and method. Participation was voluntary, and students could cease to participate at any time. All personal data were processed ensuring confidentiality and anonymity. Simulated names were generated with the aid of the tool described in Benítez et al. (2017) and used to create the network graphs [54]. This study was approved by the University of León Ethics Committee (Ref. ETICA-ULE-026-2018) and adhered to the Declaration of Helsinki, Law 15/1999 of 13 December, on Personal Data Protection, and Law 14/2007, of 3 July, on Biomedical Research.

2.7. Data Analysis

We used the Kolmogorov–Smirnov test with Lilliefors correction to determine the normality of values for all the variables analyzed except those for centrality measures. The descriptive statistics are given as means and standard deviations.

To determine differences between pre- and post-intervention variables, we used the Student's t-test when distribution was normal and Wilcoxon's t-test when it was non-normal.

To determine correlations between parametric values, we used Pearson's correlation coefficient, while for nonparametric values, we used Spearman's correlation coefficient.

Significance was set at $p < 0.05$ and $p < 0.01$. All statistical analyses were performed using SPSS v. 25.0.

2.8. Results

Universities are the institutions of higher education (HEI) whose objective is to contribute to the sustainable transformation of societies through the training of future professionals. In this sense, this sustainability and transformation project must start from the lectures at the university campus, with specific actions in the curricula and oriented towards what UNESCO calls the "Focus of the whole school". The study of collaboration networks among students of different grades constitutes a strategy for fostering the future sustainability of professionals. In our case, it was proposed that a nurse and/or an engineer could improve patient care. This is based on the fact that a nurse knows what the patient demands, and the engineer knows how to propose a technological solution and integrate the use of technologies in health systems. To achieve this, the nurse and the engineer must develop effective communication channels, have critical thinking, and know how to work in networking teams with professionals from different disciplines.

If this method of working achieves positive outcomes among the students, it will possibly be achieved also in the future, when the students become professionals. Our intervention was related to sustainability because trying to demonstrate that interdisciplinary work solves problems responds to societal demands and achieves objectives effectively.

The application of SNA helped us to deepen the structures of collaborative networks among students. The specificity of the method allowed the identification of relational roles, such as Indegree, Outdegree, Eigenvector, and Betweenness. In addition, this research related these relational roles with fundamental constructs for teamwork, such as resilience and engagement, as detailed below.

Table 4 gives the descriptive data and a comparison of normalized network centrality variables pre- and post-intervention. We found that the help and friendship networks presented variables with significant differences in InDegreeN, OutDegreeN and EigenvectorN. By contrast, no significant differences were detected for any negative network variable.

Table 4. Descriptive data and comparison of pre–post-intervention centrality variables (N = 50).

Network	Centrality Variable	Pre-Intervention	Post-Intervention	Student's *t*-Test	Sig.
Help	InDegreeN	0.10 ± 0.05	0.18 ± 0.08	−13.775	<0.001 **
	OutDegreeN	0.10 ± 0.08	0.18 ± 0.12	−4.611	<0.001 **
	EigenvectorN	13.35 ± 15.04	16.87 ± 10.85	−2.761	0.008 **
	BetweennessN	2.78 ± 3.91	2.76 ± 2.89	0.029	0.977
Friendship	InDegreeN	0.12 ± 0.06	0.18 ± 0.07	−10.326	<0.001 **
	OutDegreeN	0.12 ± 0.07	0.18 ± 0.12	−4.229	<0.001 **
	EigenvectorN	11.85 ± 16.28	17.16 ± 10.38	−3.643	0.001 **
	BetweennessN	2.78 ± 3.91	2.56 ± 2.08	0.359	0.721
Negative	InDegreeN	0.12 ± 0.05	0.12 ± 0.05	−0.01	0.992
	OutDegreeN	0.12 ± 0.23	0.12 ± 0.25	0	1
	EigenvectorN	17.31 ± 10.12	17.63 ± 9.53	−0.248	0.805
	BetweennessN	0.58 ± 2.07	0.70 ± 2.02	−0.338	0.737

Legend: InDegreeN, normalized InDegree; OutDegreeN, normalized OutDegree; BetweennessN, normalized Betweenness; EigenvectorN, normalized Eigenvector. * Correlation is significant at 0.05. ** Correlation is significant at 0.01.

Graphic representations of pre- and post-intervention networks are shown in Figures 1 and 2 (help networks), Figures 3 and 4 (friendship networks), and Figures 5 and 6 (negative networks). Blue indicates men and pink indicates women, while squares denote nursing students and circles denote computer engineering students. Tie density in the help and friendship networks showed clear changes over the course of the study. Pre-intervention, most ties in these networks occurred between students on the same degree course, with hardly any ties between degrees. In addition, more interactions were observed between nursing than between computer engineering students, with the latter presenting greater selectivity when forming friendships or seeking help. Post-intervention, the number of interactions increased in the help and friendship networks, but differences still remained. Meanwhile, the pre- and post-intervention negative networks presented a series of central nodes that connected the majority of ties, and it was more difficult to distinguish between the two degrees.

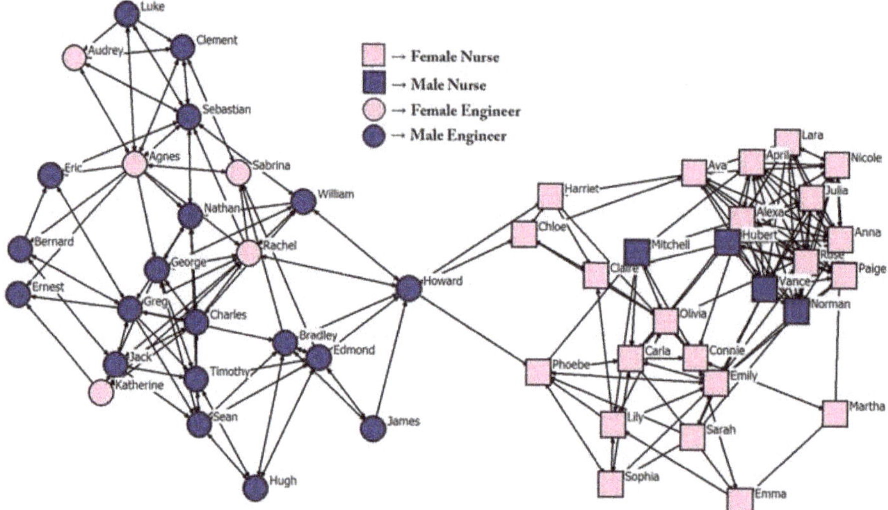

Figure 1. Graph of the pre-intervention friendship network (simulated names).

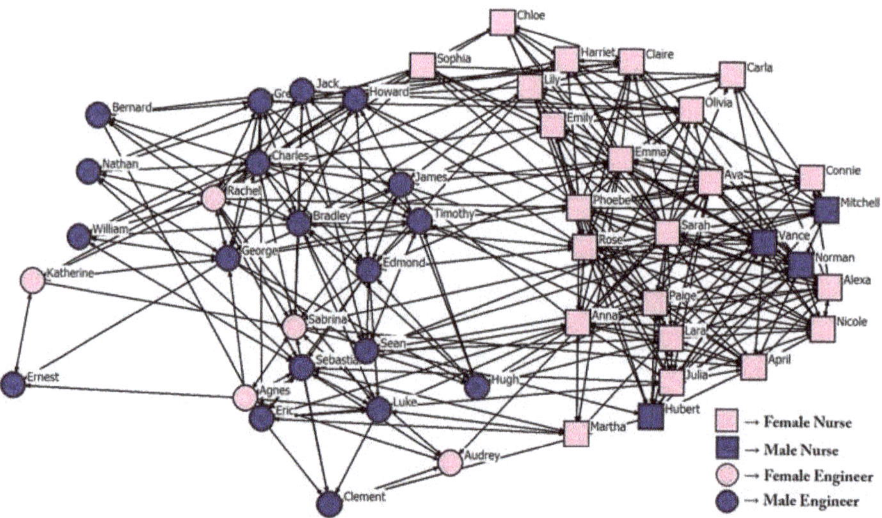

Figure 2. Graph of the post-intervention friendship network (simulated names).

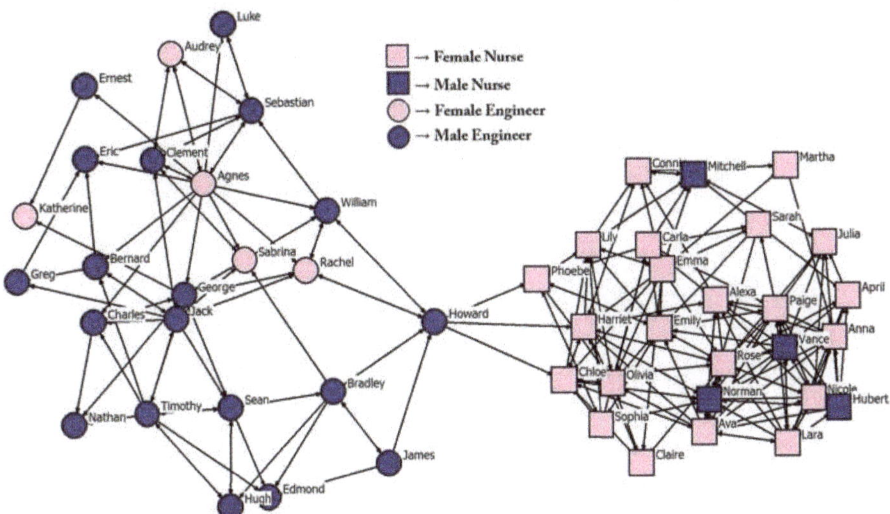

Figure 3. Graph of the pre-intervention help network (simulated names).

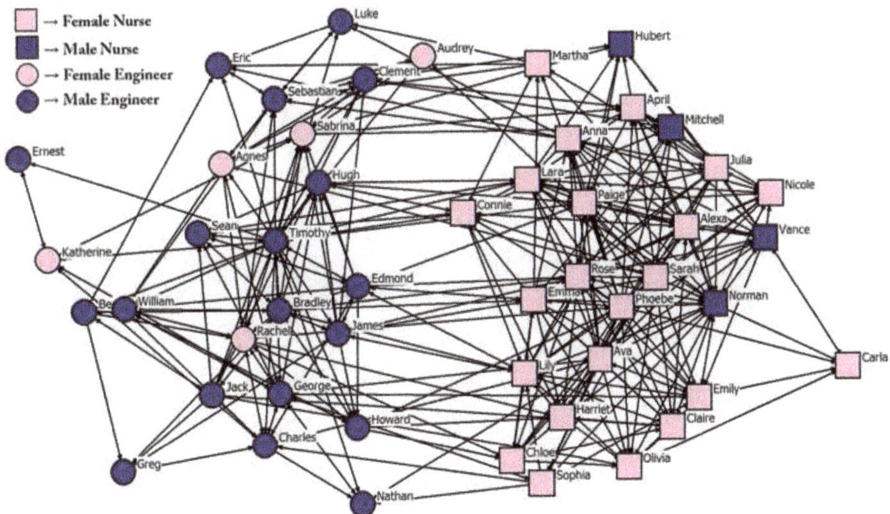

Figure 4. Graph of the post-intervention help network (simulated names).

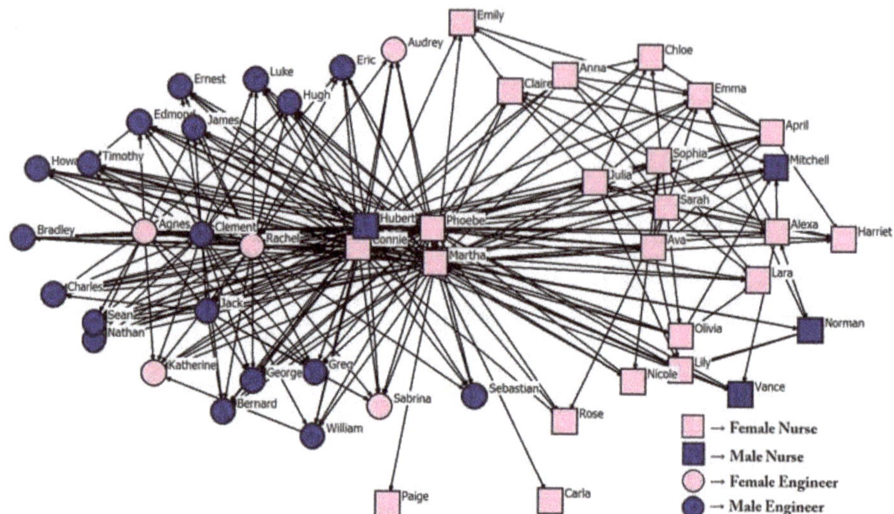

Figure 5. Graph of the pre-intervention negative network (simulated names).

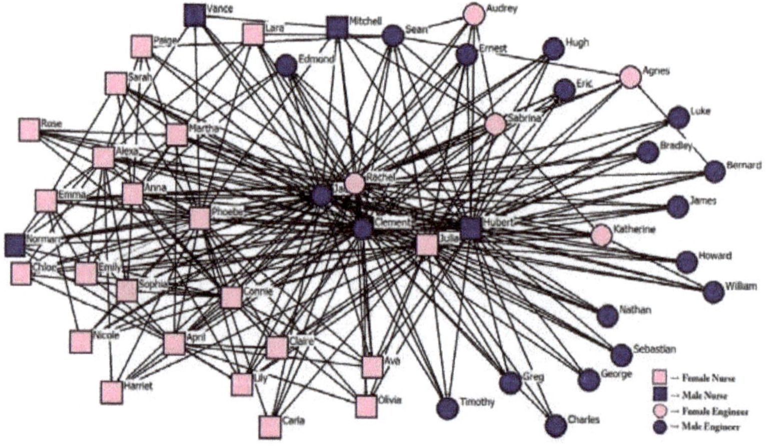

Figure 6. Graph of the post-intervention negativa network (simulated names).

Tables 5 and 6 show the behavioral change in engagement (subscales dedication, vigor, and absorption) and resilience, which was statistically significant. All the variables showed an increase following the educational intervention.

Table 5. Descriptive data and comparison of pre–post-intervention centrality variables (N = 50).

Network	Centrality Variable	Pre-Intervention	Post-Intervention	Student's *t*-Test	Sig.
	InDegreeN	0.10 ± 0.05	0.18 ± 0.08	−13.775	<0.001 **
Help	OutDegreeN	0.10 ± 0.08	0.18 ± 0.12	−4.611	<0.001 **
	EigenvectorN	13.35 ± 15.04	16.87 ± 10.85	−2.761	0.008 **
	BetweennessN	2.78 ± 3.91	2.76 ± 2.89	0.029	0.977
	InDegreeN	0.12 ± 0.06	0.18 ± 0.07	−10.326	<0.001 **
Friendship	OutDegreeN	0.12 ± 0.07	0.18 ± 0.12	−4.229	<0.001 **
	EigenvectorN	11.85 ± 16.28	17.16 ± 10.38	−3.643	0.001 **
	BetweennessN	2.78 ± 3.91	2.56 ± 2.08	0.359	0.721
	InDegreeN	0.12 ± 0.05	0.12 ± 0.05	−0.01	0.992
Negative	OutDegreeN	0.12 ± 0.23	0.12 ± 0.25	0	1
	EigenvectorN	17.31 ± 10.12	17.63 ± 9.53	−0.248	0.805
	BetweennessN	0.58 ± 2.07	0.70 ± 2.02	−0.338	0.737

Legend: InDegreeN, normalized InDegree; OutDegreeN, normalized OutDegree; BetweennessN, normalized Betweenness; EigenvectorN, normalized Eigenvector. * Correlation is significant at 0.05. ** Correlation is significant at 0.01.

Table 6. Comparison of pre–post-intervention engagement and resilence variables (N = 50).

Variable	Subscale	Pre-Intervention	Post-Intervention	Wilcoxon's *t*-Test	Sig.
	Dedication	3.0 ± 1.1	3.4 ± 1.3	−2.340	0.019 *
Engagement	Vigor	3.1 ± 1.0	3.4 ± 1.2	−2.955	0.003 **
	Absorption	4.0 ± 1.0	4.2 ± 1.2	−2.483	0.013 *
Resilience	Resilience	28.4 ± 4.9	29.8 ± 5.4	−2.239	0.025 *

* Correlation is significant at 0.05. ** Correlation is significant at 0.01.

As regards correlations between the variables of centrality and engagement, we found relationships in the help and friendship networks, but not in the negative network. The help network was statistically significantly related to the three engagement subscales and, in particular, the relationship between BetweennessN and all three engagement subscales presented a level of significance of 0.001. The same behavior was observed between EigenvectorN and the absorption subscale of engagement (Table 7).

Table 7. Correlations between network cantrality variables and engagement (N = 50).

Network	Centrality Variable	Engagement	Correlation	Sig.
	BetweennessN	Dedication	0.378 (P)	0.007 **
	InDegreeN		0.345 (S)	0.014 *
	OutDegreeN	Vigor	0.299 (S)	0.035 *
	EigenvectorN		0.289 (S)	0.042 *
Help	BetweennessN		0.451 (S)	0.001 **
	InDegreeN		0.543 (S)	<0.001 **
	OutDegreeN	Absorption	0.349 (S)	0.013 *
	EingvectorN		0.417 (S)	0.003 **
	BetweennessN		0.414 (S)	0.003 **
Friendship	EigenvectorN	Absorption	0.298 (S)	0.035 *

Legend: InDegreeN, normalized InDegree; OutDegreeN, normalized OutDegree; BetweennessN, normalized Betweenness; EigenvectorN, normalized Eigenvector; P, Pearson's correlation coefficient; S (Spearman's correlation coefficient). * Correlation is significant at 0.05. ** Correlation is significant at 0.01.

Relationships between resilience and the centrality variables of all networks were statistically significant for half of the variables in the help and friendship networks. Both networks presented an association with the normalized OutDegree and normalized Betweenness. In particular, the relationship between the help network and normalized Betweenness presented a significance of 0.001. Once again, no statistically significant relationships were observed for the negative network (Table 8).

Table 8. Correlations between network centrality variables and resilience (N = 50).

Network	Centrality Variable	Spearman's Correlation	Sig.
Help	OutDegreeN	0.297	0.036 *
	BetweennessN	0.445	0.001 **
Friendship	OutDegreeN	0.405	0.004 **
	BetweennessN	0.35	0.013 *

Legend: InDegreeN, normalized InDegree; OutDegreeN, normalized OutDegree; BetweennessN, normalized Betweenness; EigenvectorN, normalized Eigenvector; P, Pearson's correlation coefficient. * Correlation is significant at 0.05. ** Correlation is significant at 0.01.

The only statistically significant relationship observed between the sociodemographic variables (sex, degree, team) and centrality variables for the three networks and academic performance was identified between the team and academic performance (r = 0.283; sig. 0.046).

3. Discussion

In the present study, we conducted an educational intervention based on a cooperative interdisciplinary task carried out by undergraduate nursing and computer engineering students. The intervention was implemented on a management course (nursing) and a semantic modeling course (computer engineering). It was based on mixed working teams where the nursing students had to identify and explain a health demand and the computer engineering students had to formulate a technological solution. Assessment of the task formed part of each participant's final mark for academic performance.

We determined changes in the students' help, friendship, and negative networks through measures of centrality, engagement and resilience. These assessments were performed pre- and post-intervention. As regards the analysis, we correlated centrality, engagement, resilience and sociodemographic variables with the final mark for academic performance in the courses analyzed. Our main finding was that the centrality variables in the negative network did not present any statistically significant changes after the cooperative task intervention. However, we did observe statistically significant results for engagement, resilience and centrality variables in the help and friendship networks, except for Betweenness centrality. We obtained correlations between help and friendship network centrality variables and engagement and resilience. With respect to academic performance, we only found a correlation for the variable "team".

In relation to the educational intervention, we found that the cooperative task generated changes in student behavior. These findings are consistent with those reported in similar studies using team work to help students to acquire subject competencies [55], especially in health education teaching placements [56] and crossdisciplinary workshops in universities about different sustainability issues [1,2]. As regards the study population, we did not find any studies that analyzed a population as disparate as our sample (nurses and computer engineers), although some analyzed distinct but similar populations such as medical and nursing students, using a model of interprofessional problem-based learning to assess the effect on learning (mutual understanding of roles, appreciation, and interprofessional communication and collaboration) [46,57]. In this respect, our study incorporated a novel and risky interdisciplinary perspective, but with considerable success. The technological complexity of health care is rising, and this will mean that professionals in various areas of knowledge must know how to work in teams.

The main contribution of our study is the application of SNA as a quantitative methodology to determine relational behavior change. Our intervention required students to interact in order to carry out an academic task for subsequent assessment. Hence, we generated a need on the basis of which each group of students had to establish relations of interdependence to achieve a shared goal: to complete the academic task and obtain a good academic outcome. The literature indicates that relationships emerge when there is a shared goal or vision, but the results suggest that SNA applied to collaborative learning has not reflected this same diversity of actors and relational ties but instead has solely explored one-mode networks of learners connected by communication-based relational ties and has been limited to a descriptive report of SNA results [40,58]. However, SNA facilitates a structural interpretation of relationships in undergraduate learning and their impact on learning outcomes, which can inform educators in unique ways and improve educational reform [59].

We found no evidence in the literature review of any study that had conducted this type of educational intervention with the variables analyzed, and it is therefore not possible to perform a direct comparison. SNA has been used as an instrument to analyze or explain interdisciplinary behavior in the field of education [33,60]. Our research yielded associations between students and change in friendship and help but not negative networks. Both the friendship and help networks presented significant differences in the centrality variables of InDegree, OutDegree, and Eigenvector, suggesting that cooperative work with a shared goal exerts the most influence on the number of ties received and emitted and the capacity for influence in the classroom. By contrast, the negative attitude of avoiding a peer, measured by means of the negative network (which of the following classmates do you avoid interacting with? [30]), did not lead to changes following our educational intervention. One explanation for this may be that a negative perception of someone is not a barrier to working together in a team when there is a shared goal which requires interaction. Put more simply, you might not like working with someone, but when you have to in order to achieve a goal, you are capable of working efficiently with the people around you. These findings are in line with those reported by Wang et al. (2015), who proposed a machine learning algorithm for predicting positive and negative relationships in social networks. They found that both types of relationships, which include support and opposition, and trust and suspicion, are present in all networks, and that research in this context could help us to understand the formation of relationships and network structures. In turn, this would enable us to optimize them in order to achieve shared goals.

As regards the significant change in the help and friendship networks, this may be associated with two factors. First, the cooperative task intervention enhanced network contacts, because members shared experiences, frustrations, and achievements, furthering cohesion. In this respect, our findings are consistent with those of studies by Dreier-Wolfgramm et al. (2018), who showed that their problem-based learning intervention with nursing and medical students had increased each participant's level of knowledge through contact with the others and had exerted a positive effect in terms of mutual appreciation [57]. Second, cooperative work enhanced the students' self-esteem because it helped them to clarify doubts, solve problems, and achieve their goals. In other words, the intervention generated friendly sociability. An earlier study with adolescent students found that self-esteem heightened sociability, reduced symptoms of depression and sadness, and increased the number of ties with other students [37].

The graphical representations of the networks show that the number of contacts increased over the course of the intervention and that the students mainly formed groups within their discipline rather than according to sex. Thus, nursing students interacted more with each other, as did computer engineering students. This could be explained by homophily, the tendency of individuals to interact with others similar to themselves [46]. In this context, similarity would refer to the same team, race, sex or degree course. In our study, nursing students shared a common context and language in terms of health, while computer engineering students shared a context more related to technology. Previous studies of networks have found that the main variable associated with the closest ties in a class of nursing students was sex [41]. However, in this study, it was not possible to extrapolate the data in the graphics as regards sex because our sample was not homogeneous in terms of the number of men and women on each of the degree courses.

Our findings indicate that cooperative work significantly changed the psychological variables analyzed, which is consistent with the results of previous studies. With regards to university student engagement, authors such as Persky (2012) observed this change in team-based learning in a foundational pharmacokinetics course [61]. Similarly, Promo et al. (2018) suggested that incorporating tasks based on interdependence can promote engagement in small teams as well as in an entire class of undergraduate students [62]. In relation to resilience, our study highlights its role in helping university students to overcome adversity and learn from experience. Thomas and Asselin (2018) advocated strengthening resilience in order to improve clinical placements and promote support, education, and reflection in the context of university clinical education [63].

The post-intervention results obtained for the three subscales of engagement were 3.4 ± 1.3 for dedication, 3.4 ± 1.3 for vigor, and 4.2 ± 1.3 for absorption, which differ slightly from those reported in similar studies. One study conducted with 90 nursing degree students obtained means of 4.4 for dedication, 3.08 for vigor, and 3.21 for absorption [46]. Another study conducted with 134 nursing degree students obtained means of 4.82 for dedication, 3.13 for vigor, and 2.98 for absorption [41]. In both cases, it can be seen that the subscale of dedication obtained the highest means. However, in our study, the subscale of absorption (an individual's capacity to be totally focused on work) obtained the highest mean. One explanation for this finding might be that since our students were studying two different degree courses taught on different campuses, they had to overcome the obstacle of geographical distance, optimizing the time invested in work without distractions.

The result obtained for resilience post-intervention was 29.82, similarly to other studies on nursing students, with results between 28.6 and 34.7 [41,46,64]. Achieving a high degree of resilience is important because it is associated with experiencing less psychological distress and, above all, less academic burnout [65]. This may be explained by the high levels of stress experienced during university studies, which could diminish undergraduate nursing and computer engineering students' preparedness to exercise their profession. Previous studies of nursing students have found that burnout during nursing education predicts lower occupational preparedness and future clinical performance, together with high stress levels in new students [65,66].

As regards associations between the variables of centrality and engagement, we observed numerous statistically significant correlations in the help and friendship networks, with the exception of Betweenness. Similarly, Fernández-Martínez et al. (2017) observed the same behavior in 48 first-year nursing students [41]. Relationships between the variables of centrality in the friendship and help networks were associated with resilience, in contrast to the findings of previous studies with nursing students, in which resilience was only related to the friendship network [41]. In addition, it should be noted that we found relationships between the negative network and engagement or resilience, suggesting that negative ties involving enduring and recurrent negative judgements and feelings do not prevent actors from performing a task. Other researchers have claimed that workplace ties are "friendly", "positive", or at least "neutral", and that although occasional upsets may arise, creating temporary discontent with individual or team achievements, positive ties transcend negative ones, canceling the latter's effects on the actors [32].

Lastly, academic performance did not correlate with any variable except the team to which each participant belonged. One explanation for this result is that the teams obtained different marks and, therefore, students on the same team would have similar marks because the team mark contributed to individual marks. This was a surprising finding because previous studies on nursing students have reported that a better result for the three subscales of engagement was associated with better academic performance [67]. Another study found a similar result for adolescents, concluding that promoting engagement in adolescence would lead to better performance in high school [14]. The means obtained for the engagement subscales might have been affected by student motivation. Since our students were studying different degree courses, they did not know each other and, consequently, could not select their own teams, which were assigned by the researchers. This may have influenced engagement, although in a positive sense via the formation of new networks. In turn, students with greater OutDegree centrality tended to present a higher level of engagement [13]. Social prestige as measured by students' Eigenvector has been associated with better academic outcomes in some—but not all—studies [12,68]. In our study, performance in the group task did not present any statistically significant relationship [69]. The importance of the results obtained is not only in the replicability of the interventions in different educational contexts, but also in their sustainability over time, to be able to respond to the problems and demands generated by the students. These conflicting results in the literature indicate the potential importance of social prestige arising from the formation of social networks. In the recent literature, as a measure of social capital, the Eigenvector has been positively associated with academic outcome, as has InDegree [70]. Nevertheless, none of the networks obtained in our study presented any correlation with academic performance, possibly because we used the final mark as our measure of performance, which included other criteria in addition to the cooperative task, such as multiple-choice test results and classroom participation.

Our study presents several limitations. First, we did not include more variables that might have better elucidated the results obtained. Secondly, we did not include a control group undergoing a similar experience in a different form, and it was therefore not possible to compare differences between a control and experimental group. However, the inclusion of all the students and the creation of similar teams generated added value when interpreting the data. Another limitation was the correlation only between performance and equipment. This should be considered for future research so that the yield variable includes evaluations of relational competencies. This study did not aim to generalize or prove cause–effect relationships but to test some innovative and initial hypotheses that will be further tested at a larger scale in a future study with other degrees.

4. Conclusions

One of the most important premises of the context of high education is that the focus of the teaching–learning processes is oriented to the demands of society and developed in a sustainable way. In our case of health sciences and computer engineering, society increasingly demands that we complement and understand each other. That is, the nursing professional captures the demands of the patients and must be competent to evaluate the cost–benefit of the proposals to meet their needs. On the other hand, the evidence demonstrates that technological applications can be useful for patient care and save costs, but nurses are not trained to make those proposals. Computer engineers do have the core of knowledge to provide technological solutions.

This way of understanding, through collaborative work between different faculties, how the university context approaches social reality, is a useful way to propose innovative and sustainable solutions in teaching–learning.

In the intervention carried out in this research, students showed that they were able to work in a network, although their areas of knowledge were very different—nursing and computer engineering.

The behavior of nursing and computer engineering students following a cooperative task intervention changed significantly for all network centrality variables in the help and friendship, but not negative, networks.

In addition, we graphically represented these network changes, showing that ties were basically formed between students taking the same degree course.

We observed a significant change in engagement and resilience following the cooperative interdisciplinary task intervention.

We analyzed centrality, engagement, resilience, and sociodemographic variables and academic performance and found that many of the centrality variables in the help and friendship networks were associated with engagement and resilience. However, academic performance did not correlate with any variable analyzed except the team to which the students belonged.

As a practical application, teachers should identify engagement, resilience, and networks in order to improve the communication skills necessary to carry out cooperative tasks, an essential aspect of the students' future work.

Currently, the world is conceptualized as global. In this sense, we must also conceptualize the work between professionals from different fields, different organizations, and different countries as collaborative and global, given that the pace of society demands it, and technologies allow it.

We believe our analysis of an interdisciplinary teaching intervention provides some valuable evidence that can suggest future strategies for use by university teachers to develop, in students, important skills needed by professionals, in the context of a sustainable lifelong learning framework.

Author Contributions: Conceptualization, P.M.-S., I.G.-R., J.A.B.-A., J.P.-P., and M.M.R.-G.; data curation methodology, J.P.-P., M.M.R.-G., J.A.B.-A., P.M.-S., M.C.P., and I.G.-R.; software, J.A.B.-A., M.M.R.-G., P.M.-S., I.G.-R., and M.C.P.; validation, P.M.-S., I.G.-R., J.A.B.-A., M.C.P., and M.M.R.-G.; formal analysis, M.M.R.-G., P.M.-S., I.G.-R., J.A.B.-A., M.C.P., and J.P.-P.; investigation, P.M.-S., I.G.-R., J.A.B.-A., M.C.P., J.P.-P., and M.M.R.-G.; resources, P.M.-S., I.G.-R., M.M.R.-G.; J.A.B.-A., M.C.P., and J.P.-P.; data curation, J.P.-P., M.M.R.-G., P.M.-S., I.G.-R., J.A.B.-A., and M.C.P.; writing—original draft preparation, M.M.R.-G., J.P.-P., P.M.-S., I.G.-R., J.A.B.-A., and M.C.P.; writing—review and editing, J.P.-P., P.M.-S., I.G.-R., J.A.B.-A., M.C.P., and M.M.R.-G.; visualization, P.M.-S., I.G.-R., J.A.B.-A., M.C.P., J.P.-P., and M.M.R.-G.; supervision, P.M.-S., I.G.-R., M.M.R.-G., J.A.B.-A., M.C.P., and J.P.-P.

Funding: This research has had partial funding from the Master of Socio-Health Research (University of Leon).

Acknowledgments: To the students and professors of computer science and nursing of the University of León.

Conflicts of Interest: The authors declare no conflict of interest.

References

1. Albareda-Tiana, S.; García-González, E.; Jiménez-Fontana, R.; Solís-Espallargas, C. Implementing Pedagogical Approaches for ESD in Initial Teacher Training at Spanish Universities. *Sustainability* **2019**, *11*, 4927. [CrossRef]

2. Australia, S. Pacific (2017) Getting started with the SDGs in universities: A guide for universities, higher education institutions, and the academic sector. Australia, New Zealand and Pacific Edition. *Sustain. Dev. Solut. Netw. Aust. Pac. Melb.* **2018**. Available online: http://ap-unsdsn.org/wp-content/uploads/University-SDG-Guide_web.pdf (accessed on 9 November 2019).

3. Cassen, R.H. Our Common Future: Report of the World Commission on Environment and Development. *Int. Aff.* **1987**, *64*, 126. [CrossRef]

4. Boud, D. Sustainable Assessment: Rethinking assessment for the learning society. *Stud. Contin. Educ.* **2000**, *22*, 151–167. [CrossRef]

5. Boud, D.; Soler, R. Sustainable assessment revisited. *Assess. Eval. High. Educ.* **2016**, *41*, 400–413. [CrossRef]

6. Tai, J.; Ajjawi, R.; Boud, D.; Dawson, P.; Panadero, E. Developing evaluative judgement: Enabling students to make decisions about the quality of work. *High. Educ.* **2018**, *76*, 467–481. [CrossRef]

7. Clark, J.; Baker, T. Sustainable Assessment in Cooperative Learning. In Proceedings of the EDULEARN14: 6th International Conference on Education and New Learning Technologies, Barcelona, Spain, 7–9 July 2014; pp. 325–333.

8. Kearney, S.; Perkins, T.; Kennedy-Clark, S. Using self- and peer-assessments for summative purposes: Analysing the relative validity of the AASL (Authentic Assessment for Sustainable Learning) model. *Assess. Eval. High. Educ.* **2016**, *41*, 840–853. [CrossRef]

9. Crone, E.A.; Dahl, R.E. Understanding adolescence as a period of social–affective engagement and goal flexibility. *Nat. Rev. Neurosci.* **2012**, *13*, 636–650. [CrossRef] [PubMed]

10. Dishion, T.J.; Tipsord, J.M. Peer contagion in child and adolescent social and emotional development. *Annu. Rev. Psychol.* **2011**, *62*, 189–214. [CrossRef] [PubMed]

11. Wang, M.-T.; Holcombe, R. Adolescents' Perceptions of School Environment, Engagement, and Academic Achievement in Middle School. *Am. Educ. Res. J.* **2010**, *47*, 633–662. [CrossRef]

12. Rizzuto, T.E.; LeDoux, J.; Hatala, J.P. It's not just what you know, it's who you know: Testing a model of the relative importance of social networks to academic performance. *Soc. Psychol. Educ.* **2009**, *12*, 175–189. [CrossRef]

13. Liu, C.-C.; Chen, Y.-C.; Tai, S.-J.D. A social network analysis on elementary student engagement in the networked creation community. *Comput. Educ.* **2017**, *115*, 114–125. [CrossRef]

14. Wang, M.; Kiuru, N.; Degol, J.L.; Salmela-aro, K. Friends, academic achievement, and school engagement during adolescence: A social network approach to peer in fluence and selection effects. *Learn. Instr.* **2018**, *58*, 148–160. [CrossRef]

15. Li, M.-H.; Eschenauer, R.; Persaud, V. Between Avoidance and Problem Solving: Resilience, Self-Efficacy, and Social Support Seeking. *J. Couns. Dev.* **2018**, *96*, 132–143. [CrossRef]

16. Zhao, F.; Guo, Y.; Suhonen, R.; Leino-kilpi, H. Nurse Education Today Subjective well-being and its association with peer caring and resilience among nursing vs medical students: A questionnaire study. *YNEDT* **2016**, *37*, 108–113.

17. Jean, L.; Hunter, S. Nurse Education Today Resilience in nursing students: An integrative review. *YNEDT* **2016**, *36*, 457–462.

18. Lozares, C.; Sumario, R. La teoria de redes sociales. *Pap. Rev. Sociol.* **1996**, *48*, 103–126. [CrossRef]

19. Wasserman, S.; Faust, K. *Social Network Analysis: Methods and Applications*; Cambridge University Press: Cambridge, UK, 1994; ISBN 9780521387071.

20. Barnes, J.A. Class and Committees in a Norwegian Island Parish. *Hum. Relat.* **1954**, *7*, 39–58. [CrossRef]

21. Nahapiet, J.; Ghoshal, S. Social Capital, Intellectual Capital, and the Organizational Advantage. *Acad. Manag. Rev.* **1998**, *23*, 242. [CrossRef]

22. Borgatti, S.P.; Foster, P.C. The Network Paradigm in Organizational Research: A Review and Typology. *J. Manag.* **2003**, *29*, 991–1013.

23. Robins, G. *Doing Social Network Research: Network-Based Research Design for Social Scientists*; Sage: Newcastle, UK; ISBN 1446276139.

24. Zachary, W.W. An Information Flow Model for Conflict and Fission in Small Groups. *J. Anthr. Res.* **1977**, *33*, 452–473. [CrossRef]

25. Wasserman, S.; Faust, K. Social network analysis: methods and applications. *Am. Ethnol.* **1997**, *24*, 219–220.

26. Laninga-Wijnen, L.; Ryan, A.M.; Harakeh, Z.; Shin, H.; Vollebergh, W.A.M. The moderating role of popular peers' achievement goals in 5th- and 6th-graders' achievement-related friendships: A social network analysis. *J. Educ. Psychol.* **2018**, *110*, 289–307. [CrossRef]

27. Ahuja, G. Collaboration Networks, Structural Holes, and Innovation: A Longitudinal Study. *Adm. Sci. Q.* **2000**, *45*, 425. [CrossRef]

28. Hether, H.J.; Murphy, S.T.; Valente, T.W. A social network analysis of supportive interactions on prenatal sites. *Digit. Health* **2016**, *2*. Available online: https://doi.org/10.1177/2055207616628700 (accessed on 9 November 2019). [CrossRef] [PubMed]

29. Adriaensens, S.; Van Waes, S.; Struyf, E. Comparing acceptance and rejection in the classroom interaction of students who stutter and their peers: A social network analysis. *J. Fluen. Disord.* **2017**, *52*, 13–24. [CrossRef] [PubMed]

30. Marineau, J.E.; Labianca, G.J.; Kane, G.C. Direct and indirect negative ties and individual performance. *Soc. Netw.* **2016**, *44*, 238–252. [CrossRef]

31. Labianca, G.; Brass, D.J. Exploring the Social Ledger: Negative Relationships and Negative Asymmetry in Social Networks in Organizations. *Acad. Manag. Rev.* **2006**, *31*, 596–614. [CrossRef]

32. Lyons, B.J.; Scott, B.A. Integrating social exchange and affective explanations for the receipt of help and harm: A social network approach. *Organ. Behav. Hum. Decis. Process.* **2012**, *117*, 66–79. [CrossRef]

33. Brass, D.J.; Borgatti, S.P.; Halgin, D.S.; Labianca, G.; Mehra, A. *Contemporary Perspectives on Organizational Social Networks*; Emerald Group Publishing: Bingley, UK, 2014; ISBN 9781783507528.

34. Park, D.; Lee, G.; Kim, G.W.; Kim, T.T. Social Network Analysis as a Valuable Tool for Understanding Tourists' Multi-Attraction Travel Behavioral Intention to Revisit and Recommend. *Sustainability* **2019**, *11*, 2497. [CrossRef]

35. Wilkin, J.; Biggs, E.; Tatem, A.J. Measurement of Social Networks for Innovation within Community Disaster Resilience. *Sustainability* **2019**, *11*, 1943. [CrossRef]

36. Scatá, M.; Di Stefano, A.; La Corte, A.; Lió, P. Quantifying the propagation of distress and mental disorders in social networks. *Sci. Rep.* **2018**, *8*, 5005. [CrossRef] [PubMed]

37. Pachucki, M.C.; Ozer, E.J.; Barrat, A.; Cattuto, C. Mental health and social networks in early adolescence: A dynamic study of objectively-measured social interaction behaviors. *Soc. Sci. Med.* **2015**, *125*, 40–50. [CrossRef] [PubMed]

38. Jostad, J.; Sibthorp, J.; Paisley, K. Understanding groups in outdoor adventure education through social network analysis. *J. Outdoor Environ. Educ.* **2013**, *17*, 17–31. [CrossRef]

39. Mauldin, R.L.; Narendorf, S.C.; Mollhagen, A.M.; Mauldin, R.L.; Narendorf, S.C.; Mollhagen, A.M. Relationships Among Diverse Students in a Cohort-Based MSW Program: A Social Network Analysis. *J. Soc. Work Educ.* **2017**, *53*, 684–698. [CrossRef]

40. Putnik, G.; Costa, E.; Alves, C.; Castro, H.; Varela, L.; Shah, V. Analysing the correlation between social network analysis measures and performance of students in social network-based engineering education. *Int. J. Technol. Des. Educ.* **2016**, *26*, 413–437. [CrossRef]

41. Fernández-Martínez, E.; Andina-Díaz, E.; Fernández-Peña, R.; García-López, R.; Fulgueiras-Carril, I.; Liébana-Presa, C. Social Networks, Engagement and Resilience in University Students. *Int. J. Environ. Res. Public Health* **2017**, *14*, 1488. [CrossRef] [PubMed]

42. Todhunter, F. Using principal components analysis to explore competence and confidence in student nurses as users of information and communication technologies. *Nurs. Open* **2015**, *2*, 72–84. [CrossRef] [PubMed]

43. Janssen, M.; Bodin, Ö.; Anderies, J.; Elmqvist, T.; Ernstson, H.; McAllister, R.R.J.; Olsson, P.; Ryan, P. Toward a network perspective of the study of resilience in social-ecological systems. *Ecol. Soc.* 2006, 11, p. 15. Available online: http://frst411.sites.olt.ubc.ca/files/2015/01/0c96052a7b5076c538000000.pdf (accessed on 9 November 2019).

44. Castro, T.C.; Gonçalves, L.S. The use of gamification to teach in the nursing field. *Rev. Bras. Enferm.* **2018**, *71*, 1038–1045. [CrossRef] [PubMed]

45. García-cabot, A.; García-l, E.; Diez-folledo, T. Social network analysis of a gami fied e-learning course: Small-world phenomenon and network metrics as predictors of academic performance. *Comput. Hum. Behav.* **2016**, *60*, 312–321.

46. Liébana-Presa, C.; Andina-Díaz, E.; Reguera-García, M.-M.; Fulgueiras-Carril, I.; Bermejo-Martínez, D.; Fernández-Martínez, E. Social Network Analysis and Resilience in University Students: An Approach from Cohesiveness. *Int. J. Environ. Res. Public Health* **2018**, *15*, 2119. [CrossRef] [PubMed]

47. Grove, S.K.; Gray, J.R.; Faan, P.R.N. *Understanding Nursing Research: First South Asia Edition, E-Book: Building an Evidence-Based Practice*; Elsevier: New Delhi, India, 2019; ISBN 8131257118.

48. Schaufeli, W.B.; Salanova, M.; González-Romá, V.; Bakker, A.B. The Measurement of Engagement and Burnout: A Two Sample Confirmatory Factor Analytic Approach. *J. Happiness Stud.* **2002**, *3*, 71–92. [CrossRef]

49. Notario-Pacheco, B.; Solera-Martínez, M.; Serrano-Parra, M.D.; Bartolomé-Gutiérrez, R.; García-Campayo, J.; Martínez-Vizcaíno, V. Reliability and validity of the Spanish version of the 10-item Connor-Davidson Resilience Scale (10-item CD-RISC) in young adults. *Health Qual. Life Outcomes* **2011**, *9*, 63. [CrossRef] [PubMed]

50. Crespo, M.; Fernández-Lansac, V.; Soberón, C. Adaptación española de la Escala de Resiliencia de Connor-Davidson (CD-RISC) en situaciones de estrés crónico. *Behav. Psychol. Psicol. Conduct.* **2014**, 22, 219–238.

51. Rodríguez, J.J.G.; Manzanares, M.T.L.; Luque, F.F. Engagement and academics results of the grade students. *Rev. Enferm.* **2013**, *36*, 18–25.

52. Schaufeli, W.B.; Bakker, A.B. UWES–Utrecht work engagement scale: Test manual. *Unpubl. Manuscr. Dep. Psychol. Utr. Univ.* **2003**, *8*. Available online: https://www.wilmarschaufeli.nl/publications/Schaufeli/Test%20Manuals/Test_manual_UWES_English.pdf (accessed on 9 November 2019).

53. Borgatti, S.P.; Everett, M.G.; Freeman, L.C. Ucinet for Windows: Software for social network analysis. *Harv. MA Anal. Technol.* **2002**, *6*. Available online: https://www.semanticscholar.org/paper/Ucinet-for-Windows%3A-Software-for-Social-Network-Borgatti-Everett/53d2d01787895c945552d6604a30e5431bf573b3 (accessed on 9 November 2019).

54. Benítez, J.A.; Labra, J.E.; Quiroga, E.; Martín, V.; García, I.; Marqués-Sánchez, P.; Benavides, C. A Web-Based Tool for Automatic Data Collection, Curation, and Visualization of Complex Healthcare Survey Studies including Social Network Analysis. *Comput. Math. Methods Med.* **2017**, *2017*, 1–8. [CrossRef] [PubMed]

55. Webb, N.M. The teacher's role in promoting collaborative dialogue in the classroom. *Br. J. Educ. Psychol.* **2009**, *79*, 1–28. [CrossRef] [PubMed]

56. Cinelli, B.; Symons, C.W.; Bechtel, L.; Rose-Colley, M. Applying Cooperative Learning in Health Education Practice. *J. Sch. Heal.* **1994**, *64*, 99–102. [CrossRef] [PubMed]

57. Dreier-Wolfgramm, A.; Homeyer, S.; Oppermann, R.F.; Hoffmann, W. A model of interprofessional problem-based learning for medical and nursing students: Implementation, evaluation and implications for future implementation. *GMS J. Med. Educ.* **2018**, *35*, p. Doc13. Available online: https://dx.doi.org/10.3205/zma001160 (accessed on 9 November 2019).

58. Dado, M.; Bodemer, D. A review of methodological applications of social network analysis in computer-supported collaborative learning. *Educ. Res. Rev.* **2017**, *22*, 159–180. [CrossRef]

59. Grunspan, D.Z.; Wiggins, B.L.; Goodreau, S.M. Understanding Classrooms through Social Network Analysis: A Primer for Social Network Analysis in Education Research. *CBE Life Sci. Educ.* **2014**, *13*, 167–178. [CrossRef] [PubMed]

60. Carolan, B. *Social Network Analysis and Education: Theory, Methods & Applications*; SAGE Publications, Inc.: Thousand Oaks, CA, USA, 2014; ISBN 9781412999472.

61. Persky, A.M. The Impact of Team-Based Learning on a Foundational Pharmacokinetics Course. *Am. J. Pharm. Educ.* **2012**, *76*, 31. [CrossRef] [PubMed]

62. Premo, J.; Cavagnetto, A.; Davis, W.B. Promoting Collaborative Classrooms: The Impacts of Interdependent Cooperative Learning on Undergraduate Interactions and Achievement. *CBE Life Sci. Educ.* **2018**, *17*, ar32. [CrossRef] [PubMed]

63. Thomas, L.J.; Asselin, M. Promoting resilience among nursing students in clinical education. *Nurse Educ. Pr.* **2018**, *28*, 231–234. [CrossRef] [PubMed]

64. Ríos-Risquez, M.I.; García-Izquierdo, M.; Sabuco-Tebar, E.D.L.A.; Carrillo-Garcia, C.; Martinez-Roche, M.E. An exploratory study of the relationship between resilience, academic burnout and psychological health in nursing students. *Contemp. Nurse* **2016**, *52*, 1–22. [CrossRef] [PubMed]

65. García-Ros, R.; Pérez-González, F.; Pérez-Blasco, J.; Natividad, L.A. Evaluación del estrés académico en estudiantes de nueva incorporación a la universidad. *Rev. Latinoam. Psicol.* **2012**, *44*, 143–154.

66. Rudman, A.; Gustavsson, J.P. Burnout during nursing education predicts lower occupational preparedness and future clinical performance: A longitudinal study. *Int. J. Nurs. Stud.* **2012**, *49*, 988–1001. [CrossRef] [PubMed]

67. Marqués-Sánchez, P.; Alfonso-Cendón, J.; Fernández-Martínez, M.E.; Pinto-Carral, A.; Liébana-Presa, C.; Conde, M. Ángel; García-Peñalvo, F.J. Co-operative Networks and their Influence on Engagement: A Study with Students of a Degree in Nursing. *J. Med. Syst.* **2017**, *41*, 103. [CrossRef] [PubMed]

68. Gašević, D.; Zouaq, A.; Janzen, R. "Choose Your Classmates, Your GPA Is at Stake!": The Association of Cross-Class Social Ties and Academic Performance. *Am. Behav. Sci.* **2013**, *57*, 1460–1479. [CrossRef]

69. Hernández-García, Á.; González-González, I.; Jiménez-Zarco, A.I.; Chaparro-Peláez, J. Applying social learning analytics to message boards in online distance learning: A case study. *Comput. Hum. Behav.* **2015**, *47*, 68–80. [CrossRef]

70. Saqr, M.; Fors, U.; Tedre, M. How the study of online collaborative learning can guide teachers and predict students' performance in a medical course. *BMC Med. Educ.* **2018**, *18*, 24. [CrossRef] [PubMed]

 *sustainability*

Review

A Matter of Responsible Management from Higher Education Institutions

Nicolas Roos

Faculty of Business Management and Economics, Technische Universitaet Dresden, 01069 Dresden, Germany; nicolas.roos@tu-dresden.de

Received: 1 October 2019; Accepted: 13 November 2019; Published: 18 November 2019

Abstract: Higher education institutions (HEIs) are influential social institutions which disseminate knowledge, promote innovation, and educate future decision-makers. The increasing awareness of HEIs as social actors has increased the pressure on them to accept and act upon their social responsibility. Processing this responsibility requires a structured management approach. The little attention given thus far to management performance and structured steering processes of social responsibility in HEIs marks the research gap the present study is focused on. This article provides a systematic review of scientific and academic publications, applying the concept of Social Performance after Wood (1991). The study aims to combine different research and modeling approaches to examine individual elements of social performance along the dimensions of processes of social responsiveness and outcomes of institutional behavior. With this approach, the study aims to answer the question of how HEIs assume their responsibilities as social institutions. The results show that observable outcomes of social behavior in the academic environment reflect a broad understanding of different approaches. By clustering the encoded literature into processes and outcomes, the study structures the fragmented body of research reflecting the various characteristics of the higher education sector.

Keywords: responsible management; higher education institutions; social performance; institutional theory; systematic review

---

1. Introduction

Higher Education Institutions (HEIs), as influential social institutions, act as promoters of change through research and education [1]. Within this function, they create and disseminate knowledge, emphasizing their role in social capacity in society [2].

With a general dissemination of business tools in public institutions, the proliferation of responsible management practices likewise increases the awareness of effects of organizational activities within HEIs [3]. The orientation towards a higher level of awareness through improvement based mechanisms and performance enhancement leads to a professionalization of HEIs' management structures [4,5]. This constitutes an essential capacity for providing resources to process responsibility [6].

Since structured management appears to be a critical success factor for operating responsibility, the question of systematically approaching management performance and steering social sustainability activities defines a research gap currently not being addressed in the scientific literature.

Previous studies approach the implementation of responsibility [4,7] or the integration of stakeholder interests in the context of sustainability considerations [8,9]. Other authors address responsibility by examining frameworks and indicators [10–13].

Though measuring effects and outcomes is an important aspect, this forgoes the idea that controlling and managing activities require a superstructure to steer these practices. So far, operations for processing responsibility have received little attention in the literature [14–17].

This paper seeks to determine the efforts of HEIs practicing social sustainability and responsible management. For this case, the study applies a systematic literature review based on the model of social performance after Wood [18].

The review pursues the research question of how HEIs assume their responsibilities as social institutions within the dimensions of social performance of Wood (1991). The objective of this study is to apply a structured approach to HEIs' management activities concerning social sustainability, which consequently makes it feasible to apply tools for steering. Accordingly, the study makes two contributions to the literature stream of responsible management and social performance at HEIs. First, archival data research provides a welcomed opportunity to analyze current research approaches and systemize the status quo of how HEIs manage their social responsibility. In this regard, the review is useful for mapping current management performance in that research field and identifying further research areas not yet studied [19]. The review adds to the mostly qualitative and case-based research in this area, reporting on implementation approaches [19–21] or discussing factors of success [17,22,23], by proposing a pattern for the assessment and management of social performance issues in HEIs.

Second, the study expands the existing literature on sustainability in HEIs by investigating the managerial perspective on practicing social responsibility. The study highlights the role and function of social performance management after Wood [18] and applies this to the context of HEIs. This enables new perspectives on aspects of systematic steering and processing organizational responsible behavior, which contributes to a better understanding of balancing local and global demands, and to closing the gap between social values and organizational behavior in order to improve the social legitimacy of HEIs.

The remainder of the paper is structured as follows: the study continues with a more detailed theoretical background on HEIs, responsible management, and social performance after Wood in Section 2. The methodology, the research design, and measurement of constructs are explained in Section 3. In Section 4, the results of the study are presented. Finally, the study concludes with findings, and presents limitations and implications for future research, in Section 5.

2. Theoretical Background

In the last 20 years, social responsibility has been considered relevant for nearly all entities providing services to the public, including HEIs [20]. HEIs, as public institutions with a primary mission revolving around teaching and research, have an emphasized role in society, since they develop professional skills and knowledge for conscious acting and foster research on sustainable solutions and innovations. This social mission, being embedded within the responsibility to act in the general interest of society, produces social legitimacy and forms the HEI's license to operate. Creating legitimacy by fulfilling a social mission and responsible acting more or less describes a management task, with assessment being a vital component that enables its success. Anticipating this, an increasing number of coordinating bodies have emerged in HEIs [11]. Although the assessment of research and teaching performance is a field that has received significant attention, the area of managing social responsibility in the course of fulfilling social legitimacy has not [17]. The relatively low proliferation of action being taken is grounded in a diverse understanding of the term social responsibility, which goes back to the lack of a generally accepted and common definition [24–27]. Vasilescu et al. [28] approach the social responsibility of a university as "the need to strengthen civic commitment and active citizenship; it is about volunteering, about an ethical approach, developing a sense of civil citizenship by encouraging the students, the academic staff to provide social services to their local community or to promote ecological, environmental commitment for local and global sustainable development". Others define the social responsibility of HEIs from a stakeholder-oriented perspective [22,29,30]. In another approach, Reiser [31] identifies distinct types of social responsibilities: (1) organizational responsibility to lead as an ethical example, (2) cognitive responsibilities of scientific output, (3) social responsibilities towards the community on issues of social development, and (4) educational responsibilities for responsible citizens.

Since this categorization delivers a suitable approach for understanding the different types of responsibilities of institutions, the examination applies this as a working definition for the examination of responsible management from a managerial perspective, although the study also investigates the implementation of type (1) and (3) responsibilities. The study furthermore follows the assumption that defining social responsibility should deliver information on the purpose of acting, how to pursue this purpose, and the ethical obligations serving as guiding principles [32].

To avoid confusion with environmental topics, the study excludes them from the present study in order to produce a clear construct. Consequently, the study applies the term responsibility instead of sustainability [32]. Following this working definition, the study aims to seek a proper tool for the identification of certain activities. With a focus on management activities and steering mechanisms, the application of performance-driven frames focusing on social issues provides appropriate tools for the subsequent analysis.

The application of Wood's model [18] on social performance (see Table 1) enables a mapping of existing management efforts around social issues. The model serves as a framework to reveal social performance in HEIs' management, since it delivers an open structure to map the scattered field of activities in HEIs. Thereby, Wood's model is well-suited for systematically reviewing the literature along internal and external perspectives, as well as processes and outcomes. This enables a wide focus, in order to depict a broader spectrum of activities, with implications for the non-observable issues of organizations (as proposed by principles of CSR). On the other hand, the systematic review of the literature serves to check the applicability and practicability of Wood's model for theory and practice, since it demonstrates whether responsible management performance follows an underlying structure (as proposed by the model), which makes it assessable for management bodies.

In order to discuss these questions, the following section describes Wood's model of social performance in detail. Wood's model goes back to Carroll's approach [33] to mapping the social performance of organizations. In his model, Carroll describes three integral parts of social performance, which are split into the following dimensions:

(1) Corporate Social Responsibility (CSR) as legal, ethical, or economic responsibilities;

(2) Social Issues as categories CSR refers to (environment, certain interest groups/stakeholders);

(3) Corporate Social Responsiveness as action patterns dealing with CSR or social issues [33].

Complementing his thoughts, Wartick and Cochran [34] modified Carroll's model by restructuring the dimensions of social performance based on the assumption of an interaction between the principles of social responsibility, the process of social responsiveness, and the policies addressing social issues [34]. One major contribution of this modification is the recognition that social issues need to be actively managed. In response, Wood developed the model of social performance as an " ... organization's configuration of principles of social responsibility, processes of social responsiveness, and policies, programs, and observable outcomes as they relate to ... societal relationships" [18].

Table 1. Corporate social performance according to Wood (1991) (own illustration).

Corporate Social Performance		
Principles of Corporate Social Responsibility	**Processes of Corporate Social Responsiveness**	**Outcomes of Corporate Behavior**
Institutional level	Environmental assessment	Social impacts
Organizational level	Stakeholder management	Social programs
Individual level	Issues management	Social4 policies

According to this understanding, Wood defines the dimensions of social performance as follows:

(1) Principles of corporate social responsibility describe basic motivations and expectations on different levels of consideration. The principle of legitimacy justifies responsible acting on an institutional level, as there are general expectations from society for all kinds of entities.

Public responsibility describes the expectations towards distinct forms of organizations, which is exercised by managerial discretion on the individual level. At this level, responsibility is understood as a moral task, which cannot be completely determined by the organization. According to Wood, the principles are not observable features of an organization.

(2) The processes of corporate social responsiveness describe action patterns contrasting normative principles of CSR. According to Wood, processes are indirectly measurable by inference [18]. Furthermore, Wood proposes the application of a three-stage management process, consisting of environmental assessment, stakeholder management, and issues management [18]. Environmental assessment describes the examination of the organizational environment to identify and anticipate relevant influences. For the case of stakeholder management, Wood refers to the definition of Freeman as "any group or individual who can affect or is affected by the achievement of the organization's objectives." [35]. Unfortunately, further details on how to identify or manage stakes and claims are not given. Issues management is the last process and entails the monitoring and handling of relevant topics. In contrast to Wartick and Cochran, Wood assorts issues management as a process, since this topic is less oriented towards the performance itself, but rather on the outcome, as defined in the third dimension of the model [18].

(3) The outcomes of corporate behavior describe the observable/assessable part of social performance. Outcomes classify into social impacts describing their direct influence on society, social programs as processes to fulfill specific goals, and social policies as patterns of decision-making within an organization. Thereby social programs and policies are inspired by processes of corporate social responsiveness [18].

The high level of detail and its broad acceptance within the scientific community [36] make this model an appropriate basis for the case of this research. Although the concept originates from a business context, it is feasible for the HEI context as well, since the model shows a highly contextual and organization-dependent degree of integration. Practical application within the systematic screening of the literature requires a customization of Wood's model for the context of this inquiry. Though principles of corporate social responsibility describe the ethical layer behind organizational social performance, this highly contextual issue describes non-observable features [18], which, consequently, cannot be measured. Due to this, the dimension of principles will be excluded from this examination. Hence, the modified model for the examination of management performance on responsibility in HEIs comprises only the dimensions of processes of social responsiveness and outcomes of social behavior (see Table 2)

Table 2. Model of Social Performance following Wood (1991).

Social Performance	
Processes of Social Responsiveness	**Outcomes of Social Behavior**
Environmental assessment	Social impacts
Stakeholder management	Social programs
Issues management	Social policies

3. Materials and Methods

This study applies a systematic review of scientific papers and academic publications, following a structured multi-stage process after Fink and Tranfield et al. [36,37]. The study uses Wood's model for systematically scanning literature on the research question of how HEIs put social performance and responsible management into practice along the dimensions of processes and outcomes of institutional social performance (see Table 2) and follows a three-stage procedure, as described below.

Stage 1: Selecting Search Terms and Databases

As the study found no literature issuing the assessment of responsible management or social performance at HEIs on Google Scholar, the structured research process was extended on the electronic databases Ebsco (Academic Search Complete and Business Source Complete) and Web of Science for publications with topics, titles, abstracts, or keywords. A detailed description of the search and the selection of relevant literature can be found in Tables A1–A4.

The search terms and phrases were derived from the objects of investigation, resulting in three thematic blocks: (1) the social responsibility and social performance, with the phrases "sustainab*" OR "CSR" OR "soc* resp*" OR "soc* perf*" OR "CSP" OR "SR"; (2) management controls, with the phrases "manag* control*" OR "public sector accounting" OR "performance measurement" OR "Public Management" OR "Management Control System*" OR "MCS" OR "Management Accounting" OR "MAS"; and the (3) institutional background of higher education institutions, with the phrases "high* education" OR universit* OR college* OR campus OR "business school*" OR "HEI*" OR "knowledge intensive organi?ation*". The three blocks were linked with the conjunction "AND" to combine them within a query. The search terms aim to cover a vast spectrum of potentially relevant articles on issues of social performance at HEIs. For this case, the search terms in the field of management controls and institutional anchoring also contain phrases from the accounting sector. The search terms on issues of social performance focus on a more general level. The phrases were grouped around the terms of social performance and social responsibility, which were derived from the model of social performance following Wood, as well as Reiser's definition of responsible management.

The need for a wide view is even more evident as the literature is unclear about whether to apply the term sustainability or social responsibility. Consequently, the research included both keywords. Although the objective revolves around HEIs and not a corporate context, the keywords also include the phrases "CSR" (corporate social responsibility) and "CSP" (corporate social performance) to identify further articles on social sustainability.

According to Tranfield [38], the review searched the documents for full text after applying the search terms, in order to not be too restrictively focused on the bibliographic data.

Stage 2: Applying Practical Screening Criteria

To avoid bias within the results, the study included journals, academic journals, and conference contributions published in the English language between the years 1987 and 2019. The study applies the year 1987 as a starting point, since the publication of the Brundtland report marks a turning point in the global debate on responsibility and sustainability. To gain a high number of suitable results, the study uses selection requirements in the screening process to sort out cases without relevance for the research question. Therefore, publications were checked as to whether the title, topic, and abstract fit the objective of responsible management or social performance in the HE sector. Publications with no thematic affiliation, as well as articles beyond the higher education context, were not considered for further screening.

Stage 3: Application of Methodological Screening Criteria

To conclude the screening of the literature, the study checked the literature regarding availability. Moreover, the study applied snowballing to identify further articles from the bibliographies, in order to gain a wider spectrum of potentially relevant publications. To specify and structure the findings, a review protocol was generated (Table 3). The categories for the analysis include the following domains:

1. Bibliographic data: year of publication, geographic origin of research, land of publication and journal, authors, and title;

2. Approach of the publication: research design, data collection method, and data analysis method;

3. Issues of social performance: following the structure of Wood's model, focusing on observable patterns (excluding principles of social responsibility) and snowballing and synthesizing the findings.

Table 3. Review Protocol.

1. Bibliographic Data	
Author(s)	Who is/are the author(s) of the publication?
Year	In which year was the work published?
Title	What is the title of the publication?
Country	Which country does the publication focus on?
Journal name	In which journal was the publication?

Table 3. *Cont.*

2. Approach of the Publication	
Research design	Which research design was used?
Data collection method	How was the data collected?
Data analysis method	How was the data analyzed?
3. Issues of Social Performance	
Processes of Social Responsiveness	
Env. assessment	What factors influence the organization's actions?
Stakeh. management	How does the organization manage relevant interest groups?
Issues management	How does the organization manage distinct concerns?
4. Outcomes of Organizational Behavior	
Social Impacts	How do the organization's actions affect society?
Social Programs	How does the organization manage its targets?
Social Policies	How does the organization realize its goals?
5. Normative Approach	
Theoretical background	Which theory is applied in the publication?
Terminological focus	How does the publication understand responsibility/sustainability?

To be assigned to the dimensions of social performance, the statements had to fit into the definition of either the processes of social responsiveness or outcomes of social behavior.

In the case of processes, the structure splits further, into environmental assessment, stakeholder management, and issues management. Furthermore, the dimension of outcomes consists of the divisions between social impacts, social programs, and social policies. The coding was left open deliberately, to cover a large spectrum of publications on social performance and responsible management in HEIs, and to capture perspectives arguing from different theoretical foundations. The coding was undertaken by one member of the research team and double-checked for intercoder reliability within a PhD workshop to avoid bias and suggestive defamation in the results of the analysis. The application of Wood's model for the content analysis delivers useful results on managerial aspects, even with the exclusion of principles due to their non-observable character.

4. Results

The following section is structured as a bibliographic analysis and a content analysis, inspired mostly by Schaltegger and Wagner [39] and Seuring and Gold [40]. An overview of the procedure is summarized in Table 4.

4.1. Bibliographic Analysis

By screening full texts under the application of selection requirements and an availability check with the additional screening of sources within the publications, the review finally identified 50 (45 by systematic review, five by snowballing) relevant studies for inquiry. A detailed itemization is provided in the Appendix A.

Table 4. Selection process of included studies.

1st step: Applying search terms on title, topics and subject terms
26,295 results
↓
2nd step: Filtering topics, titles and abstracts on fit with research question
4327 results
↓
3rd step: Check for availability and full text screening
59 results
↓
4th step: Full text review and snowballing
50 results

Looking into the distribution of articles over time (see Figure 1), the first appearance of relevant publications on topics of social responsibility in HEIs occurred in the early 2000s, which might indicate an association with the draft of the millennium development goals. This might also be assumed for the surge of publications since the year 2016, after the inception of the sustainable development goals.

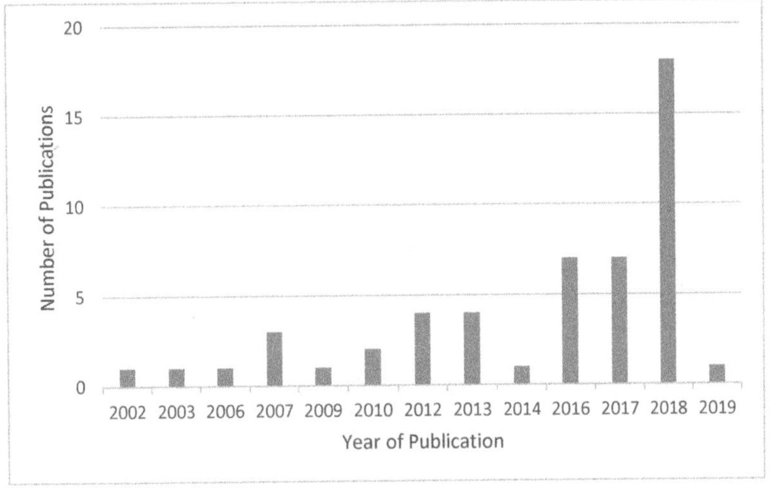

Figure 1. Distribution of Publications over time.

The majority of articles apply case studies for research design (75%), followed by systematic literature reviews (25%). Looking at the global distribution, the publications' origins crystalize around distinct geographical regions. The main focus of research is located in Europe (45%), followed by North America (27%) and Asia (17%). Looking deeper into the publications with a European origin, the study observes a publication focus in middle and eastern Europe (Germany, Austria, Poland, Latvia, Romania, Slovenia). Apart from that, Spain and the United Kingdom show the highest rate of publications.

Taking a closer look at the journals addressing social issues at HEIs, the majority of articles originate from the "Journal of Cleaner Production" (21%) and the "International Journal of Sustainability in Higher Education" (18%). The publications from both journals do not originate from a special issue. The remaining publications are scattered across various journals, thematically ranging from Management and Strategy to Public Administration and Organization topics. Table 5 provides a list of included journals and their methodological quality according to the German "Verband der Deutschen Hochschullehrer" (VHB) and the "Australian Business Deans Council" (ABDC) as valuable rankings from the Harzing Journal Quality List.

Table 5. Methodological quality of included journals.

Journal Name	№ of Results	VHB	ABDC
Abasyn Journal of Social Sciences	1	/	/
Australian Journal of Public Administration	1	/	A
Bulletin of the Transilvania University of Braşov Series V: Economic Sciences	1	/	/
Business and Professional Ethics Journal	1	/	/
Business and Society Review	1	C	C
Environment and Planning C: Politics and Space	1	/	/

Table 5. *Cont.*

Journal Name	№ of Results	VHB	ABDC
Global Business and Management Research: An International Journal	1	/	/
Higher Education	2	/	A
International Journal of Public Leadership	1	/	/
International Journal of Sustainability in Higher Education	9	/	/
Journal of Business Management	2	/	C
Journal of Cleaner Production	11	B	/
Journal of Contemporary Management Issues	1	/	/
Journal of Organizational Learning and Leadership	1	/	/
New Directions for Institutional Research	2	/	/
Procedia Environmental Sciences	1	/	/
Procedia Social and Behavioral Sciences	1	/	/
Public Administration Review	1	B	A
Research Papers of Wroclaw University of Economics	2	/	/
Resources, Conservation and Recycling	1	/	/
Social Responsibility Journal	1	/	B
Strategic Organization	1	B	A
Studies and Scientific Researches—Economic Edition	1	/	/
Studies in Higher Education	1	/	A
Sustainability	1	C	/
Sustainable Development	1	C	C
Tertiary Education and Management	1	/	/
Transylvanian Review of Administrative Sciences	1	/	/

In addition to this, the methodological quality of the journals should be examined in more detail. According to Cook and Campbell [41], methodological quality is defined along distinct criteria: statistical conclusion validity, internal validity, construct validity and external validity. The present study follows these criteria and adapts them, as Cepeda and Martin (2005) propose a similar approach for the assessment of case studies [42]. Since a large number of the literature examined is based on case studies, this provides a suitable instrument for the following investigation to assess methodological quality (see Table 6)

Following Cepeda and Martin's approach, methodological quality is composed of internal validity, construct validity, external validity and reliability. These criteria aim at capturing the conceptual framework, research cycle and theory building along the stages of planning, data collection, data analysis and critical analysis [42].

Internal validity examines descriptions of facts and findings, and their internal coherency with the intended objectives.

Construct validity refers to the explicit and detailed description of methods and procedures, including background information on presumed cause and effects.

External validity embraces descriptions, which allows for the assessment of replicability, and evaluates the appropriateness of applied theories and findings.

Reliability evaluates whether the research question is clearly outlined, and the study design is congruent with it.

According to this definition, the following table provides an overview of the methodological quality of reviewed literature.

Table 6. Methodological quality criteria in reviewed literature following Cepeda and Martin (2005).

Study	Author's Interpretation			
	Internal Validity	Construct Validity	External Validity	Reliability
(Adams, Martin, and Boom, 2018)	✓	✓	✓	✓
(Ahmad, 2012)	✓	✓	✓	✓
(Akins, Bright, Brunson, and Wortham, 2013)	unclear	✓	✓	unclear
(Anderson, Ndalamba, and Caldwell, 2017)	✓	✓	✓	✓
(Asrar-ul-Haq, Kuchinke, and Iqbal, 2017)	✓	✓	✓	✓
(Ayala-Rodríguez, Barreto, Rozas Ossandón, Castro, and Moreno, 2019)	✓	✓	✓	✓
(Bacow and Moomaw, 2007)	✓	unclear	unclear	✓
(Bice and Coates, 2016)	✓	unclear	✓	✓
(Bieler and McKenzie, 2017)	✓	✓	✓	✓
(Buchta, Jakubiak, Skiert, and Wilczewski, 2018)	✓	✓	✓	✓
(Casarejos, Frota, and Gustavson, 2017)	✓	✓	✓	✓
(Cichowicz and Nowak, 2018)	✓	unclear	✓	✓
(Comm and Mathaisel, 2003)	✓	✓	✓	✓
(Dima, Vasilache, Ghinea, and Agoston, 2013)	✓	✓	✓	✓
(Dixon and Coy, 2007)	✓	✓	✓	✓
(Ferrero-Ferrero, Fernández-Izquierdo, Muñoz-Torres, and Bellés-Colomer, 2018)	✓	✓	✓	✓
(Friman et al., 2018)	✓	✓	✓	✓
(Gamage and Sciulli, 2017)	✓	✓	✓	✓
(Gulavani, S., Nayak, N., and Nayak, M., 2016)	×	✓	✓	✓
(Hayter and Cahoy, 2018)	✓	✓	✓	✓
(Kim, Sadatsafavi, Medal, and Ostergren, 2018)	✓	✓	✓	✓
(Labanauskis, 2017)	✓	✓	✓	✓
(Leal Filho et al., 2018)	✓	✓	✓	✓
(Li, Gu, and Liu, 2018)	✓	✓	✓	✓
(Link, 2007)	✓	✓	unclear	unclear
(Lopez and Martin, 2018)	✓	✓	✓	✓
(Lozano, 2006)	✓	✓	✓	✓
(Lukman, Krajnc, and Glavič, 2009)	✓	✓	✓	✓
(Malandrakis, Panaras, and Papadopoulou, 2017)	✓	✓	✓	unclear
(Marinescu, Toma, and Constantin, 2010)	✓	✓	✓	unclear
(Mohamad et al., 2018)	✓	✓	✓	✓
(Mosier and Ruxton, 2018)	✓	✓	✓	✓
(Murray, 2018)	✓	✓	✓	✓
(Nadeem and Kakakhe, 2012)	✓	✓	✓	unclear
(Nejati and Nejati, 2013)	✓	✓	✓	✓
(Pearce, Wood, and Wassenaar, 2018)	✓	✓	✓	✓
(Popescu, M. and Beleaua, I.C., n.d.)	✓	unclear	✓	unclear
(Popović and Nedelko, 2018)	✓	✓	✓	✓
(Sassen and Azizi, 2018)	✓	✓	✓	✓
(Schaffhauser-Linzatti and Ossmann, 2018)	✓	✓	✓	✓
(Sedlacek, 2013)	✓	✓	✓	✓
(Sepasi, Rahdari, and Rexhepi, 2018)	✓	✓	✓	✓
(Shriberg, 2002)	✓	✓	✓	✓
(Turan, Cetinkaya, and Ustun, 2016)	✓	✓	✓	✓
(Vasilescu, Barna, Epure, and Baicu, 2010)	✓	✓	unclear	unclear
(Vaughter, McKenzie, Lidstone, and Wright, 2016)	✓	✓	✓	✓
(Wigmore-Álvarez and Ruiz-Lozano, 2012)	✓	unclear	✓	✓
(Yáñez, Uruburu, Moreno, and Lumbreras, 2019)	✓	✓	✓	✓
(Zahid, Ghazali, and Rahman, 2017)	✓	✓	✓	✓
(Zorio-Grima, Sierra-García, and Garcia-Benau, 2018)	✓	✓	✓	✓

4.2. Content Analysis

For the analysis of the publications' content, the study applies the model of social performance after Wood. The focus on processes of social responsiveness and outcomes of social behavior make up the core of the analysis. The presence or absence of social performance indicators, especially in the dimension of outcomes of social behavior, allows conclusions to be drawn on the general management of responsibility within HEIs.

Before looking into the dimensions of social performance, the first issue of interest was to determine the terms in use when discussing social performance. Do the publications apply the term 'sustainability' or 'responsibility' to describe the actions of the organizations in their study? The results show that a vast majority (65%) applies the label 'sustainability', whereas 25% apply 'responsibility', when referring to issues of responsible management or social performance. Ten percent of publications apply both terms interchangeably. Among them, seven articles explicitly concern university social responsibility.

For the investigation of processes of social responsiveness, the study adapts the classification process proposed by Wood.

To evaluate the organizational environment, the study distinguishes between the organizational context in which the HEI is embedded and the institutional approach, which outlines how the institution handles responsibility/sustainability.

The external field in which the higher education organization operates is strongly centered around the different responsibilities they face, ranging from a local to a global level. Casarejos et al. [43] provide an overview of the different matters an HEI has to master. A strong factor acting upon the social responsibility of a university is socio-economic pressure on different levels. This includes the influence of corporations through partnerships, legal requirements/regulation from the government, and megatrends like the demographic development or globalization, challenging the organization to act responsibly. International commitments for sustainability/responsibility, complemented by obligations for disclosure or the availability and efficient use of funding, determine the activities of HEIs.

Shifting the focus to the internal perspective of environmental assessment, the study identifies governance and leadership as core aspects forming the center of contemplation. Leadership operates the responsibility for outcomes, the society, and the world as a whole [44], with strategy, mission, vision, and values defining the point of departure for the commitment and participation of the community. This creates the social license to operate for the organization, as Ayala-Rodriguez et al. [29] point out. Management frameworks and patterns for change develop a culture and understanding for conscious acting, as with the institutional orientation between faculty and administration. Marketing and communication of engagement build a bridge from the internal to the external bonding of the institution. Communication tools (e.g., indicators) finally enable the assessment and benchmarking of efforts.

In the case of stakeholder management, the study examines the question of whether stakeholders play a certain role in the management of social responsibility at HEIs and how their interests are managed by the institution.

A central assignment follows a categorization of the groups of students, staff (academic and non-academic), and administrators [12,20,42]. Students are often perceived as the initiators of engagement, following a bottom-up approach, since they have the opportunity to " ... operate outside traditional decision-making systems and their capability to pressure their universities in ways that employees simply cannot" [45,46] or have the power to attract a certain level of attention through their actions as agents of change. However, students have only a limited time of residence at HEIs and their long-term influence is limited. Faculty and staff act as the long-term campus population and therefore dominate long-term changes [47]. A detailed examination of this interest group reveals the managing staff (presidents, faculty leaders, academic directors, or professors) to be major drivers in terms of sustainable development. This confirms the assumption that top-down approaches are decisive for the implementation of responsibility in HEIs.

In the case of managing stakeholders, HEIs apply a broad spectrum of approaches ranging from (nonfinancial) reporting frameworks on ESG providing unilateral information for the assessment of engagement [10] to the provision of action plans proposing multi-stakeholder management processes [48], shared governance [49], or participation [12,46].

The investigation of issues management, the last aspect of processes of social responsiveness, tackles the management of participatory activities within HEIs. An examination of leadership demands in order to identify relevant attributes seems useful here. Besides individual traits like integrity, respect, or courage, the strategic planning and a sustainable leadership development (embracing teaching and mentoring) which imparts self-reflection, and the empowerment of staff, are considered to be crucial for handling stakeholder demands [46,47]. On the side of procedural issues management, a strategic anchoring within long-term visions affecting the institution's policy and targets positively stimulates responsible leadership processes. As one example, the PDCA cycle is mentioned as an appropriate tool [14].

The analysis provides insights into exchange relationships between HEIs and their environment. Understanding these contextual factors creates a more tangible background from which to gather the organizational approaches to acting socially responsibly as directly observable characteristics. With the examination of perceptible outcomes, the study discloses existing practices reflecting indirectly observable processes, which moreover allows for a verification of the internal consistency of the findings.

Determining community outreach is generally perceived as a difficult task [50,51]. The analysis of impacts on social behavior requires a more detailed proceeding which enables an in-depth analysis. Therefore, the analysis of aspects of organizational philanthropy, impacts on society, and their perception/assessment make up the stages of the examination.

HEIs' impacts are strongly tied to their mission to educate responsible citizens by responding to social needs and adding value to society [4]. Communication and collaboration with the surrounding communities can be seen as an opportunity to strengthen the perception that HEIs are a valuable part of the local community, because they improve the quality of life.

Organizational philanthropy describes the institution's efforts to positively influence its environment. Although HEIs play a key role in society, evidence regarding their direct involvement in the surrounding community focuses on rather general positions. Providing social welfare and synergies with local partners from business or government, or positioning the campus as a living laboratory, describe ways in which they positively influence the local community [52].

The assessment of HEIs' social performance follows the application of measures to quantify efforts and results. Various measurement tools for the university context enable a pursuit of these objectives. A variety of tools, ranging from GRI, SSR, AASHE, STARS to the Campus Assessment Framework or the Sustainability Pathways Toolkit, provide ample opportunity for the assessment and communication of engagement and community outreach. In terms of distinct tools for the measurement of social performance, the possibilities are more limited [13]. Since most indicators are embedded in an overarching sustainability assessment scheme, environmental aspects dominate reporting standards or are prioritized. Current indicators firmly referring to social performance and responsible management include research compliance (e.g., anti-corruption, respect of privacy or outcome-responsibility), human resource management topics (health, training, diversity, satisfaction), human rights (anti-discrimination, child and forced labor), or education issues (curriculum content, teaching quality, graduation rates).

The examination shows that, so far, no reporting tool serves as a standard for assessing social performance. In the context of assessment tools, (environmental) sustainability issues seem to be a managerial fig leaf for HEIs to shrink from their (social) responsibility.

Notwithstanding, even without assessment tools of social performance, it is possible to be sustainable and act responsibly as an institution of higher education [11].

The social programs of HEIs describe the process perspective in the pursuit of certain goals. For a detailed examination, the research on this topic is divided into social (performance) objectives and management perspectives of processes, to enable a better overview of the field.

In terms of organizational responsibility, major targets include issuing holistic approaches fostering quality management or society outreach to satisfy stakeholder needs and tackle social problems [19,50]. As there is a lack of mission statements embracing responsibility issues [53,54], the use of blanket statements [51] fostering the public image of the organization [53] suggests the need for more responsible management. The evolution of mission statements simultaneously calls for an investigation of managerial processes [55] to achieve better governance and leadership [48]. As a driver of institutionalization, processes depend on the selection of objectives before initiating collaborative structures fostering commitment and comprehension [7] in order to overcome barriers and trade-offs [56]. These integrative efforts lead to an improved understanding and organizational culture of holistic and systematic change [44].

As the last aspect, social policies describe procedures concerning social issues at HEIs, as well as their implementation. The examination shows certain processes often begin in the course of informal activities [20] leading to an incremental incorporation [7], flanked by participatory processes, management frameworks, or guidelines, and driven by top level support, clear objectives, monitoring, and an institutional framing [7]. Mission statements [54] defining a certain policy [57] with transparent objectives, clear processes, monitoring, and communication patterns [58] promote a culture of social responsibility [7]. Human Resource Management functions as a critical success factor by providing the necessary resources [16] and spurring organizational change and commitment.

5. Discussion and Conclusions

The present study examines the social performance of HEIs and the operationalization of responsibility management, and answers the question of how HEIs' social performance management can be quantified within the dimensions of social performance following Wood [18]. The systematic literature review gives insight into the status quo of responsible management practices of HEIs and investigates HEIs' social performance along the dimensions of processes of social responsiveness and outcomes of social behavior. The study examines a structured approach to management activities, which makes it feasible to exercise steering and assessment processes. The results show social performance is an issue of HEIs' responsible management, though the forms and distribution across the examined dimensions vary.

The present study contributes to the discussion on management performance on social issues and responsible management at higher education institutions in three ways.

First, the study provides a systematic overview of different research approaches to responsible management affecting the social performance of HEIs. The study shows that, regarding the management of social issues, there is no common practice. It becomes evident that HEIs predominantly focus their engagement on processes of social responsiveness, especially environmental assessment and stakeholder concerns. This shows a general willingness to perceive social responsibility and act accordingly, but falls short regarding outcomes, especially in the case of determining impact or transferring commitments into actions. This can be explained by the generally low hurdles for HEI management to produce statements on voluntary obligations. Taking action, though, by implementing concrete structures and processes, appears to be of minor interest, since this step has an obligatory character as it requires the allocation of distinct resources with only loosely predictable long-term outcomes. The low proliferation of responsibility management structures and processes suggests a weak institutionalization, which might result from vague objectives, which inhibit to achieve certain goals. Since environmental assessment and stakeholder management appear to be well-elaborated topics, HEIs seem to suffer from a lack of translation mechanisms to integrate these issues into their responsible management.

Second, the study expands on the current literature on responsible management in HEIs by exploring the role of management performance for the implementation and steering of social issues. The application of social performance measures opens up new perspectives for the understanding of aims and conditions of a successful implementation of managing social issues within HEIs.

The abundance of performance information on the features of environmental assessment and stakeholder management, contrasted with the weak expressions of structures and processes, illustrate the need for a systematic management approach. Though the willingness to shoulder social responsibility values is recognizable, implementation so far is limited to a measurement of outcomes.

Third, the large variety of case-based research underlines the demand for a structured approach. The non-standardized use of the terms 'sustainability' and 'responsibility' in the examination on social performance issues opens a general discussion on whether 'responsibility' and 'sustainability' should be treated as complementary or competing terms [56]. Moreover, the subsumption of instances, and of responsibility, under the term 'sustainability', is misleading when talking only about social performance. The versatility of the topic makes it occasionally difficult to assess and manage all dimensions under one management or to supply responsibility from a single source (e.g., different stakeholder claims, functional overlaps of different institutional units and responsibilities).

Based on these findings, the study proposes implications for decision-makers and researchers. First, the assessment of the literature reveals a recent interest in social matters, responsible management issues, strategy implementation, and leadership, as well as the measurement of outcomes at HEIs. So far, ecological sustainability prevails within the scientific discourse and reporting, whereas social (performance) issues play a minor role. Following the triple bottom line, the role of social sustainability has to be emphasized and strengthened in order to be more balanced alongside environmental sustainability. Therefore, a consistent wording and a clear definition of both concepts are crucial to gaining and improving consistency. The synonymous use of the terms 'responsibility' and 'sustainability' mirrors the goal of a harmonization of both terms. The current underrepresentation of social performance issues in HEI management, caused by an unclear ex-ante assessment of consequences entailing voluntary commitment with incalculable future obligations, must be overcome by a clearly defined management construct.

Second, HEIs bear a special responsibility towards their staff members. Besides the fact that the staff constitutes the long-term campus population, their role is often given minor importance compared to other stakeholder groups. The study discovers that participatory approaches and shared governance formats can create a proper framework for the implementation of responsible management affecting social performance. Voluntary engagement has the potential to initiate activities, which can later be transferred into formalized structures and processes. This has the potential to maintain interaction with stakeholders in an ongoing process that reflexively reviews the university's efforts to improve social impacts.

Third, the examination of institutional impacts of universities' social performance provided weak evidence for the application of appropriate tools, as well as for long-term measurements, which deliver a satisfying assertion on engagement. A missing definition might be cause for the diverse understanding of responsible management, which results in the relatively weak impact of actions. So far, existing assessment tools for sustainability at HEIs provide unsatisfactory results.

As in many other surveys, the present study also shows different limitations, providing future research needs. As of now, the impact of social performance seems to be an irrelevant measure for HEIs, since existing assessment tools marginalize social topics [13]. Nevertheless, it is important to mention that a voluntary engagement for social responsibility can pursue various pathways through an organization, even without proper measures. Formal, non-formal, and informal institutionalization efforts can raise new possibilities for further research, especially by differentiating the roles of various stakeholder groups and the meaning of contextual factors for possible implementation strings. In that context, it should be questioned whether principles of social responsibility are in fact non-observable features of an organization or if it is possible to redefine and identify them based on changing practical parameters, communication, and transparency. This would also be useful for a definition of the term 'social responsibility'. Since social responsibility is a diverse field in the institutional context and finding a generally accepted definition is challenging, a proper solution finally has to emerge from the communities negotiating the field of tension between sustainability and responsibility [32]. Along these implications, the research leaves open the question as to which social responsibilities HEIs actually bear in order to fulfill their license to operate as social institutions. Since the study shows that the social impacts of organizational activities are difficult to quantify, a major task for future research will also be to find proper measurement methods with which to solve the question of how HEIs can pursue purpose in line with acting responsibly, and which values should guide them along the way. Further studies could also take up some parts of the results to provide a more significant contribution to the field of responsible management of HEIs. In the end, the major question is whether HEIs will manage to walk the talk, or continue to simply provide lip service regarding their social responsibility.

In short, this paper highlights a model that, by providing accessibility through understanding, enables the active managing of HEIs' social responsibility. The systematic review of the literature shows that HEIs are strongly determined by their mission on research and teaching, which, so far, exceeds other external demands from outside the organization. The strong orientation toward internal affairs emphasizes governance and management topics as relevant influencing variables determining the organizational focus. As a result, HEIs are currently in the process of deepening their engagement with themselves and their mission statement, and forming clearly defined goals, which reflect their social responsibility and have a meaningful effect. Due to a lack of urgency and external pressures, the search for social responsibility beyond research and teaching is difficult. Governance structures and management processes can have a steering effect, in order to proactively shape responsibility and bring it into society along transparent processes.

The implementation of performance measures on social responsibility in HEIs allows their integration into general performance measurement, which strengthens the interconnectedness between economic, environmental, and social issues. Realizing these potentials enables an effective management of social issues and encourages responsible habits, even in fields that have so far evaded measurement (e.g., principles of corporate social responsibility). To become effective operationally, responsible management has to harmonize social performance principles, processes, and outcomes, to create more rooted solutions which balance local and global demands. Managing social performance in this way can help HEIs to close the gap between social values and organizational behavior, and thereby improve their social legitimacy.

Author Contributions: Conceptualization, N.R.; methodology, N.R.; software N.R.; validation, N.R.; formal analysis N.R.; resources, N.R; data curation, N.R.; writing—original draft preparation, N.R; writing—review and editing, N.R.; visualization, N.R.; supervision, N.R.; project administration, N.R.

Acknowledgments: Open Access Funding by the Publication Fund of the TU Dresden.

Conflicts of Interest: The author declares no conflict of interest.

Appendix A

Table A1. Search on EBSCO database.

Date of Search	Source	Search-Terms <AND> Journal Names	Filter	Restrictions	Content Filter Restrictions "NOT"
31 January 2019	EBSCO Academic Search Complete EBSCO Business Source Complete	(sustainab* OR "CSR" OR "soc* resp*" OR "soc* perf*" OR "CSP" OR "SR") AND ("manag* control*" OR "public sector accounting" OR "performance measurement" OR "Public Management" OR "Management Control System*" OR "MCS" OR "Management Accounting" OR "MAS") AND ("high* education" OR universit* OR college* OR campus OR "business school*" OR "HEI*" OR "knowledge intensive organi?ation*")	Title Subject Terms Abstract	English Peer Reviewed Journals and Academic JournalsDate: 1987–2019	Education and Research Teaching and Learning Curriculum affairs Environmental and Green Issues Climate and Carbon Waste Transportation and Mobility Building and Infrastructure
		Results:	25,341	1,674	35

Table A2. Steps on filtering relevant literature on EBSCO search.

Steps		Results
1	Search yield (applying content filter restrictions and excluding duplicates)	35
2	Studies considered for practical screening (information delivered from titles, abstracts, and subject terms; excluding: not available)	27
3	Relevant references found in studies	1
4	Studies considered for methodological screening	28

Table A3. Search on Web of Science database.

Date of Search	Source	Search-Terms <AND> Journal Names	Filter	Restrictions	Content Filter Restrictions "NOT"
4 February 2019	Web of Science	(sustainab* OR "CSR" OR soc* resp* OR soc* perf* OR "CSP" OR "SR") AND (manag* control* OR "public sector accounting" OR "performance measurement" OR "Public Management" OR "Management Control" System* OR "MCS" OR "Management Accounting" OR "MAS") AND (high* "education" OR universit* OR college* OR campus OR "business" school* OR HEI* OR "knowledge intensive" organi?ation*)	Topic (Title, Abstract, Subject Terms)	English Date: 1987–2019	Education and Research Teaching and Learning Curriculum Affairs Environmental and Green Issues Climate and Carbon Waste Transportation and Mobility Building and Infrastructure
		Results:	24,621	2653	24

Table A4. Steps on filtering relevant literature on Web of Science search

Steps		Results
1	Search yield (applying content filter restrictions and excluding duplicates)	24
2	Studies considered for practical screening (information delivered from titles, abstracts, and subject terms; excluding: not available)	18
3	Relevant references found in studies	4
4	Studies considered for methodological screening (excluding: learning, teaching, training)	22

References

1. Amaral, L.P.; Martins, N.; Gouveia, J.B. Quest for a sustainable university: A review. *Int. J. Sustain. High. Educ.* **2015**, *16*, 155–172. [CrossRef]
2. Kneer, G.; Schroer, M. *Handbuch Soziologische Theorien*, 1st ed.; Verlag für Sozialwissenschaften: Wiesbaden, Germany, 2009.
3. Matten, D.; Moon, J. "Implicit" and "Explicit" CSR: A Conceptual Framework for a Comparative erstanding of Corporate Social Responsibility. *Acad. Manag. Rev.* **2008**, *33*, 404–424. [CrossRef]
4. Dima, A.; Vasilache, S.; Ghinea, V.; Agoston, S. A model of academic social responsibility. *Transylv. Rev. Adm. Sci.* **2013**, *38*, 23–43.
5. Schmidt, U.; Günther, T. Public sector accounting research in the higher education sector: A systematic literature review. *Manag. Rev. Q.* **2016**, *66*, 235–265. [CrossRef]
6. Brinkhurst, M.; Rose, P.; Maurice, G.; Ackerman, J.D. Achieving campus sustainability: Top-down, bottom-up, or neither? *Int. J. Sustain. High. Educ.* **2011**, *12*, 338–354. [CrossRef]
7. Lozano, R. Incorporation and institutionalization of SD into universities: Breaking through barriers to change. *J. Clean. Prod.* **2006**, *14*, 787–796. [CrossRef]
8. Nadeem, A.; Kakakhe, S.J. An investigation into Corporate Social Responsibility (CSR) of public sector universities in KPK. *Abasyn J. Soc. Sci.* **2012**, *5*, 14–27.

9. Wigmore-Álvarez, A.; Ruiz-Lozano, M. University Social Responsibility (USR) in the global context: An overview of literature. *Bus. Prof. Ethics J.* **2012**, *31*, 475–498. [CrossRef]

10. Comm, C.L.; Mathaisel, D.F.X. Less is more: A framework for a sustainable university. *Int. J. Sustain. High. Educ.* **2003**, *4*, 314–323. [CrossRef]

11. Leal Filho, W.; Londero Brandli, L.; Becker, D.; Skanavis, C.; Kounani, A.; Sardi, C.; Papaioannidou, D.; Paço, A.; Azeiteiro, U.; de Sousa, L.O.; et al. Sustainable development policies as indicators and pre-conditions for sustainability efforts at universities: Fact or fiction? *Int. J. Sustain. High. Educ.* **2018**, *19*, 85–113. [CrossRef]

12. Link, T. Social indicators of sustainable progress for higher education. *New Dir. Inst. Res.* **2007**, *134*, 71–82. [CrossRef]

13. Sassen, R.; Azizi, L. Assessing sustainability reports of US universities. *Int. J. Sustain. High. Educ.* **2018**, *19*, 1158–1184. [CrossRef]

14. Labanauskis, R. Key features of sustainable universities: A literature review. *J. Bus. Manag.* **2017**, *2017*, 56–69.

15. Li, Y.; Gu, Y.; Liu, C. Prioritising performance indicators for sustainable construction and development of university campuses using an integrated assessment approach. *J. Clean. Prod.* **2018**, *202*, 959–968. [CrossRef]

16. Popescu, M.; Beleaua, I.C. Improving management of sustainable development in universities. *Bull. Transilv. Univ. Brasov Ser. V Econ. Sci.* **2014**, *7*, 97–106.

17. Popović, T.; Nedelko, Z. Social responsibility and strategic orientation of higher education—Cases of Croatia and Bosnia and Herzegovina. *Manag. J. Contemp. Manag. Issues* **2018**, *23*, 123–140. [CrossRef]

18. Wood, D.J. Corporate social performance revisited. *Acad. Manage. Rev.* **1991**, *16*, 691–718. [CrossRef]

19. Speklé, R.F.; Widener, S.K. Challenging issues in survey research: Discussion and suggestions. *J. Manag. Account. Res.* **2018**, *30*, 3–21. [CrossRef]

20. Cichowicz, E.; Nowak, A.K. Social responsibility of an organizationfrom the perspective of public universities. *Pr. Nauk. Uniw. Ekon. Wrocławiu* **2018**, *520*, 34–45. [CrossRef]

21. Gamage, P.; Sciulli, N. Sustainability reporting by Australian universities. *Aust. J. Public Adm.* **2017**, *76*, 187–203. [CrossRef]

22. Marinescu, P.; Toma, S.; Constantin, I. Social responsibility at the academic level. Study case: The University of Bucharest. *Stud. Sci. Res. Econ. Ed.* **2010**, *15*, 404–410. [CrossRef]

23. Buchta, K.; Jakubiak, M.; Skiert, M.; Wilczewski, A. University social responsibility—Theory vs. practice. *Pr. Nauk. Uniw. Ekon. Wrocławiu* **2018**, *520*, 22–33. [CrossRef]

24. Disterheft, A.; Caeiro, S.; Azeiteiro, U.M. Sustainable universities – a study of critical success factors for participatory approaches. *J. Clean. Prod.* **2015**, *106*, 11–21. [CrossRef]

25. Dahlsrud, A. How corporate social responsibility is defined: An analysis of 37 definitions. *Corp. Soc. Responsib. Environ. Manag.* **2008**, *15*, 1–13. [CrossRef]

26. Taneja, S.S.; Taneja, P.K.; Gupta, R.K. Researches in corporate social responsibility: A review of shifting focus, paradigms, and methodologies. *J. Bus. Ethics* **2011**, *101*, 343–364. [CrossRef]

27. Turker, D. Measuring corporate social responsibility: A scale development study. *J. Bus. Ethics* **2009**, *85*, 411–427. [CrossRef]

28. Vasilescu, R.; Barna, C.; Epure, M.; Baicu, C. Developing university social responsibility: A model for the challenges of the new civil society. *Procedia—Soc. Behav. Sci.* **2010**, *2*, 4177–4182. [CrossRef]

29. Ayala-Rodríguez, N.; Barreto, I.; Ossandón, G.R.; Castro, A.; Moreno, S. Social transcultural representations about the concept of university social responsibility. *Stud. High. Educ.* **2019**, *44*, 245–259. [CrossRef]

30. Othman, R. Higher education institutions and social performance: Evidence from public and private universities. *Int. J. Bus. Soc.* **2014**, *15*, 1–18.

31. Reiser, J. Managing University Social Responsibility (USR). In Proceedings of the International Sustainable Campus Network: Best Practices—Future Challenges, Zurich, Switzerland, 25–27 April 2007; p. 2.

32. Bansal, P.; Song, H.-C. Similar but not the same: Differentiating corporate sustainability from corporate responsibility. *Ann. Acad. Manag.* **2017**, *11*, 105–149. [CrossRef]

33. Carroll, A.B. A three-dimensional conceptual model of corporate performance. *Acad. Manage. Rev.* **1979**, *4*, 497–505. [CrossRef]

34. Wartick, S.L.; Cochran, P.L. The evolution of the corporate social performance model. *Acad. Manage. Rev.* **1985**, *10*, 758–769. [CrossRef]

35. Freeman, R.E. *Strategic Management: A Stakeholder Approach, Reissue*; Cambridge University Press: Cambridge, UK, 2010.

36. Wood, D.J. Measuring corporate social performance: A review. *Int. J. Manag. Rev.* **2010**, *12*, 50–84. [CrossRef]

37. Fink, A. *Conducting Research Literature Reviews: From the Internet to Paper*; SAGE: Thousand Oaks, CA, USA, 2005.

38. Tranfield, D.; Denyer, D.; Smart, P. Towards a methodology for developing evidence-informed management knowledge by means of systematic review. *Br. J. Manag.* **2003**, *14*, 207–222. [CrossRef]

39. Schaltegger, S.; Wagner, M. *Managing the Business Case for Sustainability: The Integration of Social, Environmental and Economic Performance*, 1st ed.; Routledge: Abingdon, UK, 2017.

40. Seuring, S.; Gold, S. Conducting content-analysis based literature reviews in supply chain management. *Supply Chain Manag. Int. J.* **2012**, *17*, 544–555. [CrossRef]

41. Hyman, R. Quasi-experimentation: Design and analysis issues for field settings (book). *J. Personal. Assess.* **1982**, *46*, 96–97. [CrossRef]

42. Cepeda, G.; Martin, D. A review of case studies publishing in Management Decision 2003–2004: Guides and criteria for achieving quality in qualitative research. *Manag. Decis.* **2005**, *43*, 851–876. [CrossRef]

43. Casarejos, F.; Frota, M.N.; Gustavson, L.M. Higher education institutions: A strategy towards sustainability. *Int. J. Sustain. High. Educ.* **2017**, *18*, 995–1017. [CrossRef]

44. Shriberg, M. Toward sustainable management: The University of Michigan Housing Division's approach. *J. Clean. Prod.* **2002**, *10*, 41–45. [CrossRef]

45. Atherton, A.; Giurco, D. Campus sustainability: Climate change, transport and paper reduction. *Int. J. Sustain. High. Educ.* **2011**, *12*, 269–279. [CrossRef]

46. Murray, J. Student-led action for sustainability in higher education: A literature review. *Int. J. Sustain. High. Educ.* **2018**, *19*, 1095–1110. [CrossRef]

47. Mohamad, Z.F.; Abd Kadir, S.N.; Nasaruddin, A.; Sakai, N.; Zukia, F.M.; Husseinae, H.; Sulaiman, A.H.; Salleha, M.S.A.M. Heartware as a driver for campus sustainability: Insights from an action-oriented exploratory case study. *J. Clean. Prod.* **2018**, *196*, 1086–1096. [CrossRef]

48. Sedlacek, S. The role of universities in fostering sustainable development at the regional level. *J. Clean. Prod.* **2013**, *48*, 74–84. [CrossRef]

49. Pearce, C.L.; Wood, B.G.; Wassenaar, C.L. The Future of Leadership in Public Universities: Is Shared Leadership the Answer? *Public Adm. Rev.* **2018**, *78*, 640–644. [CrossRef]

50. Akins, R.; Bright, B.; Brunson, T.; Wortham, W. Effective Leadership for Sustainable Development. *E J. Organ. Learn. Leadersh.* **2013**, *11*, 29–36.

51. Bieler, A.; McKenzie, M. Strategic planning for sustainability in Canadian higher education. *Sustainability* **2017**, *9*, 161. [CrossRef]

52. Mosier, S.; Ruxton, M. Sustainability university–community partnerships: Lessons for practitioners and scholars from highly sustainable communities. *Environ. Plan. C Polit. Space* **2018**, *36*, 479–495. [CrossRef]

53. Ahmad, J. Can a university act as a corporate social responsibility (CSR) driver? An analysis. *Soc. Responsib. J.* **2012**, *8*, 77–86. [CrossRef]

54. Lopez, Y.P.; Martin, W.F. University Mission Statements and Sustainability Performance: Business and society review. *Bus. Soc. Rev.* **2018**, *123*, 341–368. [CrossRef]

55. Lukman, R.; Krajnc, D.; Glavič, P. Fostering collaboration between universities regarding regional sustainability initiatives—The University of Maribor. *J. Clean. Prod.* **2009**, *17*, 1143–1153. [CrossRef]

56. Ferrero-Ferrero, I.; Fernández-Izquierdo, M.Á.; Muñoz-Torres, M.J.; Bellés-Colomer, L. Stakeholder engagement in sustainability reporting in higher education: An analysis of key internal stakeholders' expectations. *Int. J. Sustain. High. Educ.* **2018**, *19*, 313–336. [CrossRef]

57. Vaughter, P.; McKenzie, M.; Lidstone, L.; Wright, T. Campus sustainability governance in Canada: A content analysis of post-secondary institutions' sustainability policies. *Int. J. Sustain. High. Educ.* **2016**, *17*, 16–39. [CrossRef]

58. Kim, A.A.; Sadatsafavi, H.; Medal, L.; Ostergren, M.J. Impact of communication sources for achieving campus sustainability. *Resour. Conserv. Recycl.* **2018**, *139*, 366–376. [CrossRef]

 *sustainability*

Article

Sustainability Assessment and Benchmarking in Higher Education Institutions—A Critical Reflection

Sandra Caeiro [1,2,*], **Leyla Angélica Sandoval Hamón** [3], **Rute Martins** [1] **and Cecilia Elizabeth Bayas Aldaz** [3]

1   Department of Science and Technology, Universidade Aberta, 1269-001 Lisbon, Portugal; amrfilipe@fmh.ulisboa.pt
2   Center for Environmental and Sustainability Research, NOVA School of Science and Technology, NOVA University Lisbon, 2829-516 Caparica, Portugal
3   Department of Business Organization, Universidad Autónoma de Madrid, 28049 Madrid, Spain; angelica.sandoval@uam.es (L.A.S.H.); cecilia.bayas@uam.es (C.E.B.A.)
*   Correspondence: scaeiro@uab.pt

Received: 13 November 2019; Accepted: 7 January 2020; Published: 10 January 2020

Abstract: Higher Education Institutions (HEIs) play a crucial role in implementing practices for Education for Sustainable Development (ESD). This implementation should be done in different dimensions according to a holistic and whole-school approach. Different tools have been adapted and developed to assess this integrated approach. The aim of this research is to critically reflect the existing tools to assess and benchmark ESD implementation and to discuss their applicability in two case studies. Two public Universities in Southern Europe, with headquarters in the capitals of Portugal and Spain were selected to assess and compare the integration of ESD according to a whole-school approach—Universidade Aberta in Portugal and Universidad Autónoma de Madrid in Spain. After a critical analysis of the existing tools based on literature review and a list of criteria classified by experts, two tools were selected to be applied in the case studies. The online Sustainability Tracking, Assessment & Rating System Reporting Tool was used in Universidade Aberta and Green Metrics tool was used in Universidad Autónoma de Madrid. The tools were complemented with focus group with key-actors in both universities. The results obtained allowed to identify the need to define a common objective of the assessment tools and limitations they still have. The tools need improvements on their development namely to integrate the external impact of Higher Education Institutions on sustainability, to integrate participatory processes and to assess non-traditional aspects of sustainability. This research hopes to contribute to the continuous research about the usefulness of these assessment and benchmarking tools as drivers to HEIs improve their sustainability performance and their role as agents of changes.

Keywords: higher education institutions; sustainability benchmarking; sustainability tracking; assessment and rating systemTM; green metrics; education for sustainability

---

1. Introduction

Higher Education Institutions (HEIs) have a critical responsibility in education for sustainable development (ESD) due to "the main component for raising awareness of SD among the population" [1]. In other words, ESD provides the knowledge and skills for students to start to create SD initiatives. Furthermore, HEIs should "lead by example" [2].

Sustainable development as a development model integrates environmental, social and economic considerations [3] and so HEIs must assume a holistic focus in all activities. Indeed, HEI with or without governmental policies and recommendations are moving towards holistic and systemic approaches when addressing Education for Sustainable Development. The Sustainable Development Goals (SDG), recently adopted by United Nations in 2015 [4] and in particular, SDG 4, are an additional driver for the implementation of sustainability at HEI in an integrative way.

According to the United Nations guidelines followed by several researchers the integrative approach to implement sustainability in HEIs includes six major ESD dimensions to allow a whole-school approach—(i) Facilities or Operations; (ii) Teaching and Curriculum; (iii) Organizational Management; (iv) External Community; (v) Research; (vi) Assessment and Communication (based on [5,6]).

Based on early work [7,8], some authors stress that the integration of sustainability in HEIs can be done at various levels, from national to the institution level [9]. The institutions that integrate sustainability according to the whole-school approach (in terms of dimensions and involvement of the whole HEIs community) can achieve the sustainability maturity curve, thus enabling them to be agents of change and transformation. In this context, Kapitulcinová [9] introduced the model from a "business-as-usual university" to a "sustainable university" where sustainability has been fully integrated under three degrees, initiation/awakening, implementation/pioneering and institutionalization/transformation, including in this last stage a consolidation of changes. Thus, the HEIs are under mounting pressure to partner with societal stakeholders and organizations to collaboratively create and implement sustainability-advancing knowledge, tools and societal transformations [10]. So, in recent years, an increasing number of institutions have begun the adjustment and restructuring of education, research, campus operations and community outreach towards sustainability [8,11,12]. However, deep and full integration of sustainable development at all of dimensions and community of HEIs is still lacking [13,14].

Different tools have been developed to assess and benchmark ESD implementation at HEIs but how well are being performed in case studies and how well they evaluate sustainability and its impact is still an open question. This study attempts to provide a holistic sustainability maturation path reflecting essential dimensions that HEIs need to approach and assess as whole-school integration towards a potential sustainability management change. The aim of this research is, in the first part, to critically analyze the existing tools to assess and benchmark ESD implementation in HEIs. In the second part of the article, the aim is to assess the integration of ESD according to a whole-school approach through the use of those tools in the case studies and discuss their applicability. Two of the analyzed tools were selected and applied in two HEIs in Portugal and Spain. The assessment was complemented with stakeholder's engagement. European HEIs have been ahead in the implementation of sustainability at the different dimensions [13,15]. In particular, in southern Europe neighbor countries like Portugal and Spain, despite the lack of national policies, HEIs are working towards ESD implementation through the development of plans and actions in sectorial areas. The two case studies are public Universities (one smaller and in distance learning regime and others in traditional learning and bigger) engaged in the last years in integrating sustainability in different dimensions. They have their headquarters in the capitals of Portugal and Spain and are classified in levels of excellence in their teaching and learning methods: Universidade Aberta, Portugal and Universidad Autónoma de Madrid in Spain.

This article starts after this Introduction with a literature review about the tools to assess sustainability in HEI (Section 2). Section 3 presents the methods for the tool's assessment and case studies evaluation. Section 4 describes the case studies, Universidade Aberta and Universidad Autónoma de Madrid. Section 5 discuss the results and the final section is dedicated to final conclusions and contribution to new knowledge and research implications.

2. Sustainability Assessment in Higher Education

Given the rapid growth of ESD initiatives at HEIs, the measures, assessment and reports of HEIs progress toward SDG has become increasingly important [16]. The assessment of sustainability in HEIs is one of the most important dimensions of EDS implementation in HEIs and could be conducted base on specific tools, that allows to assess whether all possible dimensions to the implementation of sustainability are being implemented and whether they are doing so holistically [9]. Also, when well-developed these tools can be used as benchmarking practices, comparing HEIs processes and performance metrics. Nevertheless, these tools must be able to uniformly evaluate the implementation of sustainability in HEIs without necessarily placing them in a ranking or competition. So, the tools must identify the important themes, be measurable and comparable, go beyond eco-efficiency, measure progress and motivations and be understandable to a broad set of key actors [17]. According to other authors [18], these tools can be based on indicators and conceptual models that support sustainability decisions, as well as facilitating communication efficiently and for a wide audience, knowing how to respond to complex processes capable of assessing the transformation for sustainability.

Much work has been done on the development of tools to assess sustainability specifically in HEIs, showing the importance of the theme. Several articles have reviewed these tools from different perspectives (e.g., [7,9,14,16,17,19–33]). However, it is observed that in these various reviews, the systematization is not homogeneous, noting a lack of common designations and objectives and including tools that do not have as main objective the assessment of the implementation of ESD or are not per se an assessment tool. So, for example, the reviewed tools:

(i)   assess only one of the dimensions of implementing sustainability in HEIs such as Campus operations (e.g., Campus Sustainability Assessment Framework-CSAF, [20]; National Wildlife Federation's State of the Campus, [34]) or Curricula (e.g., Sustainability Tool For Assessing Universities' Curricula Holistically, STAUNCH, [11]);

(ii)   evaluate only one pillar of sustainability, namely Environmental (e.g., Campus Ecology, [35], Environmental Performance Survey, [34]);

(iii)   only serve as manuals or supportive conceptual models (e.g., Greening Campuses, [36]);

(iv)   assess only the level of literacy and knowledge of population sustainability (e.g., SULITEST, [37]);

(v)   adapt by an Institution but based on other existing metrics (e.g., UNI-Metrics—Value Metrics and Policies for a Sustainable University Campus, [31]);

(vi)   specific to a type of HEI (e.g., Business School Impact System BSIS, see Reference [14]);

(vii)   serve only as guidelines for supporting communication of performance for sustainability but are not themselves an assessment tool. This is the case of the guidelines developed by the International Campus Sustainability Network and Global University Leader Forum [38], two international networks that suggest the organization of sustainability reports for HEIs, based on GRI indicators and the Sustainability Tracking tool, Assessment & Rating System—STARS [38]. This is also the case of the work developed by Nixon [19] who developed other guidelines for the implementation of sustainability in HEIs (under the CSARP project "Campus Sustainability Assessment Review Project" developed at Western Michigan University), in its different dimensions and practices, based on a literature review of existing tools, proposing no new tool.

Thus, for this study a systematic review of the tool for sustainability implementation assessment in HEIs was conducted, based on the following conditions:

(a)   developed specifically for assessing the performance of sustainability implementation in HEIs;

(b)   covering at least two of the various dimensions of sustainability implementation in HEIs ((i) facilities or campus operations; (ii) teaching and curriculum; (iii) organizational management; (iv) external community; (v) research; (vi) assessment and communication;

(c)   covering at least two of the sustainability pillars (environmental, social. and economic), to guarantee that the tools in some way are based on a holistic and whole-school approach and since most are based on only two of the pillars.

Based on these criteria 27 tools were searched on google scholar using the key-words: "Sustainability assessment" and "Higher Education," from October 2018 and March 2019. Each tool was then characterized (see Table 1).

Table 1. Sustainability Assessment Tools in Higher Education (HE) and brief description.

Tool and Source	Brief Description and Region Where Applicable
AISHE 2.0 Assessment Instrument for Sustainability in Higher Education. (Latest version) AISHE 2.0 Assessment Instrument for Sustainability in Higher Education [39,40]	Based on narrative and indicators: 30 indicators, 5 dimensions (Operations, Education, Research, Society, Identity), less emphasis on the environmental component (just 1 indicator); incorporating the Deming cycle approach; the intended target is the university system; with a wide world application across the universities; the application domain adapts according to the university structure as an entire university, campus, buildings or research institute; AISHE first version was developed in 2000 and 2001 only focused in the educational role of universities, however, AISHE 2.0 has a wider scope in terms of the research, operations and relation with the society; Developed by a researcher in Europe; Tool not available online, only the manual. The Netherlands / international
AMAS—Adaptable Model for Assessing, Sustainability in Higher Education [29]	Based on 3 domains (Institutional commitment, Leadership, Advanced sustainability); with 4 levels of application hierarchy that lead to the use of standardized indicators (based on other existing tools), with different weights and key actors' participation, allowing to be adapted by each institution but comparable in the same country; With an expert consultation system; Developed by a researcher in Chile; Tool not available online. Chile
ASSC—Assessment System for Sustainable Campus [41]	Based on a questionnaire and reported on graphical form; 26 indicators; 4 dimensions (Management, Education and Research, Environment, Local Community); based on other tools (STARS, Uni-metrics, GM, BIQ - AUA). Rating system with 4 levels, allowing to obtain a certification: platinum, gold, silver and bronze; It provides information on the strengths and weaknesses of implementing sustainability in HEIs and helps them decide future strategies; includes specificities of the country where it was developed (e.g., natural disasters); Developed by Hokkaido University in 2013, within CAS-NET JAPAN (Campus Sustainability Network in Japan) but used in other universities in Japan; Tool available online, https://www.osc.hokudai.ac.jp/en/action/assc (see Reference [41]). Japan
AUSP—Evaluación de las políticas universitarias de sostenibilidad como facilitadoras para el desarrollo de los campus de excelencia internacional [29,42,43]	Based on 4 areas (Organization, Teaching, Research, Environmental management); less emphasis on the social component; 176 indicators; data collection by questionnaire and interviews (self-assessment) and reviewed by an external organization; the purpose is to improve performance and policies in terms of social responsibility, environment (including public procurement), economy and implementation of the Sustainable Development Goals; graphical representation of indicators; with several updates (last in 2018); Developed by the De La Crue Sectoral Commission for Environmental Quality, Sustainable Development and Risk Prevention of CRUE - Conference of Rectors of Spanish Universities specifically for HEIs in Spain and tested in several Spanish Universities; Questionnaire available online, https://goo.gl/forms/Fol9qwVvYF2juTbC2 (see Reference [42]). Spain

Table 1. *Cont.*

Tool and Source	Brief Description and Region Where Applicable
BIQ-AUA—Alternative University Appraisal [29,44,45]	Based on self-assessment with questions to calculate indicators for benchmark (BIQ) and Dialogue. BIQ has a special focus on governance, education, research and communication. It is divided into 4 categories, 15 sub categories (with equal weight) and includes 30 indicators and 50 questions; does not include environmental management and social responsibility indicators; the highest rating is 100, thus allowing comparison; dialogue is the component that enables institutions to share their concerns, best practices and learning about ESD. Applied to 28 universities in Asia and the Pacific; Developed by ProSPER (Promotion of Sustainability in Postgraduate Education and Research Network), an academic alliance between Asia and the Pacific. It is composed of three components based on the evaluation according to the United Nations Decade for Sustainable Development; Tool not available online Ásian-Pacific
CITE AMB—Red de Ciencia, Tecnologia, Innovacion y Educación Ambienal em Iberoamerica [46]	Based on a questionnaire with 27 questions (Yes/No answers); 4 areas (Management, Research, Education, Community), without focusing on the environmental component of campus infrastructure and social component; Developed by the Network of Science, Technology, Innovation and Environmental Education in Iberoamerica; Not available online (existence of a link in the report for "google docs" but is no longer available), developed in 2014 but with no updates available. Colombia
DUK—German Commission for UNESCO AG HS (2011) fidé [14,28]	Based on indicators in 4 areas (Operations, Research, Education, Community); With a strong focus on the institutional part, the tool operates as moderator in the whole-school approach. It contains 10 action fields and each one offers 5 stages of implementation, allowing the HEIs option; Developed by the German Commission for UNESCO in 2011 for the German context; Tool not available online; only a report about the tool in German is available. Germany
ESDGC—Education for Sustainable Development and Global Citizenship [9,47]	Based on a ranking system with 5 areas (Commitment and leadership, Teaching and learning, Institutional Management, Partnerships, Research and Monitoring) and the categorization of 4 levels; adaptation of a maturity model and training usually applied to companies and the industrial sector; results with a semaphore system; Developed specifically for HEIs in Wales, UK and outlined by the Government of Wales to enable an assessment of implementation of ESD in Universities; Tool not available online. Wales / United Kingdom
GASU—Graphical Assessment of Sustainability in Universities tool [7]	Based on GRI report with adaptations to HEIs; applied in many universities, 8 dimensions (Direct economic impact, Environmental, Labor practices and decent work, Human rights, Society, Product liability, Curricula, Research), up to 126 indicators; graphical presentation of results; Developed by a researcher in Europe and marketed through a company; Tool not available online for free, only with a fee payment. United Kingdom / international

Table 1. *Cont.*

Tool and Source	Brief Description and Region Where Applicable
GC—Good Company's Sustainable Pathways Toolkit [24,48]	Based on 20 key performance indicators plus an additional 10 indicators; more emphasis on campus operations; without focusing 2 categories of sustainability implementation in HEIs, namely, Research and Stakeholder involvement; the purpose is to aid decision support / management and benchmarking; Developed by a US company (Good company) and without support to key experts / actors; Tool not available online, neither report nor update. USA / international
GM—Green Metrics University Ranking [49,50]	Based on 6 domains (Scenario and infrastructure, Energy and climate change, Waste, Water, Transport, Education & Research); 33 indicators, two focus on the environment, no community involvement or other social components; ranking point system allowing benchmarking and comparison; with a wide world application across the universities; Tool available online, http://greenmetric.ui.ac.id (see Reference [50]. Indonesia/international
GMID—Graz Model for Integrative Development [51]	Based on narrative and domains: 5 domains (Leadership, Social Networks, Participation, Education and Learning, Research); applicable but not specific to HEIs; applied to the RCE-an international network of formal, non-formal and informal education organizations -, mobilized to provide ESD to the local and regional community at 3 levels; Developed by a researcher in Europe; Tool not available online. Austria / international
GP—Green Plan and the Label DD&RS ou Plan Vert [52]	Based on 5 domains (Strategy governance, Education and training research, Environmental management, Social policy, Regional presence); 44 indicators; can be audited and certified by internal and external stakeholders concerning the ISO 26000; purpose of assisting in the elaboration of sustainability plans/policies; Developed by Conférence des Grandes Ecoles, Conference of University Presidents, French Government and Non-Governmental Organizations within the Grenelle Environment Roundtable; Tool not available online at the present
HE 21—Higher Education 21's sustainability Indicators or HEPS Higher Education Partnership for Sustainability [24,53]	Based on indicators (12 key indicators and 8 strategic management indicators); focusing mainly on parameters of organizational management change; less emphasis on social indicators and does not encompass in a balanced way all the dimensions of ESD in HEIs (more emphasis on governance); difficult to benchmarking; latest version and network activity in 2003; Developed for 18 universities in the UK that have partnered to support English universities and their monitoring in the implementation of sustainability-HEPS Higher Education Partnership for Sustainability; Tool not available online. United Kingdom
PSIR—Penn State Indicator Report [24,54]	Based on 33 indicators, covering the environmental dimensions of the campus, transport, decision support, research and community; results of each indicator reported in 4 levels of implementation and with proposals for improvement; less emphasis on social indicators and without teaching and curriculum components; last version available in 2000; Developed by the Penn State Green Destiny Council to be applied at US universities, in the State of Pennsylvania and to be communicated to the general public how sustainability is being implemented; Tool not available online, only on the report, http://www.willamette.edu/~{}nboyce/assessment/PennState.pdf (see Reference [54]). USA, Pennsylvania State

Table 1. *Cont.*

Tool and Source	Brief Description and Region Where Applicable
P&P—People & Planet University League [55]	Based on 13 indicators (not divided into dimensions), greater focus on environmental operations and less on community; graphical presentation of results; in operation for several years allowing the annual comparison and an annual ranking; data collection is carried out in the universities' webpages and the UK Higher Education Statistics Agency; Developed by a network of UK students - People & Planet for universities in the UK and tested at various UK universities; Tool available online. https://peopleandplanet.org/university-league (see Reference [55]). United Kingdom
SAQ—Sustainability Assessment Questionnaire [56]	Based on narrative and indicators: 35 indicators, 8 dimensions (Curriculum, Research and scholarship, Operations, Faculty and staff, Extension and services, Student opportunities, Administration, Mission and planning); with greater emphasis on campus operations; presented through a questionnaire addressed to various internal stakeholders; Developed by the secretariat of the signatories of the Tailloires Declaration - Association of University Leaders for a Sustainable Future; Tool available online, http://ulsf.org/sustainability-assessment-questionnaire/ (see Reference [56]). International
SRC—Sustainability Report Card [57]	Based on narrative and indicators: 52 indicators, 5 dimensions (Campus operations, Meal service, Donation investment, Transportation, Involvement of key stakeholders); more focus on energy saving and less emphasis on education; presented through a questionnaire with a final grade from A to D; suspended in 2012; Developed by a North American Non-Governmental Institution - Sustainable Endowments Institute; Tool not available online. USA / Canada
STARS—Sustainability Tracking, Assessment & Rating System [58]	Based on narrative and indicators: 74 indicators, 5 dimensions (Academic, Involvement of key actors, Operations, Planning and Administration, Innovation and leadership); 5 levels of final classification, allowing the ranking (reporter, bronze, silver, gold, platinum); one of the most used tools internationally; updated every year; developed by a North American Non-Governmental Institution-Association for the Advancement of Sustainability in Higher Education and initially developed for HEIs in the US and Canada but applicable to any region; tool available online, https://reports.aashe.org/accounts/login/?next=/tool/ (see Reference [58]). USA / international
SUM—Sustainable University Model [8]	Based on narrative and indicators: 23 indicators, 4 dimensions (Education, Research, Dissemination and partnership, Campus sustainability); divided in 4 phases (Vision development, Mission, Sustainable committee, Audit of sustainability strategies) incorporating the Deming cycle approach; tested at various world universities; without updates; Developed by a researcher in Mexico; Tool not available online. Mexico / international

Table 1. *Cont.*

Tool and Source	Brief Description and Region Where Applicable
SLS—Sustainability Leadership Scorecard [59]	Based on performance indicators; 4 domains (leadership and governance, learning, teaching and research, operations); self-assessment developed specifically for colleges and universities to improve social responsibility and environmental performance through a whole-school approach; final scores with a range from 0–4; no weights in the indicators and final result in a dashboard index; adapted from the Green Scorecard and linked to Sustainable Development Objectives standards; Developed by a Non-Governmental Association of the United Kingdom and Ireland—The Alliance for Sustainability Leadership in Education; Tool available online for free to United Kingdom and Ireland, https://www.sustainabilityleadershipscorecard.org.uk/#!/login (see Reference [59]). United Kingdom / Ireland
SustainTool—Program Sustainable Assessment Tool [60]	Based on indicators, focused on areas / programs or at the institution level; 8 dimensions (Environmental support, Funding stability, Partnership, Organizational capacity, Program, Evaluation, Program adaptation, Communications, Strategic planning) with a low weight in the environmental component; presented in a 40 multiple-choice questions in self-assessment questionnaire, with answers being given individually or in a group; allows the communication, review and development of an action plan; available for several years with updates; Developed by a north American university—Washington University for any university, particularly in the North American context but especially directed to the health area; Tool available online, https://sustaintool.org/assess/ (see Reference [60]). USA / international
THE—Times Higher Education Impact University Ranking [61]	Based on the evaluation of the implementation of the Sustainable Development Objectives (ODS) in HEIs: 11 ODSs: 3,4,5,8,9,10,11,12,13,16, 17; each ODS has a small number of indicators associated with it; equal weight is given to each ODS; first version available for 7 ODS but still in development (1st version April 2019); Developed by the Times; Tool available online, requesting by email, https://www.timeshighereducation.com/how-participate-times-higher-education-rankings (see Reference [61]). International
TUR—Three Dimensional University Ranking [62]	Based on indicators: 15 indicators, weighted based on a participatory process and Analytical Hierarchical Process (AHP), 3 dimensions (Research, Education, Environment); less holistic approach; graphical presentation of results; allows ranking based on rankings of world universities, simplified sustainability only in 5 indicators; tested in the best universities but without updates; Developed by researchers in Europe; Tool not available online. International
UEMS—University Environmental Management System [21]	Based on EMAS / ISO14001 with a social responsibility component and indicators: 27 indicators, 3 dimensions (University EMS, Public participation and social responsibility, Teaching and research in sustainability); greater focus on the environment and campus areas; Developed by researchers in Saudi Arabia; Tool not available online. International

Table 1. *Cont.*

Tool and Source	Brief Description and Region Where Applicable
USAT—Unit-Based Sustainability Assessment tool [63]	Based on indicators: 75 indicators, 4 domains (Teaching, Research and community services, Operation and management, Student involvement, Written policy and statement); score of 1 to 4 indicators; adapted from SAQ, AISHE and GASU; can be used in the department, college or HE unit; without updates; Developed by the United Nations Environment Program (UNEP) for the African context; Tool not available online but questionnaire available online on report, https://www.ru.ac.za/elrc/publicationsandresources/unit-basedsustainabilityassessmenttoolusattool/ (see Reference [63]). Africa
uD-SiM model—Uncertainty-based quantitative assessment of sustainability for HEIs [9,64]	Based on indicators and the Models of Pressure, Exposure, Effects, Action (DPSEEA) and a multicriteria decision process (applying Fuzzy logic). Aggregate score in a final index that integrates the non-linear effects of the indicators, with different weights and normalized indicators. Indicators based on the GASU model; 4 areas (Environmental, Economic, Social and Education); applied to Canadian Universities but its implementation is international; calculation method is complex; Developed by researchers; Tool not available online. Canada/International

According to several authors, the overall implementation of these tools is still low and its development is still at an early stage [14,28,32].

The various tools for assessing the sustainability of HEIs are mostly based on indicators, using graphs or final rankings to communicate the results. Indicator-based tools have the advantage of being potentially more transparent, consistent and comparable, thus useful for monitoring and decision support [14,18], although support for decision making is not yet fully demonstrated [65].

Another common characteristic of these tools listed in Table 1 is the fact that they are filled out by self-assessment, requiring only a leader or researcher to complete them. In order to create a sustainable university, it is important not only to use assessment tools for a real application as well as integrate on the process different agents of Higher Education Institutions [32]. Stakeholder participatory approaches can be seen as a requirement, as well as a benefit towards the integration of SD into the university culture [66]. Furthermore, active stakeholder participation is essential to grow the model's level of complexity, promote model ownership and use it in the organizational strategic planning process in a higher education organization [67]. Stakeholder engagement is also crucial to achieve the visions and goals for a Green university [68] and current SDG. The second draft of the People's Sustainability Treaty on Higher Education [69] divided higher education stakeholders into three broad categories: (1) those engaged in the activities of higher education institutions: executive, academic managers, educators, researchers and students. (2) those engaged in the higher education system: administrative officers, ministries, assessment bodies, international organizations. (3) those forming part of the communities, which the HEI system serves: local communities, professional bodies, companies, among others. Hence, a socially responsible HEI considers stakeholder behavior and perception to better understand their expectations and priorities and use these to define the strategy and goals, to monitor the objectives in view of promoting activities and accountability and to enhance a community-university engagement. At the end it contributes to change management and to a more mature sustainable university [9]. Thus, sustainability reflects a condition based on the relationship between stakeholders and HEI [70].

Based on the characteristics of these tools (see Table 1), the 27 tools were critically analyzed to evaluate their real assessment of the sustainability implementation and integration in HEIs (see the sections of the methods and results).

3. Cases Studies

The Universidade Aberta (UAb) was founded in 1988 with the distinctive feature of being the only Portuguese public distance education university, a distinction that still remains today. Filling the Universidade Aberta vision and mission of being a global university, the Universidade Aberta offers undergraduate and graduate higher education courses and Lifelong Learning courses, which are especially dedicated to the whole Portuguese speaking country community. The Universidade Aberta campus corresponds to four facilities, namely, its headquarters in Lisbon and two other support buildings in Coimbra and Porto (regional offices). The Universidade Aberta also has other facilities in Lisbon and Local Learning Centers spread throughout the country but since they are not owned but rented, they were not considered in the sustainability evaluation. In 2018 Universidade Aberta community comprise the rounded numbers of 6000, which 5000 were full-time students and exclusively engaged in distance education and 340 employees, of whom 150 belongs to the academic and research staff and 190 to the administrative staff. It is structured in 5 academic departments: Science and Technology, Social and Management Sciences, Distance Learning and Humanities, Lifelong Learning, with an educational offer of 10 graduate Ph.D. programs; 22 graduated master programs, 11 under graduated programs and 9 post-graduated programs. UAb has 2 research institutes and 5 more research institutes with branches in this university.

The Universidade Aberta has focused on the quality of its service and has been distinguished by several national and international entities: (a) EFQUEL Award—European Foundation for Quality in E-learning in 2010; (b) the UNIQUe—The Quality Label for the use of ICT in Higher Education (Universities and Institutes) in 2010; (c) the 1st Level of Excellence Committed to Excellence (c2e2) of the European Foundation for Quality Management (EFQM) in 2011; (d) 2nd Level of Excellence Recognized for Excellence (R4E) of the European Foundation for Quality Management (EFQM) in 2016; (e) the certification of the International Standard Organization (ISO) 27001 by the Portuguese Association of Certification attesting the security of its platform of e-learning and ISO 9001 quality, in 2017. UAb has been applied sustainability in different ways, namely: (i) curricula (through a e-learning three cycle degrees system from undergraduate, to master and Ph.D. aiming to actively promote education for sustainable development, along with an increase in transdisciplinary across subjects and also through non formal courses about Climate Change, Education for Sustainability and Environment awareness), (ii) application of a quality management policy with a Recognition of Excellence, (iii) a specific inclusion program for students with disabilities. UAb formal compromise to Sustainability was achieved by being an institutional member of the Association for the Advancement of Sustainability in Higher Education (AASHE) and a signatory member of the Letter of Commitment for Sustainable Campus in Portugal, both in 2019. UAb has no formal office for Sustainability due to its small size but sustainability issues are informally addressed within the Quality Office.

The Universidad Autónoma de Madrid (UAM) is a Spanish public university established in 1968. In 2018, 26.733 students of all levels were enrolled at the UAM. This university has a teaching staff and/or researcher around 3.141 and an administrative and services staff around 1.053. The university is organized into eight schools: Sciences (biology, mathematics, physics and chemistry), Economics and Business, Law, Computer Science and Engineering, Arts (philosophy, history, philology, translation and interpretation), Education and Psychology, offering a wide range of programs in different scientific and technical fields and in the Humanities. UAM has 11 research institutes and these are located on campus, as well as the Madrid Science Park, with growing university-business collaboration (contracts, internships and sponsored chairs). In 2009, UAM was declared International Excellence Campus, at the same time with the Spanish Research Council (CSIC). Nowadays, UAM has been placed among the top universities for its levels of excellence in national and international rankings. In the QS World University Ranking 2019, the UAM has managed to locate itself in the first position in Spain and in the 159th place in the world.

Most of the faculties and specialized institutes are on the Cantoblanco campus, 15 km North of Madrid. The Faculty of Medicine is on another campus (near La Paz Hospital). These two campuses

(Cantoblanco and La Paz) were considered in the sustainability evaluation. In the field of sustainability, this university, from the Rio de Janeiro Summit (1992) (where a global action plan for Sustainable Development was approved: Agenda 21), formalized its commitment to Agenda 21 through the ECOCAMPUS project. In 1997, the Ecocampus office was created, which has a special involvement in the maintenance of the campus as a sustainable territory and at the same time promotes activities related generally to sustainability. Social commitment and sustainability continue to be part of the frame of reference of the different lines of action for the entire institution in the 2025 strategy. UAM has also an SDG Lab that is a multi-stakeholder initiative that contributes to the implementation of the Sustainable Development Goals at the UAM.

4. Materials and Methods

4.1. Tools Assessment

The critical analysis of the tools listed in Table 1 was based on earlier reviews and according to the following criteria (based on the research of [9,18,24,28,65]: (i) comprehensibility; (ii) comparability; (iii) availability of baseline data; (iv) assessment of progress over time; (v) comprehensiveness and integration of sustainability dimensions in HEIs; (vi) usefulness for decision-making and communication; (vii) level of participation of the public or key actors and (viii) tool accessibility on the internet. The tools were then classified from 1 to 3 for each of the criteria (1. Low, 2. Medium and 3. High), based on documentary analysis of the tools and expert knowledge (according to the methodology referred in Reference [71]).

The classification of each tool was conducted independently by four judges (authors of this paper) using the defined criteria. The procedure proposed in Reference [72] has been followed in order to determine the level of agreement between the judges (values between 0 and 1 for each agreement between two judges):

Index of Agreement = (C1,2 + C1,3 + C1,4 + C2,3 + C2,4 + C3,4)/6

For data analysis, the mode and relative frequencies were calculated for each criterion. An average of the Index of the agreement was calculated for each criteria and tool. Two of the listed tools were not classified since their information was not available in English (DUK was only accessible in German and ASSC only accessible in Japanese).

Validity, reliability and generalizability are limitations associated with this type of qualitative approach [71] and were weighed up in the qualitative assessment and discussion of the results and when drawing the conclusions.

4.2. Sustainability Assessment in the Case Studies

The sustainability assessment process at the Universidade Aberta and the Universidad Autónoma de Madrid was carried out using different tools, STARS and GM, respectively, complemented with stakeholder's engagement (focus groups/workshops) in each University. The focus groups and workshops were used for engagement and awareness of the assessment process and also to find paths of improvement (Table 2). The STARS and GM tools were selected since they were well classified according to the main criteria and also in terms of free access on the internet (see Section 5.1 heading the in results section).

For UAb, the implementation of the sustainability assessment was conducted across the year 2018, corresponding to a 3-year assessment analysis (2015 to 2017). STARS 2.1 version tool is based 5 dimensions with different weights each: Academic and Research (20%), Involvement of Key Actors (20%), Campus Operations (35%), Planning and Administration (15%), Innovation and Leadership (2%) and a total 74 indicators divided in each category. The indicators are quantified and filled in a web application with a written justification or document upload. The different indicators and methodological procedures are explained in AASHE [58]. The raw data collection was the first procedure, involving not only a web search as also the requirement of specific informants for the technical information, reach by a face to face interview or information request by email. Possessing the requested data, the STARS assessment was fulfilled, resulting in a diagnostic report on the implementation of sustainability at the Universidade Aberta, where the most and least punctuated indicators were identified, revealing the weaknesses and the strengths to the implementation of sustainability. Since UAb is an institutional member of AASHE, a final assessment of the STARS scores was conducted, through an internal (by the rectorate) and external validation (by the AASHE technical staff) to allow a final awarded rating between bronze, silver or gold label.

With the final STARS report information, two focus groups (one with university experts and others with different stakeholders from the all University—see Table 2) were held in June and July of 2018. The main objective of the participatory moments was to show and discuss the STARS results and development of proposals for improvement for the implementation of sustainability at the Universidade Aberta. In a first part of the focus groups the information provided in the STARS report was presented and improvements were then proposed according to STARS dimensions. The first focus group was conducted with university experts that have been working in UAb sustainability implementation, including the Vice-rector for quality (since the University does not have any green sustainability office and three researchers or teachers). The other focus group was organized with different stakeholders from the all University from the different departments (one professor per department chosen according to a convenience sampling), administrative staff (leader of each administrative service) and two students from environmental and sustainability graduation programs (chosen according to a convenience sampling). Detailed information about all the methods applied in UAb available in Reference [73]. These discussion groups were the first time the university stakeholders participated in discussions related to sustainability.

Table 2. Operationalization design of the sustainability assessment in the case studies.

University	Data Collection	Method	Participants
Universidade Aberta	Raw data	Documents Analysis: Activity plans and reports, Strategic Plan and programs and courses study guides, databases.	Academics, Administrative services, Campus Operations and Rectorate
	Data standardization	Search in databases /websites	-
	Improvement Proposals	1st Focus group	Vice-rector for quality and 3 researchers (experts in sustainability assessment)
		2nd Focus group	16 (researchers, students, teachers, administrative staff)
Universidad Autónoma de Madrid	Raw data	Documents Analysis: Reports and University web page information	Administration and services staff, teacher
		1st Focus group	4 members of Eco-campus team (Eco-campus Manager, Infrastructure Manager, Environmental Participation Officer, Electric Cars on Campus Officer)
		2nd Focus group	3 Vice-Chancellors (Undergraduate Studies, Sustainability and Campus and Strategy and Planning)
	Improvement Proposals	3rd Focus group	6 students' leaders from Faculty of Business and Economic Sciences
		UAM SDGs LAB Sustainable Development Goals Workshop	22 Professors and researchers 6 External experts

The process of evaluating the implementation of sustainability in the UAM has begun in 2013 and annually assessed (from 2014 to 2018), through collecting the data according to the six dimensions stipulated by GM with different weights each: Setting and Infrastructure (SI) (15%), Setting and Infrastructure (SI) (15%), Waste (WS) (18%), Water (WR) (10%), Transportation (TR) (18%), Education and Research (ED) (18%). All information about the indicators and methodologies are available at Green Metric [50]. The indicators are quantified and filled in a web application (survey type). Each year in a predefined calendar, Green Metrics validates the submitted universities and publishes the international rating scores.

Representatives of the various groups of the university have participated in the contribution of these data: administration and services staff, teacher and manager (see Table 2). They participated through face-to-face interviews and email, together with collecting the available information, for example in corporate reports and on the web page of the University. The GM assessment along the years allowed to observe the evolution of the university in each of the dimensions and to compare with the scores of participating universities worldwide in GM.

For a review of the sustainability implementation and improvement proposals of the university sustainability implementation, two participatory techniques were carried out, focus group and workshop) (see Table 2). The focus group and workshop were then conducted using semi-structured questions based on the four dimensions of STARS tools to allow better comparison with UAb case study (Academics; Engagement, Operations and Planning & Administration). The collection process was in May 2018. Participants were selected based on their crucial role in the university management system, specifically on the sustainability activities at the campus. In the 1st focus group, 4 members managers of the Eco-campus team participated and in the 2nd workshop the 3 Vice-Chancellors of UAM with responsibilities in the sustainability issues were called to participate. In the 3rd focus group, 6 students' leaders from the Faculty of Business and Economic Sciences participated in the participative process (according to convenience sampling).

In a fourth participatory phase, SDGs LAB—Sustainable Development Goals Workshop was held. The UAM SDGs LAB was organized by the vice-chancellor of Sustainability and Campus and coordinated with professors and researchers in the field and 6 external experts from the town hall staff, already used to collaborate with the university, in a total number of 22. It took place in the UAM Campus for three days. It was the first participatory workshop to design a roadmap to improve the contribution of the university towards SDGs. This workshop was organized in three stages: (I) Inspiration: review of previous experiences and new inspiring ideas, (II) Observation: potential contribution to the campus in situ and proposals, (III) Implementation: discussion, analysis and outcome.

The results of the discussion groups in UAb and UAM were transcribed and then manually analyzed to identify recurring topics in the responses according to Reference [71]. The analyses were validated by the participants.

5. Results and Discussion

5.1. Tools Assessment

In an overall performance, Comprehensibility and Comparability were the criteria higher ranked by the judges (72% and 56% respectively classified as 3, see Table 3). Since the majority of the assessed tools are based on indicators (see Table 1) it is expected that they fulfil these criteria, as well as being a support tool for communication and decision-maker. Nevertheless, is still in doubt of real demonstration if indicators can indeed support decision-maker (only 48% were classified with the higher score in this criterion and also the index of agreement between judges was low—43%). This doubt is in accordance to Ramos [65] in his reflection article. Many of the assessed tools are adaptations to existing ones, often being used to contain geographic specificities (namely from North America, Latina America, Asia and Europe), what make the comparison at the national level easier but more difficult at the international level.

Table 3. Mode and relative frequencies for each criterion within 25 HE sustainability assessment tools. Sum and Level of agreement for each tool and criteria.

Tool		Comprehensibility	Comparability	Data Access	Progress over Time	Sustainability Broadness	Support to Decision	Participation	Accessibility	Sum	Index of Agreement
1. AISHE	Mode	3	3	2	3	2	2	3	2	20	70.8%
	R.F. (1)										
	R.F. (2)			1	0.5	0.75	0.75	0.5	1		
	R.F. (3)	1	1		0.5	0.25	0.25	0.5			
2. AMAS	Mode	2	2	2	2	2	2	2	1	15	52.1%
	R.F. (1)			0.25	0.25		0.25		1		
	R.F. (2)	0.5	1	0.75	0.5	0.75	0.5	0.75			
	R.F. (3)	0.5			0.25	0.25	0.25	0.25			
3. AUSP	Mode	3	2	2	1	2	3	1	2	16	54.1%
	R.F. (1)				0.5			0.75	0.25		
	R.F. (2)	0.5	1	1	0.25	0.75	0.5	0.25	0.75		
	R.F. (3)	0.5			0.25	0.25	0.5				
4. BIQ AUA	Mode	2	2	2	2	2	2	2	1	15	60.5%
	R.F. (1)				0.25		0.25		1		
	R.F. (2)	0.75	0.75	1	0.5	1	0.5	0.75			
	R.F. (3)	0.25	0.25		0.25		0.25	0.25			
5. CITE AMB	Mode	3	2	3	1	2	2	2	1	16	48.0%
	R.F. (1)	0.25		0.25	0.75	0.25		0.25	1		
	R.F. (2)	0.25	0.75	0.25	0.25	0.75	0.75	0.75			
	R.F. (3)	0.5	0.25	0.5			0.25				
6. ESDGC	Mode	3	2	2	3	3	3	1	1	18	43.8%
	R.F. (1)	0.25			0.25		0.25	0.5	1		
	R.F. (2)	0.25	0.5	1	0.25	0.5	0.25	0.5			
	R.F. (3)	0.5	0.5		0.5	0.5	0.5				
7. GASU	Mode	3	3	2	3	3	3	2	1	20	56.1%
	R.F. (1)			0.5				0.25	0.75		
	R.F. (2)	0.5	0.25	0.5	0.5			0.75	0.25		
	R.F. (3)	0.5	0.75		0.5	1	1				

Table 3. *Cont.*

Tool		Comprehensibility	Comparability	Data Access	Progress over Time	Sustainability Broadness	Support to Decision	Participation	Accessibility	Sum	Index of Agreement
8. GC	Mode	2	3	2	2	2	2	1	1	15	45.9%
	R.F. (1)	0.25			0.25	0.25		0.25	1		
	R.F. (2)	0.5	0.5	0.75	0.5	0.75	0.75	0.75			
	R.F. (3)	0.25	0.5	0.25	0.25		0.25				
9. GM	Mode	3	3	3	1	2	3	1	3	19	56.1%
	R.F. (1)				0.5		0.5	0.5			
	R.F. (2)		0.25	0.25	0.25	0.75		0.5			
	R.F. (3)	1	0.75	0.75	0.25	0.25	0.5		1		
10. GMID	Mode	2	1	1	2	2	2	2	1	13	35.4%
	R.F. (1)		0.75	0.5	0.25	0.25			0.75		
	R.F. (2)	0.5	0.25	0.5	0.5	0.5	0.75	0.5			
	R.F. (3)	0.5			0.25	0.25	0.25	0.5	0.25		
11. GP	Mode	3	3	2	3	3	3	2	1	20	41.6%
	R.F. (1)		0.25					0.25	0.5		
	R.F. (2)	0.25	0.25	1	0.5	0.5	0.5	0.75	0.25		
	R.F. (3)	0.75	0.5		0.5	0.5	0.5		0.25		
12. HE21	Mode	1	1	2	2	2	2	2	1	13	52%
	R.F. (1)	0.5	0.75		0.25	0.5	0.5	0.5	1		
	R.F. (2)	0.25	0.25	1	0.75	0.5	0.5	0.5			
	R.F. (3)	0.25									
13. P&P People & Planet	Mode	3	3	3	1	2	2	1	3	18	50.1%
	R.F. (1)	0.25	0.25		0.5		0.25	1			
	R.F. (2)	0.25	0.25	0.5	0.25	1	0.5				
	R.F. (3)	0.5	0.5	0.5	0.25		0.25		1		
14. PSIR	Mode	3	3	3	2	2	3	1	1	18	56.3%
	R.F. (1)			0.25	0.25			1	1		
	R.F. (2)	0.25	0.5	0.25	0.5	1	0.5				
	R.F. (3)	0.75	0.5	0.5	0.25		0.5				

Table 3. Cont.

Tool		Comprehensibility	Comparability	Data Access	Progress over Time	Sustainability Broadness	Support to Decision	Participation	Accessibility	Sum	Index of Agreement
15. SAQ	Mode	3	1	3	2	2	1	2	3	17	54.1%
	R.F. (1)		0.5		0.5		0.75	0.25			
	R.F. (2)		0.25		0.5	0.75	0.25	0.75	0.5		
	R.F. (3)	1	0.25	1		0.25			0.5		
16. SRC	Mode	3	2	2	1	3	3	2	1	17	39.5%
	R.F. (1)		0.25	0.5	0.5			0.25	1		
	R.F. (2)	0.5	0.5	0.5	.5	0.25	0.5	0.75			
	R.F. (3)	0.5	0.25			0.75	0.5				
17. STARS	Mode	3	3	2	3	3	3	2	3	22	70.8%
	R.F. (1)			0.25							
	R.F. (2)			0.75	0.25			0.5	0.5		
	Freq. Relative (3)	1	1		0.75	1	1	0.5	0.5		
18. SUM	Mode	3	3	2	3	3	3	2	1	20	68.8%
	R.F. (1)							0.25	1		
	R.F. (2)	0.25	0.25	0.75			0.25	0.75			
	R.F. (3)	0.75	0.75	0.25	1	1	0.75				
19. SLS	Mode	3	3	2	3	3	3	1	2	20	54.1%
	R.F. (1)							0.5			
	R.F. (2)		0.25	0.75	0.25		0.5	0.25	0.5		
	R.F. (3)	1	0.75	0.25	0.75	1	0.5	0.25	0.5		
20. SustainTool	Mode	3	3	3	1	2	2	2	3	19	64.5%
	R.F. (1)				0.75						
	R.F. (2)	0.5	0.5		0.25	1	0.75	0.75			
	R.F. (3)	0.5	0.5	1			0.25	0.25	1		
21. THE	Mode	3	3	2	2	2	3	1	3	19	49.9%
	R.F. (1)		0.25					0.5			
	R.F.(2)	0.25		0.5	0.5	0.75	0.25				
	R.F. (3)	0.75	0.75	0.5	0.5	0.25	0.75	0.5	1		

Table 3. *Cont.*

Tool		Comprehensibility	Comparability	Data Access	Progress over Time	Sustainability Broadness	Support to Decision	Participation	Accessibility	Sum	Index of Agreement
22. TUR	Mode	3	1	3	1	2	2	3	1	16	41.8%
	R.F. (1)		0.5		0.5		0.25	0.25	0.75		
	R.F. (2)	0.25	0.25	0.25	0.25	1	0.5	0.25	0.25		
	R.F. (3)	0.75	0.25	0.75	0.25		0.25	0.5			
23. UEMS	Mode	2	2	2	2	2	2	3	1	16	54.1%
	R.F. (1)		0.25						1		
	R.F. (2)	0.75	0.75	0.75	0.75	0.5	0.75	0.25			
	R.F. (3)	0.25		0.25	0.25	0.5	0.25	0.75			
24. USAT	Mode	3	3	3	3	3	3	2	2	22	54.1%
	R.F. (1)								0.25		
	R.F. (2)		0.5	0.5	0.25	0.25	0.25	1	0.5		
	R.F. (3)	1	0.5	0.5	0.75	0.75	0.75		0.25		
25. uD-SiM model	Mode	2	3	2	3	3	2	2	1	18	58.3%
	R.F. (1)	0.25		0.25				0.25	1		
	R.F. (2)	0.75	0.5	0.75	0.5		0.75	0.75			
	R.F. (3)		0.5		0.5	1	0.25				
Index of Agreement		51.3%	47.3%	59.3%	35.3%	70.0%	4.6%	47.3%	77.3%	-	-

Legend: R.F. (1)—Relative Frequency in (1) low; R.F. (2)—Relative Frequency in (2) medium; R.F. (3) —Relative Frequency in (3) high.

The more complex and complete the tools become, the more complex become to fill them and so the access to the basic data needed (only 32% of the analyzed tools had a maximum data availability rating, such as tools where is only need to answer "yes" or "no" or in the case of multiple-choice closed answers).

The heterogeneity of dimensions of sustainability implementation in HEIs covered by the tools is still remarkable, with the teaching and curriculum dimension and campus operations remaining the best addressed. Also, the environmental pillar is the more addressed, neglecting the social and economic pillars. These results are in accordance with [74], who also found in a study conducted in Spanish Universities that more attention is given to the environmental pillar and that is still necessary to achieve an integrated perspective of sustainability in universities. Those results justify the maximum ranking of only 36% of the tools in the criterion of coverage of sustainability implementation dimension and the same percentage in the criterion of measuring the progress of sustainability implementation in HEIs (see Table 3). Assessment of the progress over time was also the criterion with a higher level of disagreement between the judges (35%). According to Reference [75], variability between judges' classification is usually expected. Nevertheless, this criterion can be subjective since the assessment along the years can also depend on the HEIs and not only the tool himself. These results are in line with recent studies (e.g., [14,16,32]). Also related to this criterion, the state of development of the tools is revealed by the lack of assessment of impact outside the institution, that means on society in general and long-term impact [76].

The Participation and Accessibility criteria were the ones that have the lowest classification (12% and 24% respectively classified as 3, see Table 3). Concerning the level of agreement between the judges, Accessibility was the criterion that reunited more consensus what is easily explained by the fact that the tool is or not available on the internet to be used (Index of agreement of 77%). Indeed, only a small number of tools have an easily accessible application on the internet to fill in the data and obtain the final result. Participatory approaches have gained increasing attention in the implementation of sustainability in higher education but often remain vague and less addressed in sustainability assessment procedures, as Disterhelft et al. [66] also defended. The policy agenda of ESD calls for innovative and more transformative approaches than reductionist practices, in order to respond better to the need for an institutional learning culture that envisions dialogue and change as stressed by the same authors.

The results of the classification of the tools by the judges (Table 3) highlighted that STARS and USAT have good performance in 6 criteria ranked with maximum classification (both had a sum of 22 points in a maximum of 24). More specifically, STARS ranked medium criteria only in Data access and Participation categories and USAT in Participation and Accessibility criteria. Additionally, from Berzosa et al. [32] experience of "USAT is simple to apply, however, complicates comparations and benchmarking, as there are not any mechanisms to standardize the interviewee and the evaluator."

Following these tools were GASU, GM, SUM, SLS and GP (a sum of 20 points, see Table 3) with 5 criteria with the highest classification. For example, GASU has being pointed out by other authors (e.g., [7,16,29]) has having the advantages of giving institution visual illustration of sustainability, turning easy to compare and contracts the universities efforts towards sustainability within and among other universities. Also, AISHE was well classified with a sum of 20 points, with 4 criteria ranked with maximum classification (and a high index of agreement between the judges—71%). As discussed by other authors like Berzosa et al. [32] "AISHE score depends on stages of development of sustainability policies and actions within the organization and varies from an activity oriented (1) to a society oriented (5) approach." In addition, according to Alghamdi et al. [16], AISHE was designed to incorporate only the most significant criteria and not necessarily the whole framework.

Regarding the tools that scored the lowest on the ranking, TUR, HE21 and GMID were the ones with 3 criteria with the lowest classification, where Comparability and Accessibility had poor scores in all three (see Table 3). HE21 [31] is difficult to measure and compare, also indicators may not cover most important issues. ESDGD, P&P and TUR are the tools that had a high divergence in the judgment agreement (44%, 50% and 42% respectively), evidencing a great variability in Understanding, Comparability, Progress Over Time and Support to Decision criteria. TUR and GMID are old tools only available in research papers. P&P, HE21 and ESDGD are tools adapted specifically to be applied in Universities in a geographical region (UK).

Considering the confounding results, THE and GM evidenced a disperse result for the Participation and the Support to Decision criteria, respectively, because the score was divided in the lowest and highest ranking (also with a low index of agreement between judges 50 and 56%, respectively). In GMID, there is a clear outlier in the Accessibility category (since the tool is not available online) and in THE, an outlier for the Comparability category (since it is an international ranking). These facts highlight the importance of the classification being made by several judges and reducing the bias of the results.

Overall, the results confirmed STARS as the tool that collected a high level of agreement among the judges (71%), with the four criteria expressing the same score, Understanding, Comparability, Sustainability Broadness, Support to Decision. Nonetheless, STARS major strength according to Sonetti et al. [31] is its preference for performance over strategy. Earlier studies that reviewed these kinds of tools highlighted that STARS, AISHE and SAQ have a higher incidence on the percentage of indicators for the Governance (in accordance with reference [25]) and Operations, as highlighted by Reference [28] dimensions. However, STARS has the widest coverage across all indicators, capturing a little of all areas compared, for example, with AISHE and SAQ [as also defended by other authors [24,28], SRC (in accordance with Shi and Lai [77]) or with GM [also as highlighted by Lauder et al. [49]. In addition, as authors like Berzosa et al. [32] claimed "the main weaknesses of SAQ are those related with open-ended questions, not establishing a final score so it is difficult to apply it as a tracking tool."

In a more recent study, STARS appear to be comparable to AISHE and BIQ-AUA, considering the availability of academic and management staff as agents of change in the implementation of sustainability within universities [9]. Analyzing the filing process between STARS and AISHE, it can be seen that the former has a higher percentage of closed questions [24], so it is possible to deduce a greater ease of completeness and greater reliability when comparing the results, thus proving to be more efficient for a regular implementation [14,26,73]. This information is also underlined in a comparative study of STARS with other tools [78]. Authors such as Stough et al. [79] highlight various strengths in STARS namely innovation, understanding, popularity and be based on a holistic and integrative approach to sustainability, while also considering the United Nations Development Goals [79].

GM tool was also overall well classified, besides worst classification in terms of Participation and Progress Over Time criteria but has been also widely used since is free, easy to fill and good as a benchmark between universities worldwide, as stressed by many other authors (e.g., [16,49,74]). As Lauder et al. [49] stressed under scientific analysis, no ranking can be free from at least some limitations, resulting in unavoidable practical considerations, such as the need for the ranking to be at a level of complexity that can appeal to a wide audience. In addition, Sonetti et al. [31] mentioned one of the major weakness of GM is the use of generic quantitative indicators which does not underpin local dimensions as well as lack of the social dimension.

5.2. Application of the Tools in the Case Studies

Universidade Aberta scores for each STARS dimensions are shown in Table 4 and Figure 1 for the period of 2015–2017. The overall scores allowed to awarded the University with a Bronze label on April 5th of 2019 after a three-time external review process by STARS technical staff where some scores rectifications were made after documentation and numbers checked. UAb STARS report is now listed on the website of AASHE. Academic and Research dimensions are the ones where UAb is better classified leaving much room for implementing sustainability in its Planning and Management dimension, unlike in other universities where the focus is more on the Campus Operations [80]. Since UAb is a distance learning institution there is no formal campus for students' classes and the resources demanding are low (in particular in terms of energy) what justifies that difference. Nevertheless, Campus Operations improvements can be put in place as discussed in the focus groups (see Table 4). However, it is worth mentioning the importance of this approach for recognizing the advantages of the UAb teaching model for sustainability, from the perspective of its social dimension, namely that it facilitates education for all and at all stages. This is evidenced as an innovation according to the STARS criteria [73]. Another strong point that the assessment highlighted is the university bet in sustainability teaching not only at formal programs from 1st, 2nd and 3rd level (bachelor to Ph.D.) but also at non-formal programs like open and massive on-line courses (see Table 4).

Through the assessment of the sustainability in UAb using STARS the key-actors were able to discuss ways of improvement namely new paths to implement sustainability practices (see Table 4). The improvements found are feasible and inexpensive but UAb should integrate sustainability into organizational practices and allocate financial and human resources in the next strategic plan. In particular, the application of STARS can be an important basis for the definition of a currently non-existent University sustainability strategy/policy, one of the fundamental pillars for the whole-school approach, which agrees with other authors [13,32]. In addition, the participatory process that took place during the completion and evaluation of the tool alerted and sensitized the focus group participants on sustainability issues at UAb, contributing to the fact that some of the proposals are already being implemented at the moment. Examples are the registration of the SDG in the resources uploaded in the Open Repository of the University, the development of more moments of joy and sociability (e.g., Christmas lunch) and ongoing process of the dematerialization of the administrative process.

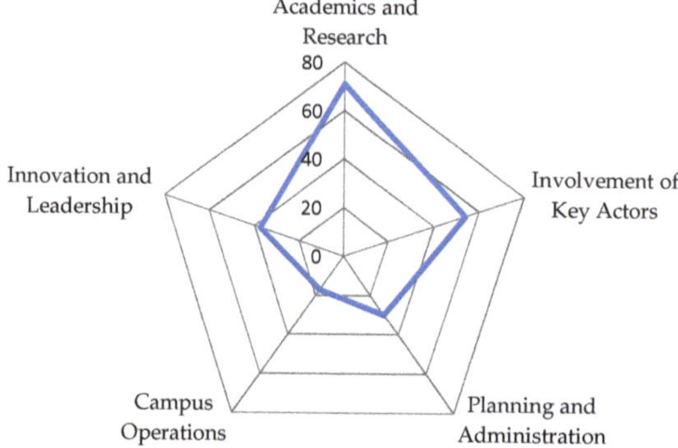

Figure 1. Results of the Sustainability Tracking tool, Assessment & Rating System (STARS) in the six dimensions for Universidade Aberta (UAb) in the period of 2015–2017. Maximum scores calculated based on version of STARS 2.1 [58].

Table 4. Summary of the results of Assessment and Improvements for Universidade Aberta regarding the STARS dimensions.

Academic

Assessment	Improvement
• Comprehensive formal cycle: BSc Environmental science, MSc Environmental Citizenship and Participation, PhD Social Sustainability and Development • Most of undergraduate programs with a least a module about Sustainability • Non formal courses (Open classes, Massive Open Online Courses, MOOCS, related with Environment) • Strategic research line about "Sustainability and Environment" (but with no financial support from University) • There is no available place to register the SDG activities of the university.	• Institution specific sustainability learning outcomes for all students • All undergraduate students should enroll at least a sustainability course (3 ETCS) • Enlarge research on sustainability linking students with the labor market, according to transdisciplinary research

Table 4. *Cont.*

Engagement, Planning and Administration	
Assessment	Improvement
• Staff training about better sustainable practices at work • Open courses on Sustainability and Environment • PhD thesis aiming at solving local problems related with SDG (action research, transdisciplinary) • No formal Institutional sustainability policy/strategy • Formal support to Students with disabilities • Assessed employee satisfaction • Participation of community members in the Institution governance (General Council) • Engagement of UAb community in DREAMLAB for Sustainability (Dragon Dreaming technic)	• Develop a University policy for sustainability • Integrated in the Quality office the Sustainability Practices • Inclusion within the program of Welcome of new employees (already existing) a sustainability performance kit • Promote awareness within academic community for students to work with the local community problems in the realm of Local Learning Centers; • Give more emphasis to the sustainability academic offer of the University

Operations and Innovation	
Assessment	Improvement
• No formal campus; e-learning regime (low ecological footprint) • LED lighting has reduced consumption • Use of local food/resources/services for events • Videoconference as prime communication service to all events • Decentralized Local Learning Centers in areas of low population density/close contact with society	• Disseminate/Monitor the Ecological footprint – CO_2 equivalent, water, waste (improvement of the GEE inventory) • Sustainable procurement practices (also within national policies): e.g., recycle paper, hybrid cars, cleaning material) • Separate bins in all facilities • Engage all university community in an online collaborative platform for sustainable ideas

This tool was first used in Portugal and at a distance learning university. Its application in the UAb has also identified some points of better adaptation to European reality (in particular related to units of measurement and benchmarks) and distance learning universities (since these institutions do not have a formal campus with students). These adaptations were communicated to the tool implementation support services of STARS. Earlier studies indicate that e-learning has a lower impact on greenhouse gas emissions and climate change, as observed in the UAb assessment and according to other studies (e.g., [81]) but its direct and indirect impacts on sustainability need to be better studied, as also advocated by Findler et al. [14]. Indeed, the long-term impact of the practices being implemented and the impact of UAb on a more sustainable society are issues that the tool has not been able to assess on its own.

The scores result of the GM assessment from 2014 to 2018 at Universidad Autónoma de Madrid are available in Figure 2 and Table 5. The results showed that UAM is in a good position compared with other worldwide universities, namely a 55th position in a total number of 719 in 2018. In 2018 UAM reached the highest scores due to considerable improvement in the Education and Infrastructures dimension but in 2014 its performance was better in terms of Water, Waste, Energy and Transportation. These results do not exactly mean the worst performance in the following years since the number of indicators related to these dimensions changed in the GM tool. An overall 34th position in the ranking of 2014 was obtained in a total of 361 universities ranked (corresponding to half of the universities compared to 2018).

According to the Vice-chancellors focus group (2nd focus group—Table 2), one critical area where the university should make an extra effort is the outreach, highlighting that university is a place to develop sciences, knowledge and new solutions able to contribute and affect to the society, local and global community. This factor is not considered in this tool, although considered in STARS. Participants from this 2nd focus group also highlighted the importance of undertaking the SDGs agenda in the university activities towards sustainability.

For the student focus group, a key factor to be included in the future is "the communication strategy" to foster the environmental message in the university community. This item is partly included in UI Green Metrics, in the Education and Research criteria, which requires the existence of a university-run sustainability website and the existence of a published sustainability report (Green Metrics [50]). This improvement was also highlighted in the UAb case study where it was proposed for better engagement of all university communities the development of an online collaborative platform for sustainable ideas (see Table 4).

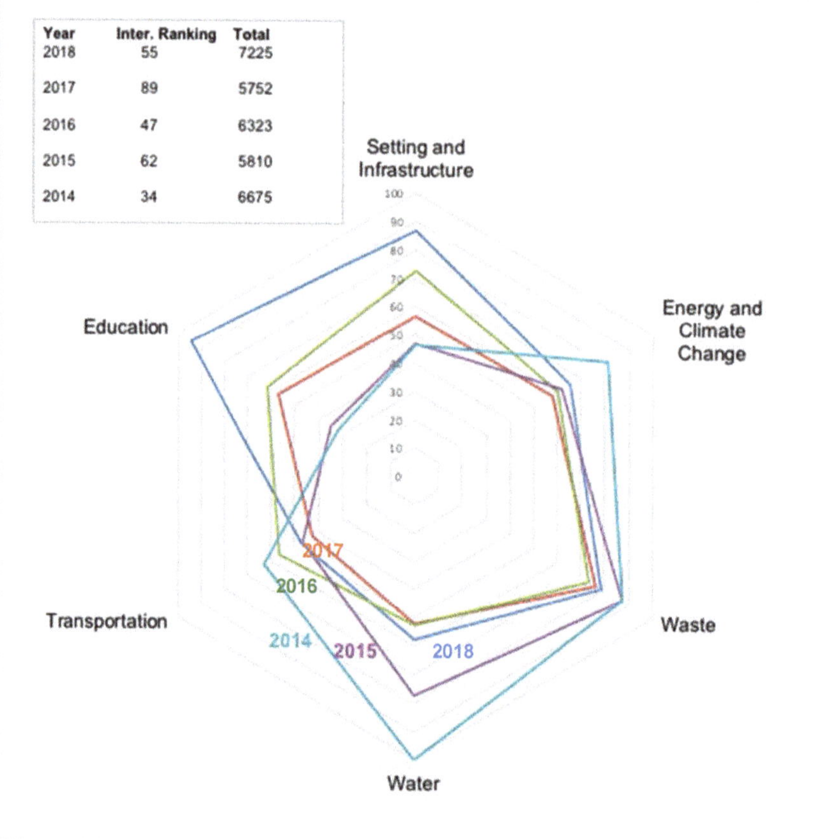

Figure 2. Scores of GreenMentrics for Universidad Autónoma de Madrid from 2014 to 2018 in total and in each dimension and also international ranking compare to total of other universities. Maximum scores calculated based on version of Green Metrics 2018 [50].

Common barriers for sustainability implementation in both Southern European institutions are related to low levels of community participation and sustainability awareness, financial constraints and a lack of HEIs legislative framework. Those barriers are also found in earlier studies (e.g., [76]).

Some weaknesses found in UAb are also found in UAM, namely in terms of SDG disclosure and low sustainability awareness from the university community and lack of financial support. Of course, in terms of campus operations, both universities' reality is quite different due to their different sizes and regime of teaching. Nevertheless, they can learn from each other. For example, UAM, in terms of the Academic dimension can use teaching methods more based on new technologies like the ones used in distance learning at UAb, allowing also more transdisciplinary research and less theory (in accordance with authors like Lozano et al. [15]). UAM can also develop more lifelong learning programs or open courses to increase society sustainability awareness, like MOOCs, learning from the experience of UAb. In addition, UAb can learn from the experiences in terms of the SDG LAB and Eco campus initiatives that UAM uses and improve their worst performance in terms of Campus Operation. Even that UAb is a distance learning university without face to face students at the campus it has facilities where energy, water and waste measures can be implemented, learning from UAM experiences. More recently, at both Universities questionnaires are being developed to the students to understand their sustainability perceptions, engagement and motivations. With the questionnaires results new strategies can be put in place.

Looking into Figure 2 it can be seen that GM dimensions are mainly focused on environmental initiatives and actions on campus, hence, there are scarce indicators of policies, management, diversity, equity and community participation, what is in accordance with several authors (e.g., [17,29]). Also, this ranking's main weakness could be considered that the information provided by the universities many times do not include evidence or in-situ verification. According to the experiences of the Eco campus focus group, the information sometimes can be considered subjective according to the understanding of the person in charge to provide the data. In the case of STARS, the data is validated by the rector team and also externally certified (like in UAb case study). However, the main strength is the accessibility and comparability of the outcome among the years (also in accordance with other authors like Lauder et al., [49]). The results from all university participants are available in the Green Metrics, easily accessible to compare by regions, countries and universities. From the benchmarking point of view, it could also facilitate the decision-making process for managers based on the potential analysis of the progress on these factors. The level of understanding is high and the authors of the GreenMetrics have provided the criteria, indicators and methodology used on their website (see for example the 2018 guide [50]).

Table 5. Summary of the results of Assessment and Improvements for Universidad Autónoma de Madrid regarding the STARS dimensions.

Academic	
Assessment	Improvement
<ul><li>Sustainability plays a role in many subjects on the majority of the faculties, 632 courses are connected to environment and sustainability.</li><li>The university is attempting to promote a participatory approach to sustainability activities together with the campus community, integrating an overall vision to a local impact</li><li>There is no available tracking system to register and follow the SDG activities of the university. The SDG approach is implicit in the academic programs. Also, the research activities are scattered.</li><li>When it comes to research and teaching, there are leading teams in different aspects of environmental, economic and social sustainability.</li><li>UAM was one of the first universities in Spain to offer a degree in the environmental science and incorporate environment into other degrees.</li><li>However, the learning method for these topics should consider a more practical method than only theory</li></ul>	<ul><li>Potential outreach in the society through scientific contribution including SDG areas in research and academic content</li><li>Use the campus as an environmental classroom for practicum, thesis, etc.</li><li>Collaborative online platform to show publications, projects, thesis, events connected with the SDGs</li><li>Observatory for dissemination - Inventory research team SDG (dynamic, active)</li><li>Involvement of teachers to motivate student participation</li></ul>

Table 5. *Cont.*

Engagement, Planning and Administration	
Assessment	Improvement
• There is a lack of awareness from the university community and society, as well as a low level of involvement and participation. • UAM must manage energy and water consumption according to the restrictions defined by the City Council and National Government. • UAM is working on a list of rigorous requirements for the process for suppliers in 2019. • Authorities are aware of the university's great expenses related to unsustainable actions on the campus. • There is no evolution on the metrics or rankings due to a lack of financial support. • Very high engagement of the UAM to sustainability: "SDG Lab Campus"; the presence of the UAM in Sustainability Conference of Rectors of Spanish Universities about Sustainability	• Social center for sustainability reference: store, organic coffee shop, distribution point Association of Parents and Friends of the Disabled of the UAM (APADUAM) Social and community involvement • Guide the UAM towards the circular economy • Implementation of a SDGs Road Map 2018

Operations and Innovation	
Assessment	Improvement
• A good connection between Madrid and UAM campus by public transportation, a daily average of 8k12k cars on the campus • Recycling initiatives by the implementation of technological resources. • Green areas have a great impact on students' attitudes toward the university. • UAM nature protection areas are a great improvement in air quality. • LED lighting has reduced consumption from 100% to 10%. • There is not an evolution of the environmental image of the UAM. • ECOCAMPUS is a leader of environmental programs at HEIs since 1997	• Agro-ecological use: fruit trees, orchards, livestock, beekeeping. • Analyze waste production and improve separation • Pedestrianize the historic core area. • Improve the treatment plant, analysis of water quality (contaminants); geofilter design • Metro minute (Distance in meters and time on feet) located at access points (train station, bus stop) and central (Rectorate, Plaza Mayor)

5.3. Overall Discussion

The tools critical analysis on the tools and cases studies application, still raise the question of the effect of implementing these tools versus the actual integration of sustainability into an HEI. The integration of sustainability in HEIs should come along with the modification of existing structures and habits, which creates many challenges related to the involved actors, the available resources, values and strategic choices to be made [82,83]. Also, based on Kapitulcinova et al. [9], the "transformational change" should occur at the level of the entire HE. It is therefore essential that a critical mass of units comprising the institution adopt sustainable development principles in their respective tasks and duties. In addition, according to Alonso-Almeida et al. [84] and Beringer et al. [85], to achieve the sustainability maturation, the sustainability integration at HEIs should involve all dimensions into a whole-school approach.

For all those reasons it is considered that the implementation of integrative approaches and models still needs to be encouraged and further research is needed [9,14,31,76]. Regardless of this fact, the case studies demonstrated that the implementation of the tools make possible to assess the state of implementation of sustainability in HEIs, monitor it, communicate it, share it within and outside the organization and improve and stimulate change, often enabling low-cost measures to be implemented. These statements are also advocated by other authors (e.g., [16,24,28,32]).

A change process enforces an overall vision, an increasing need for change that is experienced by the stakeholders, resources to support the process and short-term gains that can be communicated [86].

In this research the application of the assessment tools was complemented with participatory processes. While participatory methods are not commonly considered in these tools, they are central support for more holistic implementation allowing for the best man-nature link and a reflection that can better respond to the transformation of the institution and individual towards sustainability [66]. These tools must have a stronger component of student community participation and involvement, as this community is a major agent of change. Given the characteristics of experimentation and research of HEIs, indicators that can be incorporated into these tools or that can be used independently, should be tested to allow a long-term evaluation of whether the transformation process has been successfully achieved. Disterheft et al. [66] suggested examples of these types of indicators, based on perspectives such as the whole-school approach, interconnection between man and nature, community cohesion, celebration and happiness and principles of democracy. As argued by Ramos [65] the challenges in the area of sustainability assessment indicators should be based on transdisciplinary, collaborative and innovative scientific development where communities and the individual play a central role. Greater emphasis should be given to the development of indicators to assess non-traditional aspects of sustainability, such as ethics, culture and art, aesthetics, governance efficiency, spirituality, solidarity, compassion and trust, which represent fewer tangible dimensions of society [65,87].

These tools are also too operational not evaluating the strategic processes, as also stressed by Arroyo [33]. being able to incorporate the unpredictable and not only knowing how to deal with linear problems but also being able to assess what external impact HEIs have in practice on sustainability and going beyond the limits of HEIs [14]. As an example, the impact assessment of the research that is developed in HEIs on ESD should be carried out. Since this cannot be done based on a simple citation counting and bibliometric analysis because they do not allow to accurately define the result of this same investigation in the SD, more qualitative and documentary analyses are necessary for a more robust evaluation [14].

Few studies have explored and evaluated the role of HEIs as agents of change [31]. Also, there is a weak link between HEIs and external networks and key actors and with local and regional policies, what does not contribute to change in organizational management [82], holistic integration of sustainability in HEIs and their impact abroad [14,88]. The external impact of implementing sustainable development in HEIs can be measured in the local economy and culture, in challenges in society, in the natural environment and in the policies. Impacts can be short-term and direct, such as by training skilled workers or reducing greenhouse gases in the facilities but also indirect and long-term, such as changing graduates' sustainable lifestyles or in the implementation of Sustainable Development Goals (based on [14] suggestions).

6. Conclusions

A high increase has been seen in the implementation of sustainable development principles into Higher Education Institutions and in the research to assess its performance. In this article, a critical analysis of the existing tools to assess and benchmark ESD implementation was conducted by four judges according to a list of criteria. There are limitations associated with these kinds of critical review and qualitative evaluation but those limitations were weighed up in the mains findings. The tools applicability was then discussed in two case studies. The tool STARS—Sustainability Tracking, Assessment & Rating System was the one that was better classified in terms of understanding, comparability, sustainability broadness and support to decision, being one of the tools more worldwide used by different HEIs. The tool STARs was also used in one of the case studies. So, the use of STARS could be a good choice for HEIs that would like to assess and benchmark their sustainability performance according to a holistic and integrated approach.

The tools currently available for evaluating sustainability initiatives in HEIs do not all have the same objective or do not homogeneously evaluate the implementation of sustainability. As a consequence, some ambiguity is translated to its actual implementation and real contribution to the transformation for change. It is therefore recommended to set common general sustainability objectives in HEIs considering the integrative whole-school approach, regardless of some regions of institutions specificities (which may be indicated by specific objectives). Some improvements to the tools were discussed in this research and suggestions for future research. Sustainability assessment in HEIs should be viewed as a social construction, emerging from the different partners involved and according to mixed, bottom-up, top-down approaches, where the various actors, internal and external, contribute positively to the implementation of the sustainability assessment tool.

The implementation of Sustainability assessment tools in the HEIs case studies has shown that its use is an important driving force for the first diagnosis, a source for defining ways of improvement and also for future changes in organizational management. Also, the application of the tools in two HEIs of neighbor countries complemented with participation activities with key-actors, allowed to show the weakness of these tools namely the lack of the assessment of the impact of the sustainability performance of the university in the society and their real contribution for a sustainability transition. Bearing in mind that both countries have some similarities, namely cultural, this collaboration also brought up difficulties that need to be addressed in order to apply and implement EDS successfully. Barriers were identified and examples of improvement were given. Nevertheless, both universities learned from each other besides their differences and future developments for their sustainable performance improvement are already in place.

This article adds new insights about the main characteristic, common designations and objectives this type of tools should have and defined a list of criteria for their evaluation, trying to uniformize their meaning. Also, this research gives suggestions about improvements that are still needed on these tools so they can more fully answer their main purpose. Improvements can be in terms of incorporating in the tools an integrated process of stakeholder's participation, adding of indicators to assess non-traditional aspects of sustainability and being able to assess what external impact HEIs have in practice on sustainability.

Author Contributions: S.C. led the process of the tools review and assessment and overall discussion of the results. R.M. and S.C. developed the case study in UAb and wrote the respective methods and results. L.A.S.H. and C.E.B.A. developed the literature review and case study in UAM and wrote the respective methods and results. All the authors classified the tools according to the listed criteria and contribute for the article writing, revision and approval of the manuscript. All authors have read and agreed to the published version of the manuscript.

Funding: CENSE is financed by Fundação para a Ciência e Tecnologia, I.P., Portugal (UID/AMB/04085/2019). The funding sources played no part in the design, analysis, interpretation or writing-up of the paper or in the decision to publish.

Acknowledgments: The authors wish to thank all participants in this research in the different focus groups and workshops held in UAb and UAM, in particular Ana Paula Martinho and Teresa Teixeira in the data collection at UAb. The authors want also to thanks the rectorate support on the assessment processes in each university, in particular to Vice-cancellor for Qualaty Carla Oliveira in UAb and Santiago Atrio Cerezo Vice-cancellor for Sustainability and Campus and Jesús Rodríguez Pomeda, Vice-cancellor of Strategy and Planning in UAM.

Conflicts of Interest: The authors declare no conflicts of interest.

References

1. Mora, H.; Pujol-López, F.A.; Mendoza-Tello, J.C.; Morales-Morales, M.R. An education-based approach for enabling the sustainable development gear. *Comput. Hum. Behav.* **2018**. [CrossRef]

2. Wheeler, S.M. State and Municipal Climate Change Plans: The First Generation. *J. Am. Plan. Assoc.* **2008**, *74*, 481–496. [CrossRef]

3. United Nations. World Commission on Environment and Development. In *Our Common Future*; A/42/427; United Nations: New York, NY, USA, 1987.

4. U.N. *Transforming Our World: The 2030 Agenda for Sustainable Development*; A/RES/70/1; United Nations: New York, NY, USA, 2016.

5. UNESCO. *Education for Sustainable Development. Source Book*; United Nations Educational, Scientific and Cultural Organization: Paris, France, 2012.

6. Lozano, R.; Ceulemans, K.; Seatter, C.S. Teaching organisational change management for sustainability: Designing and delivering a course at the University of Leeds to better prepare future sustainability change agents. *J. Clean. Prod.* **2015**, *106*, 205–215. [CrossRef]

7. Lozano, R. Incorporation and institutionalization of SD into universities: Breaking through barriers to change. *J. Clean. Prod.* **2006**, *14*, 787–796. [CrossRef]

8. Velazquez, L.; Munguia, N.; Platt, A.; Taddei, J. Sustainable university: What can be the matter? *J. Clean. Prod.* **2006**, *14*, 810–819. [CrossRef]

9. Kapitulcinova, D.; AtKisson, A.; Perdue, J.; Will, M. Towards integrated sustainability in higher education—Mapping the use of the Accelerator toolset in all dimensions of university practice. *J. Clean. Prod.* **2018**, *172*, 4367–4382. [CrossRef]

10. Trencher, G.; Nagao, M.; Chen, C.; Ichiki, K.; Sadayoshi, T.; Kinai, M.; Kamitani, M.; Nakamura, S.; Yamauchi, A.; Yarime, M.; et al. Implementing Sustainability Co-Creation between Universities and Society: A Typology-Based Understanding. *Sustainability* **2017**, *9*, 594. [CrossRef]

11. Lozano, R. Diffusion of sustainable development in universities' curricula: An empirical example from Cardiff University. *J. Clean. Prod.* **2010**, *18*, 637–644. [CrossRef]

12. Lozano, R.; Lukman, R.; Lozano, F.J.; Huisingh, D.; Lambrechts, W. Declarations for sustainability in higher education: Becoming better leaders, through addressing the university system. *J. Clean. Prod.* **2013**, *48*, 10–19. [CrossRef]

13. Lozano, R.; Ceulemans, K.; Alonso-Almeida, M.; Huisingh, D.; Lozano, F.J.; Waas, T.; Lambrechts, W.; Lukman, R.; Hugé, J. A review of commitment and implementation of sustainable development in higher education: Results from a worldwide survey. *J. Clean. Prod.* **2015**, *108*, 1–18. [CrossRef]

14. Findler, F.; Schönherr, N.; Lozano, R.; Stacherl, B. Assessing the Impacts of Higher Education Institutions on Sustainable Development—An Analysis of Tools and Indicators. *Sustainability* **2018**, *11*, 59. [CrossRef]

15. Lozano, R.; Barreiro-Gen, M.; Lozano, F.J.; Sammalisto, K. Teaching Sustainability in European Higher Education Institutions: Assessing the Connections between Competences and Pedagogical Approaches. *Sustainability* **2019**, *11*, 1602. [CrossRef]

16. Alghamdi, N.; den Heijer, A.; de Jonge, H. Assessment tools' indicators for sustainability in universities: An analytical overview. *Int. J. Sustain. High. Educ.* **2017**, *18*, 84–115. [CrossRef]

17. Shriberg, M. Institutional assessment tools for sustainability in higher education. *Int. J. Sustain. High. Educ.* **2002**, *3*, 254–270. [CrossRef]

18. Ramos, T.; Pires, M.S. Sustainability Assessment: The Role of Indicators. In *Sustainability Assessment Tools in Higher Education—Mapping Trends and Good Practices at Universities around the World*; Caeiro, S., Leal Filho, W., Jabbour, C., Azeiteiro, U., Eds.; Springer: Cham, Switzerland, 2013; pp. 81–100.

19. Nixon, A. Improving the Campus Sustainability Assessment Process. Bachelor's Thesis, Western Michigan University, Kalamazoo, MI, USA, 2002.

20. Cole, L. Assessing Sustainability on Canadian University Campuses: Development of a Campus Sustainability Assessment Framework. Master's Thesis, Royal Roads University, Victoria, BC, Canada, 2003.

21. Alshuwaikhat, H.M.; Abubakar, I. An integrated approach to achieving campus sustainability: Assessment of the current campus environmental management practices. *J. Clean. Prod.* **2008**, *16*, 1777–1785. [CrossRef]

22. Larouce, D.C. *Tracking progress: Development and Use of Sustainability Indicators in Campus Planning and Management*; Arizona State University: Phoenix, AZ, USA, 2009.

23. *Implementing Sustainability at the Campus—Towards a Better Understanding of Participation Processes within Sustainability Initiatives; Sustainable Development at Universities: New Horizons. Umweltbildung, Umweltkommunikation und Nachhaltigkeit—Environmental Education, Communication and Sustainability*; Peter Lang: Frankfurt, Germany, 2012; Volume 34, pp. 345–361.

24. Yarime, M.; Tanaka, Y. The Issues and Methodologies in Sustainability Assessment Tools for Higher Education Institutions: A Review of Recent Trends and Future Challenges. *J. Educ. Sustain. Dev.* **2012**, *6*, 63–77. [CrossRef]

25. Kosta, K. Institutional Sustainability Assessment. In *Encyclopedia of Sustainability in Higher Education*; Leal Filho, W., Ed.; Springer Nature Switzerland AG: Cham, Switzerland, 2019; pp. 1–7. [CrossRef]

26. Sayed, A.; Kamal, M.; Asmuss, M. Benchmarking tools for assessing and tracking sustainability in higher educational institutions. *Int. J. Sustain. High. Educ.* **2013**, *14*, 449–465. [CrossRef]

27. Ceulemans, K.; Molderez, I.; Van Liedekerke, L. Sustainability reporting in higher education: A comprehensive review of the recent literature and paths for further research. *J. Clean. Prod.* **2015**, *106*, 127–143. [CrossRef]

28. Fischer, D.; Jenssen, S.; Tappeser, V. Getting an empirical hold of the sustainable university: A comparative analysis of evaluation frameworks across 12 contemporary sustainability assessment tools. *Assess. Eval. High. Educ.* **2015**, *40*, 785–800. [CrossRef]

29. Gómez, F.; Sáez-Navarrete, C.; Lioi, S.; Marzuca, V. Adaptable Model for Assessing Sustainability in Higher Education. *J. Clean. Prod.* **2015**, *107*, 475–485. [CrossRef]

30. Bullock, G. The comprehensiveness of competing higher education sustainability assessments. *Int. J. Sustain. High. Educ.* **2016**, *17*, 282–304. [CrossRef]

31. Sonetti, G.; Lombardi, P.; Chelleri, L. True Green and Sustainable University Campuses? Toward a Clusters Approach. *Sustainability* **2016**, *8*, 83. [CrossRef]

32. Berzosa, A.; Bernaldo, M.O.; Fernández-Sanchez, G. Sustainability assessment tools for higher education: An empirical comparative analysis. *J. Clean. Prod.* **2017**, *161*, 812–820. [CrossRef]

33. Arroyo, P. A new taxonomy for examining the multi-role of campus sustainability assessments in organizational change. *J. Clean. Prod.* **2017**, *140*, 1763–1774. [CrossRef]

34. National Wildlife Federation. Campus Environmental Report Card. Available online: https://www.nwf.org/EcoLeaders/Campus-Ecology-Resource-Center/Reports/Campus-Report-Card/index.html (accessed on 5 October 2018).

35. Smith, A.A. The Student Environmental Action Coalition. In *Campus Ecology: A Guide to Assessing Environmental Quality and Creating Strategies for Change*; Living Planet Press: Los Angeles, CA, USA, 1993.

36. Chernushenko, D. *Greening Campuses: Environmental Citizenship for Colleges and Universities*; International Institute for Sustainable Development: Winnipeg, MB, Canada, 1996.

37. Decamps, A.; Barbat, G.; Carteron, J.-C.; Hands, V.; Parkes, C. Sulitest: A collaborative initiative to support and assess sustainability literacy in higher education. *Int. J. Manag. Educ.* **2017**, *15*, 138–152. [CrossRef]

38. GULF; ISCN. *Implementation Guidelines to the ISCN-GULF Sustainable Campus Charter. Suggested Reporting Contents and Format*; Global University Leader Forum (GULF); International Campus Sustainability Network (ICSN): Lausanne, Switzerland, 2016.

39. Roorda, N. *AISHE, Assessment Instrument for Sustainability in Higher Education*; Atichting Duurzaam Hoger Onderwijs (DHO): Amsterdam, The Netherlands, 2001.

40. Roorda, N.; Rammel, C.; Waara, S.; Fra Paleo, U. AISHE 2.0 Manual: Assessment Instrument for Sustainability in Higher Education. Available online: http://www.eauc.org.uk/theplatform/aishe (accessed on 5 October 2018).

41. CAS-NET Japan. Good Practices on Campus Sustainability in Japan. Available online: https://www.osc.hokudai.ac.jp/en/action/assc (accessed on 5 October 2018).

42. CRUE; CSC. *Evaluación de las Políticas Universitarias de Sostenibilidad como Facilitadoras Para el Desarrollo de los Campus de Excelencia Internacional*; Conferencia de Rectores de las Universidades Españolas; Comisión Sectorial de Calidad: Madrid, Spain, 2011.

43. CRUE; CSC. *Diagnóstico De La Sostenibilidad Ambiental En Las Universidades Españolas*; Conferencia de Rectores de las Universidades Españolas; Comisión Sectorial de Calidad: Madrid, Spain, 2018.

44. ProSPER.Net. Alternative University Appraisal Model for ESD in Higher Education Institutions. Available online: http://prospernet.ias.unu.edu/projects/upcoming-projects/alternative-university-appraisal-aua (accessed on 5 October 2018).

45. Razak, D.; Sanusi, Z.; Jegatesen, G.; Khelghat-Doost, H. Alternative University Appraisal (AUA). Reconstructing Universities' Ranking and Rating Toward a Sustainable Future. In *Sustainability Assessment Tolls in Higher Education in Higher Education Institutions: Mapping Trends and Good Practices Around the World*; Caeiro, S., Leal Filho, W., Jabbour, C., Azeiteiro, U.M., Eds.; Springer: Cham, Switzerland, 2013; pp. 139–158.

46. CITE-AMB M; ASCUN; RISU; RAUS; PNUMA; RFA-ALC; GUPES-LA; ARIUSA. *Encuesta para el Diagnóstico de la Institucionalización Del Compromiso Ambiental En Las Universidades Colombianas*; Red de Ciencia, Tecnología, Innovación y Educación Ambiental en Iberoamérica, Ministerio de Ambiente y Desarrollo Sostenible, Asociación Colombiana de Universidades, Red de Indicadores de Sostenibilidad en las Universidades, Red Ambiental de Universidades Colombianas, Programa de las Naciones Unidas para el Medio Ambiente, Red de Formación Ambiental para América Latina y el Caribe, Alianza Mundial de Universidades por el Ambiente y la Sostenibilidad and Alianza de Redes Iberoamericanas de Universidades por la Sustentabilidad y el Ambiente: New York, NY, USA, 2014.

47. Glover, A.; Jones, Y.; Claricoates, J.; Morgan, J.; Peters, C. Developing and Piloting a Baselining Tool for Education for Sustainable Development and Global Citizenship (ESDGC) in Welsh Higher Education. *Innov. High. Educ.* **2012**, *38*. [CrossRef]

48. Good Company. *Sustainable Pathways Toolkit for Universities and Colleges: Social and Environmental Indicators for Campuses. Part II: Toolkit Technical Manual*; Good Company: Eugene, OR, USA, 2002.

49. Lauder, A.; Sari, R.F.; Suwartha, N.; Tjahjono, G. Critical review of a global campus sustainability ranking: GreenMetric. *J. Clean. Prod.* **2015**, *108*, 852–863. [CrossRef]

50. UI Green Metric. Available online: http://greenmetric.ui.ac.id/ (accessed on 5 October 2018).

51. Mader, C. Sustainability process assessment on transformative potentials: The Graz Model for Integrative Development. *J. Clean. Prod.* **2013**, *49*, 54–63. [CrossRef]

52. CdGE & CdPdU, Plan Vert—Version 2012. Available online: http://www.developpement-durable.gouv.fr/Green-Plan.html (accessed on 5 October 2018).

53. Buckland, H.; Brookes, F.; Seddon, D.; Johnston, A.; Parkin, S. The UK Higher Education Partnership for Sustainability (HEPS). 2001. Available online: https://ulsf.org/the-uk-higher-education-partnership-for-sustainability-heps/ (accessed on 5 October 2018).

54. Penn State Green Destiny Council. *Penn State Indicators Report 2000: Steps toward A Sustainable University*; Penn State Green Destiny Council: State College, PA, USA, 2000.

55. People & Planet University League. How Sustainable Is Your University? Available online: https://peopleandplanet.org/university-league (accessed on 5 October 2018).

56. ULSF. Sustainability Assessment Questionnaire (SAQ). Available online: http://ulsf.org/sustainability-assessment-questionnaire/ (accessed on 5 October 2018).

57. Sustainable Endowments Institute. The College Sustainability Report Card. Available online: http://www.greenreportcard.org/index.html (accessed on 5 October 2018).

58. AASHE. *STARS Technical Manual—Version 2.1 Administrative Update Three*; Association for the Advancement of Sustainability in Higher Education: Philadelphia, PA, USA, 2017.

59. EUAC. Sustainability Leadership Scorecard—Launching a Transformational Tool for the Education Sector—The Sustainability Leadership Scorecard. Available online: http://www.eauc.org.uk/sustainability_leadership_scorecard (accessed on 5 October 2018).

60. Washington, U. The Program Sustainability Assessment Tool v2. Available online: https://sustaintool.org/ (accessed on 5 October 2018).

61. Times, The World University Ranking. The University Impact Ranking. Available online: https://www.timeshighereducation.com/how-participate-times-higher-education-rankings (accessed on 5 October 2018).

62. Lukman, R.; Krajnc, D.; Glavič, P. University ranking using research, educational and environmental indicators. *J. Clean. Prod.* **2010**, *18*, 619–628. [CrossRef]

63. Togo, M.; Lotz-Sisitka, H. *Unit Based Sustainability Assessment Tool*; A Resource Book to Complement the UNEP Mainstreaming Environment and Sustainability in African Universities Partnership; UNEP: Howick, South Africa, 2009.

64. Waheed, B.; Khan, F.; Veitch, B.; Hawboldt, K. Uncertainty-based quantitative assessment of sustainability for higher education institutions. *J. Clean. Prod.* **2011**, *19*, 720–732. [CrossRef]

65. Ramos, T. Sustainability Assessment: Exploring the Frontiers and Paradigms of Indicator Approaches. *Sustainability* **2019**, *11*, 824. [CrossRef]

66. Disterheft, A.; Caeiro, S.S.; Leal Filho, W.; Azeiteiro, U.M. The INDICARE-model—Measuring and caring about participation in higher education's sustainability assessment. *Ecol. Indic.* **2016**, *63*, 172–186. [CrossRef]

67. Zaini, R.M.; Pavlov, O.V.; Saeed, K.; Radzicki, M.J.; Hoffman, A.H.; Tichenor, K.R. Let's Talk Change in a University: A Simple Model for Addressing a Complex Agenda. *Syst. Res. Behav. Sci.* **2017**, *34*, 250–266. [CrossRef]

68. Yuan, X.; Zuo, J.; Huisingh, D. Green Universities in China—What matters? *J. Clean. Prod.* **2013**, *61*, 36–45. [CrossRef]

69. UN. *People's Sustainability Treaty on Higher Education*; United Nations: New York, NY, USA, 2012.

70. Salvioni, D.M.; Franzoni, S.; Cassano, R. Sustainability in the Higher Education System: An Opportunity to Improve Quality and Image. *Sustainability* **2017**, *9*, 914. [CrossRef]

71. Saunders, M.; Lewis, P.; Thornhill, A. *Research Methods for Business Students*, 6th ed.; Pearson Education Limited: England, UK, 2012.

72. Hernández Sampieri, R.; Férnandez Collado, C.; Baptista Lucio, P. *Metodologia De Pesquisa [Research Methodology]*; McGraw-Hill: São Paulo, Brazil, 2006.
73. Martins, R. Ferramentas de Avaliação Da Sustentabilidade no Ensino Superior: A adequação ao Contexto de uma Universidade a Distância. Master's Thesis, Universidade Aberta, Lisboa, Portugal, 2019.
74. Blasco, N.; Brusca, I.; Labrador, M. Assessing Sustainability and Its Performance Implications: An Empirical Analysis in Spanish Public Universities. *Sustainability* **2019**, *11*, 5302. [CrossRef]
75. Krippendorff, K. *Content Analysis. An Introduction to Its Methodology*; Sage: Thousand Oaks, CA, USA, 2004.
76. Leal Filho, W.; Wu, Y.-C.J.; Brandli, L.L.; Avila, L.V.; Azeiteiro, U.M.; Caeiro, S.; Madruga, L.R.D.R.G. Identifying and overcoming obstacles to the implementation of sustainable development at universities. *J. Integr. Environ. Sci.* **2017**, *14*, 93–108. [CrossRef]
77. Shi, H.; Lai, E. An alternative university sustainability rating framework with a structured criteria tree. *J. Clean. Prod.* **2013**, *61*, 59–69. [CrossRef]
78. Maragakis, A.; van den Dobbelsteen, A. Sustainability in Higher Education: Analysis and Selection of Assessment Systems. *J. Sustain. Dev.* **2015**, *8*, 1–9. [CrossRef]
79. Stough, T.; Ceulemans, K.; Lambrechts, W.; Cappuyns, V. Assessing sustainability in higher education curricula: A critical reflection on validity issues. *J. Clean. Prod.* **2018**, *172*, 4456–4466. [CrossRef]
80. Lidstone, L.; Wright, T.; Sherren, K. An analysis of Canadian STARS-rated higher education sustainability policies. *Environ. Dev. Sustain.* **2014**, *17*, 259–278. [CrossRef]
81. Versteijlen, M.; Perez Salgado, F.; Janssen Groesbeek, M.; Counotte, A. Pros and cons of online education as a measure to reduce carbon emissions in higher education in the Netherlands. *Curr. Opin. Environ. Sustain.* **2017**, *28*, 80–89. [CrossRef]
82. Verhulst, E.; Lambrechts, W. Fostering the incorporation of sustainable development in higher education. Lessons learned from a change management perspective. *J. Clean. Prod.* **2015**, *106*, 189–204. [CrossRef]
83. Hugé, J.; Mac-Lean, C.; Vargas, L. Maturation of sustainability in engineering faculties—From emerging issue to strategy? *J. Clean. Prod.* **2018**, *172*, 4277–4285. [CrossRef]
84. Alonso-Almeida, M.; Marimon, F.; Casani, F.; Rodriguez-Pomeda, J. Diffusion of sustainability reporting in universities: Current situation and future perspectives. *J. Clean. Prod.* **2015**, *106*, 144–154. [CrossRef]
85. Beringer, A.; Adomßent, M. Sustainable university research and development: Inspecting sustainability in higher education research. *Environ. Educ. Res.* **2008**, *14*, 607–623. [CrossRef]
86. Barth, M. Many roads lead to sustainability: A process-oriented analysis of change in higher education. *Int. J. Sustain. High. Educ.* **2013**, *14*, 160–175. [CrossRef]
87. Viegas, O.; Caeiro, S.; Ramos, T. A Conceptual Model for the integration of Non-Material components in Sustainability Assessment. *Ambiente Soc.* **2018**, *21*, 1–20. [CrossRef]
88. Vargas, V.R.; Lawthom, R.; Prowse, A.; Randles, S.; Tzoulas, K. Sustainable development stakeholder networks for organisational change in higher education institutions: A case study from the UK. *J. Clean. Prod.* **2019**, *208*, 470–478. [CrossRef]

MDPI
St. Alban-Anlage 66
4052 Basel
Switzerland
Tel. +41 61 683 77 34
Fax +41 61 302 89 18
www.mdpi.com

Sustainability Editorial Office
E-mail: sustainability@mdpi.com
www.mdpi.com/journal/sustainability